T0387959

Bioremediation Technology

Bioremediation Technology

Hazardous Waste Management

Edited by

M.H. Fulekar and Bhawana Pathak

CRC Press is an imprint of the
Taylor & Francis Group, an **informa** business

CRC Press
Taylor & Francis Group
6000 Broken Sound Parkway NW, Suite 300
Boca Raton, FL 33487-2742

Printed on acid-free paper

International Standard Book Number-13 978-0-367-27310-1 (Hardback)

Library of Congress Cataloging-in-Publication Data

Names: Fulekar, M. H., editor. | Pathak, Bhawana, editor.
Title: Bioremediation technology : hazardous waste management / edited by M.H. Fulekar and Bhawana Pathak.
Description: Boca Raton : CRC Press, [2020] | Includes bibliographical references and index. | Summary: "The book describes hazardous waste industries, sources of waste generation, characterization and treatment processes/ methods, technique and technology to deal with the treated waste as per the prescribed standard. Advanced treatment based on the microbial remediation, plant-based decontamination, rhizoremediation and nano-based remediation is also explained. Advances in treatment technology using biotechnological tools/bionanotechnology for removal of contaminants are described. This volume will help readers to develop of bio-nano-based technology, which can be transferred from lab to land, pilot to commercial scenario"-- Provided by publisher.
Identifiers: LCCN 2019029239 | ISBN 9780367273101 (hardback ; acid-free paper) | ISBN 9780429296031 (ebook)
Subjects: LCSH: Hazardous wastes--Biodegradation.
Classification: LCC TD1061 .B536 2020 | DDC 628.4/2--dc23
LC record available at https://lccn.loc.gov/2019029239

Visit the Taylor & Francis Web site at
http://www.taylorandfrancis.com

and the CRC Press Web site at
http://www.crcpress.com

Contents

Preface

Environmental pollution has become a global concern due to rapid industrialization, urbanization and modern technological development in chemical processes, operation and production. Technological innovations in industries have given rise to new products and new pollutants in abundant levels, which are above the self-cleaning capacity of our environment. The present treatment technology involving physicochemical and biological methods are not efficient/effective to treat the contaminant in compliance with standards. Today environmental biotechnology is considered an emerging science for the treatment of hazardous waste. The technology involves the use of microorganisms and plants for the biological treatment of pollutants. Biotechnological treatment is carried out at lower temperature and pressure that requires less energy than the conventional physicochemical treatment technology and has been found safer for environmental protection.

Biotechnological innovation for treatment of hazardous waste under controlled environmental condition have been found to be cost-effective in reducing the pollution potential of toxic contaminants to enhance public acceptance and compliance with environmental legislation. The contaminants present in soil or surface/ground water can be remediated by bioremediation and/or phytoremediation and use of advance green and clean technology. The research case studies pertaining to detoxification and decontamination of hazardous waste by advance technology in compliance with standards will help to clean up the environment.

This book covers hazardous waste generation from different industrial sources; characterization and treatment processes/methods; technique and technology to comply the treated waste with the prescribed standard. The advanced treatment is based on the microbial remediation, plant-based decontamination, rhizoremediation and nano-based remediation. Advances in treatment technology using biotechnological tools and bionanotechnology for the removal of contaminants are described in this book. It will benefit the academician, industries and research scholars in the development of bio-nano-based technology. This technology can be transferred from laboratory to land—pilot to commercial-based technology.

Editors

M. H. Fulekar (MSc, MPhil, PhD, LLB, MBA, DSc (submitted)) is a professor and joint director (R&D) at Centre of Research for Development, Paul University, Vadodara. He was a senior professor and dean at the School of Environment and Sustainable Development, and director, Central University of Gujarat. He was professor and head at the University Department of Life Sciences, University of Mumbai. He has to his credit more than 350 research papers and articles published in international and national journals of repute. He also has to this credit 200 NCBI, USA Gene Bank submissions. He authored 15 books published by CRC, Springer IK International; Oxford, etc. He has guided 22 PhD and 21 MPhil students in Environmental Science, Life Sciences, and Nano Sciences and at present guiding PhD students in Environmental Biotechnology and Environmental Nanotechnology. He has to his credit five patents (1 awarded, 2 published and 2 in progress).

As principal investigator, Professor Fulekar has completed research projects with the UGC, CSIR, BRNS, and DBT R&D, among other organizations. He was awarded the International Labor Organization (ILO) fellowship ILO/UNDP1985: ILO/FINNIDA; ILO/JAPAN. UGC awarded him made him BSR fellow in 2018.

Professor Fulekar was named in the "Who's Who" in Science and Engineering, United States, in 1998, and was awarded "Outstanding Scientist of the 20th Century" in 2000 by the International Biographical Centre, Cambridge, England. He is also a member of the New York Academy of Sciences, New York. He is a recipient of the Education Leadership Award and International Award for Environmental Biotechnology. Recently he has been awarded the title of Eminent Educationist.

Professor Fulekar has worked in various capacities in university administration. Areas of research interest are Environmental Science, Environmental Biotechnology, Environmental Nanotechnology. As an expert, he visited United States, Australia, Singapore, Bangkok, Thailand, Hong Kong, Nepal etc.

Bhawana Pathak is associate professor and dean at the School of Environment and Sustainable Development, Central University of Gujarat. Earlier she has worked as a pool officer at the University Department of Life Sciences, University of Mumbai. She was awarded the CSIR Research Associateship and worked on plant behavior ecology and biodiversity conservation at the G. B. Pant Institute of Himalayan Environment and Development, Almora, India. She was a gold medallist in MSc Botany from the Kumaun University, Nainital, and did her PhD from G. B. Pant Institute of Himalayan Environment and Development. She has more than 15 years of research and teaching experience in the specialized areas of Environmental Ecology, Biodiversity Conservation, Environmental Biotechnology and Environmental Nanotechnology. Fourteen MPhil and 6 PhD degrees have been awarded under her guidance and at present she is guiding 6 PhD students. Dr. Pathak also contributes to the development of curriculum for different courses; has published research papers, scientific research articles, chapters in books and also contributed to innovative research work that has been instrumental in shaping future policies. She has also presented scientific papers at national and international conferences and seminars. She is a recipient of the Outstanding Faculty Award in bio science and technology. Dr. Pathak's biographical note was also published in *Asia Pacific—Learned India: Educationist* "Who Is Who." Presently her major focus is on transferring the laboratory technology to the land.

Contributors

Nikhil Bhatt
P.G. Department of Microbiology and Biogas Research Centre
Gujarat Vidyapith
Gandhinagar, India

Harish Chandra
Department of Botany and Microbiology
Gurukula Kangri Vishwavidyalaya
Haridwar, India

Smita Chaudhry
Institute of Environmental Studies
Kurukshetra University
Kurukshetra, India

Rafael Alberto Fonseca-Correa
Civil Engineering Program
Universidad Piloto de Colombia
Bogotá, Colombia

Ramesh Chandra Dubey
Department of Botany and Microbiology
Gurukula Kangri Vishwavidyalaya
Haridwar, India

Dimple P. Dutta
Chemistry Division
Bhabha Atomic Research Centre
Mumbai, India

Jyoti Fulekar
Parul Univerity
Varodara, India

M. H. Fulekar
Centre of Research for Development
Parul University
Varodara, India

Pooja Ghosh
School of Environmental Sciences
Jawaharlal Nehru University
New Delhi, India

Liliana Giraldo
Departamento de Química
Universidad Nacional de Colombia-Sede Bogotá
Bogotá, Colombia

Hiral Gohil
Soil and Water Science Department
University of Florida
Gainesville, Florida

Juhi Gupta
School of Environmental Sciences
Jawaharlal Nehru University
New Delhi, India

Shalini Gupta
School of Environment and Sustainable Development
Central University of Gujarat
Gujarat, India

Bhawana Pathak
School of Environment and Sustainable Development
Central University of Gujarat
Gujarat, India

Pooja Hirpara
P.G. Department of Microbiology & Biogas Research Centre
Faculty of Sciences & Applied Sciences
Gujarat Vidyapith
Gandhinagar, India

Chandrahas N. Khobragade
School of Life Sciences
Swami Ramanand Teerth Marathwada University
Nanded, India

P. Mahla
P.G. Department of Microbiology and Biogas Research Centre
Gujarat Vidyapith
Gandhinagar, India

Juan Carlos Moreno-Piraján
Departamento de Química
Universidad de los Andes
Bogotá, Colombia

Andrew Ogram
Soil and Water Science Department
University of Florida
Gainesville, Florida

Rashmi Paliwal
Institute of Environmental Studies
Kurukshetra University
Kurukshetra, India

Janmejay Pandey
Laboratory of Environmental Biotechnology
Institute of Microbial Technology (IMTECH-CSIR)
Chandigarh, India

Department of Biotechnology
School of Life Sciences
Central University of Rajasthan
Ajmer, India

Ashita Rai
School of Environment & Sustainable Development
Central University of Gujarat
Gandhinagar, India

Manviri Rani
Department of Chemistry
Malaviya National Institute of Technology
Jaipur, India

Ronal Orlando Serrano-Romero
Civil Engineering Program
Universidad Piloto de Colombia
Bogotá, Colombia

Madhushree M. Routh
School of Life Sciences
Swami Ramanand Teerth Marathwada University
Nanded, India

Abdul Barey Shah
Department of Environmental Science
Babasaheb Bhimrao Ambedkar University
Lucknow, India

Uma Shanker
Department of Chemistry
Dr. B. R. Ambedkar National Institute of Technology
Jalandhar, India

Rana Pratap Singh
Department of Environmental Science
Babasaheb Bhimrao Ambedkar University
Lucknow, India

Indu Shekhar Thakur
School of Environmental Sciences
Jawaharlal Nehru University
New Delhi, India

Suchita C. Warangkar
School of Life Sciences
Swami Ramanand Teerth Marathwada University
Nanded, India

1 The Microbial Degradation of DDT and Potential Remediation Strategies

Hiral Gohil and Andrew Ogram

CONTENTS

1.1 INTRODUCTION

DDT (1,1,1-trichloro-2,2-di(4-chlorophenyl)ethane) was first synthesized in 1874, and large-scale manufacturing began during World War II to protect troops from insect-mediated diseases. Postwar, it was released for the use of civilians (EPA, 1975), and by the 1950s it had become the highest-selling insecticide in the United States (Smith, 1991).

* Corresponding author: Email: aogram@ufl.edu

By the late 1960s, bioaccumulation of DDT and its primary metabolites, DDD (1-chloro-4-[2,2-dichloro-1-(4-chlorophenyl)ethyl]benzene) and DDE (1,1-*bis*-(4-chlorophenyl)-2,2-dichloroethene) (collectively known as DDx), in wildlife became a concern (Ratcliffe, 1967, 1979; Turusov, 2002). This created concerns about human safety because measurable quantities were detected not only in wildlife such as birds, fish, and mammals but also in soil and water (Carson, 1962; EPA, 1975). In the early 1970s, DDT was partially banned for routine use in many developed countries and only permitted for the control of emergency public health problems (Ratcliffe, 1967; WHO, 1979; Turusov, 2002). Even though it has been banned in the United States since the early 1970s, it is still nearly ubiquitous in the environment. Over 30 years after its ban, DDx were found at 305 of 441 hazardous waste sites in the United States (HazDat, 2002).

The occurrence of DDx in the environment continues to pose a threat, as they may bioaccumulate in fatty tissues of all life forms, resulting in higher concentrations in higher trophic levels (LeBlanc, 1995). DDx's recalcitrant nature, persistence, and toxicity have been recognized as a serious environmental and ecological threat (Dimond and Owen, 1996; Kunisue, 2003). Ecological and environmental concerns prompted inclusion of DDx in the U.S. EPA's National Priority List (NPL) (U.S. EPA, 2000) as one of the most recalcitrant and environmentally significant pollutants (EPA, 2002). This review will focus on the biochemistry and microbiology of DDx degradation and their application to bioremediation.

1.2 STRATEGIES FOR DDx ENVIRONMENTAL REMEDIATION

Various chemical, physical, and biological methods have been employed to remediate soils that have been contaminated with DDx. Conventional abiotic methods, such as excavation, solvent washing, soil inversion, incineration, use of surfactants, thermal desorption, microwave enhanced treatment, ultraviolet (UV) irradiation, and sulfuric acid treatment (Foght et al., 2001), have been used. Although such chemical and physical methods have been applied, they commonly damage the health of the soils and are expensive and labor and energy intensive. Bioremediation is an alternative approach that uses biological agents to remove or transform pollutants to a less harmful form. Although bioremediation may include a wide array of biological systems, we will restrict our focus to the roles played by microbes in the treatment of DDx.

The metabolic diversity of microorganisms empowers them to use an extensive array of compounds for growth. We can exploit this capability to promote biologically mediated remediation. Such methods include the use of simple methods, such as the addition of various amendments to promote cometabolization and stimulation of the indigenous degrading consortia, if present. Amendments ranging from simple carbon compounds, such as glucose, yeast extract, peptone (Chacko et al., 1966), glucose, diphenylmethane (Pfaender and Alexander, 1972), octane, hexadecane or glycerol (Golovleva and Skryabin, 1981), and cellulose (Castro and Yoshida, 1974), to complex amendments, such as alfalfa (Guenzi and Beard, 1968; Burge, 1971), rice straw, and cellulose (Castro and Yoshida, 1974), have been shown to increase the bioremediation potential for DDx. Since soil microbes have been previously shown to cometabolize DDx (Focht and Alexander, 1971; Subba-Rao and Alexander, 1985; Bumpus and Aust, 1987), an alternative carbon source may be important to determine the biodegradation potential.

One approach to bioremediation is bioaugmentation. This method requires large numbers of an active population to create a measurable change, and may require multiple inoculations (Morgan and Watkinson, 1989). Methods have been developed to improvise and ensure survival of the active degraders, which include encapsulating bacteria (Chen and Mulchandani, 1998) or fungi (Leštan et al., 1996). An additional concern, as suggested by Lindqvist and Enfield (1992), is adsorption of DDx to the introduced microbes, which could promote mobility of the contaminants. An additional concern is that the active population must be in sufficient amounts to survive potentially adverse environmental conditions and compete with the indigenous soil population. Research is also required to select potential degraders, such that this area requires extensive research before application.

Analog induction is another approach to remediation in which either nontoxic chemicals or natural amendments with structural similarity to the contaminant are added to the system to increase biodegradation potential. This approach is based on the growth of degraders in the presence of structurally similar chemicals that could induce production of DDx-degrading enzymes. Pfaender and Alexander (1972) reported increases in DDT metabolism with the use of diphenylmethane as an analog. Another group (Beunink and Rehm, 1988) employed the same analog to enrich DDx-metabolizing microorganisms in sewage sludge. Biphenyl was used to induce the production of DDx-degrading enzymes by Hay and Focht (1998) and Aislabie and coworkers (1999), whereas another group (Nadeau et al., 1994) used chlorobiphenyl as an analog. A major concern with using analogs is the potential toxicity. Using natural analogs is an alternative to chemical analog induction. Natural substances with complex polyaromatic structures, including terpenes, are found in orange peels and pine needles and have been used for inducing biphenyl degradation (Hernandez, 1997).

According to Sayles and coworkers (1997), reducing agents, including zero valent iron, may be used as an adjunct to biodegradation to increase the conversion of DDT to DDD. Cysteine or sodium sulfide may be added as reducing agents to anoxic microcosms to produce redox conditions suitable for DDT degradation (You et al., 1995).

The great versatility of microbial metabolism makes bioremediation simpler, more environmentally friendly, and more cost-effective for environmental remediation (Jacques et al., 2008) compared with conventional physicochemical methods.

1.3 REASONS FOR RECALCITRANCE OF DDxs

A major reason for the recalcitrance of DDT is the chlorine substitutions, since its nonhalogenated analog diphenylmethane is readily degradable (Focht and Alexander, 1971; Subba-Rao and Alexander, 1985; Hay and Focht, 1998; Juhasz and Naidu, 2000). Halogenated aromatic compounds such as DDx are typically resistant to microbial attack because the halogen atom is larger than the carbon and hydrogen atoms on the DDx molecule and withdraws more electrons. Chlorine is an electronegative substituent, and when present on an aromatic ring lowers the aromaticity by shifting the electron density toward itself and hence alters resonance across the ring. As a result, chlorine makes the aromatic ring less susceptible to oxidative attack. Furthermore, because of the electronegativity and the size of chlorine atom, it can have a detrimental electronic effect and steric hindrances for enzyme-mediated degradation reactions (Crooks and Copley, 1993; Sylvestre and Sandossi, 1994). However, the great metabolic diversity of soil microorganisms enables them to overcome many of the obstacles and degrade haloaromatics such as DDxs.

1.4 BIOAVAILABILITY

The term "bioavailability" refers to the available or accessible fraction of a substrate to biological processes (Juhasz et al., 1999; Foght et al., 2001). Although bioremediation is a promising approach, accessibility of the contaminant to microbes may be a major limiting factor contributing to the recalcitrance of most hydrophobic organic compounds (HOCs) such as DDx (Hunt and Sitar, 1988). Bioavailability may limit remediation by physically sequestering and hence protecting the pollutant from microbial attack (Alexander, 1995; 1997). After entry into soils, HOCs may rapidly combine with the organic matter, and a combined effect of diffusion and sequestration to inaccessible sites may lead to tightly binding to and within soil particles (Weissenfels et al., 1992).

With time, the sorbed residues may become resistant to chemical extraction, such that they are considered to be tightly bound (Alexander, 2000). Bioavailability of a contaminant depends on a range of factors, such as the chemical and physical properties of the soil, chemical nature of the contaminant, duration of contact, and microbes present (Juhasz et al., 1999). Soils containing high amounts of organic matter therefore have a higher proportion of sorbed DDxs as compared to sandy soils (Peterson et al., 1971; Castro and Yoshida, 1974; Vollner and Klotz, 1994).

The accessibility of a contaminant to degrading microbial populations may be increased by physical dispersion of soils and employing various techniques, such as the use of surface active agents, cosolvents, surfactants, and monovalent cations, which have all been successfully employed (Keller and Rickabaugh, 1992; You et al., 1995; Sayles et al., 1997; Juhasz et al., 1999; Kantachote et al., 2001; 2004). Approaches such as the addition of surfactants have been employed to increase desorption of DDx from soil particles, thereby increasing bioavailability (Keller and Rickabaugh, 1992).

1.5 AEROBIC DEHALOGENATION AND RING CLEAVAGE

Aerobic microorganisms use oxidative catabolic reactions to break down halogenated aromatic compounds (Nadeau et al., 1994; Aislabie et al., 1999; Hay and Focht, 1998), whereas anaerobic microbes use reductive dehalogenation (Suflita et al., 1982; Quensen et al., 1990). Oxidative attack is usually a two-step process—although the intermediates may vary depending on the parent compounds, the general strategy for oxidative attack remains similar. The first step typically involves activation of the aromatic ring, and the second step is ring fission (Dagley, 1971).

In the case of oxidative ring cleavage, the initial strategy is replacement of substituents from the aromatic ring with hydroxyl groups, which facilitates oxidative cleavage of the aromatic ring. Molecular oxygen serves as a reactant, and mono- or dioxygenases introduce either one or both atoms from molecular oxygen into the substrate. Such enzymatic oxidation reactions introduce oxygen in the ring of aromatic substrates to form dihydrodiols. The continued oxidation of dihydrodiols leads to the formation of catechols, which may serve as substrates for other dioxygenases to form ring cleavage products (Juhasz and Naidu, 2000) by either entering the gentisate pathway or ortho- or meta-cleavage degradation pathways (Dagley, 1977; Harayama and Timmis, 1989; Schink et al., 1992, Fuchs et al., 1994). For example, Nadeau and colleagues (1994) demonstrated aerobic degradation of DDT through a meta-cleavage pathway.

1.6 ANAEROBIC DEHALOGENATION AND RING CLEAVAGE

Anaerobic systems cannot rely on oxygenases, such that anoxic degradation frequently involves reductive dechlorination, a process that involves replacement of a chlorine with a hydrogen atom (Suflita et al., 1982; Quensen et al., 1990). Successive reductive dechlorination reactions decrease the number of chlorines on the aromatic molecule, decreasing resonance, and hence making it easier to degrade. The primary reaction for most aromatic ring cleavage under anaerobic conditions is reduction of the ring into different central metabolic pathway intermediates, e.g. CoA thioesters (e.g. benzoyl CoA). This leads to ring saturation and loss of resonance, followed by ring cleavage, which is eventually incorporated into the tricarboxylic acid cycle (Heider and Fuchs, 1997).

1.7 PROPOSED DDx DEGRADATION PATHWAYS

The initial attack on DDT is on the trichloroalkyl backbone, after which the molecule is typically converted to DDD under anaerobic, and DDE under aerobic, conditions. Although DDT degradation has occurred at sites contaminated with DDT, the subsequent metabolites (DDD and DDE) accumulate and are limited to further degradation (Aislabie, 1997). Degradation rates and concentrations of transformation products depend on the soil conditions, water content, and the microbes present in the soil (Aislabie, 1997). Under anaerobic conditions, DDD is produced from DDT by reductive dechlorination. This process may proceed by either chemical or biological reduction (Zoro et al., 1974; Baxter, 1990). Under aerobic conditions, DDE is formed from DDT via dehydrodechlorination (Pfaender and Alexander, 1972). It should be noted that it is difficult to account for all DDT metabolites in complex environmental systems (Guenzi and Beard, 1967; Burge, 1971; Jensen et al., 1972). The major microbial biotransformation products of DDT are presented in Table 1.1.

TABLE 1.1
Biotransformation Pathways and Products of DDT

Organism	Transformation Products	Mechanism	Source	Reference
Proteus vulgaris	DDT → DDD	Reductive dechlorination	Mouse intestine	Barker et al., 1965
Escherichia coli *Enterobacter*	DDT → DDD	Reductive dechlorination	Rat feces	Mendel and Walton, 1966
Pseudomonas aeruginosa *Bacillus* sp. *Flavobacterium*	DDT → DDD	Reductive dechlorination	Activated sludge	Sharma et al., 1987
Enterobacter cloacae	DDT → DDD	Reductive dechlorination	Sewage sludge	Beunink and Rehm, 1988
Bacillus sp.	DDT → DDD	Reductive dechlorination	Soil	Katayama et al., 1993
Cyanobacteria	DDT → DDD	Reductive dechlorination	Soil	Megharaj et al., 2000
Enterobacter aerogenes	1. DDT → DDD, DDMU, DDMS, DDNU, DDOH, DDA, and DBP 2. DDT → DDE	Anaerobic pathway	Not mentioned	Wedemeyer, 1967
Bacillus sp., *E. coli*, *E. aerogenes*	1. DDT → DDD, DDMU, DDMS, DDNU, DDOH, DDA, and DBP 2. DDT → DDE	Anaerobic pathway	Not mentioned	Langlois et al., 1970
Pseudomonas isolated as Hydrogenomonas	^{14}C DDT → DDD, DDMS, DDNU, and DBP	Anaerobic pathway by cell-free extracts	Sewage	Pfaender and Alexander, 1972
P. aeruginosa 640X	DDT → ring cleavage metabolites	Reductive dechlorination	Soil	Golovleva and Skryabin, 1981
Strain B-206	DDT → DDE, DDD, DDMU to hydroxylated metabolites	Transformation of DDT to hydroxylated metabolites	Activated sludge	Masse et al., 1989
Ralstonia eutropha	Meta cleavage	Meta cleavage	Soil	Nadeau et al., 1994; 1998
Alcaligens sp. *JB1*	Meta cleavage	Meta cleavage	Soil	Parsons et al., 1995
Pseudomonas acidovorans M3GY	Meta cleavage of DDE	Meta cleavage of DDE	Genetically engineered	Hay and Focht, 1998
Terrabacter sp. Strain DDE1	Meta cleavage of DDE	Meta cleavage of DDE	Soil	Aislabie et al., 1999
R. eutropha A5	Meta cleavage of DDD	Meta cleavage of DDD	Soil	Hay and Focht, 2000

It was believed that, although anoxic conditions contribute to the successive reductive dechlorination of the aliphatic moiety of DDT, the presence of oxygen is required for efficient ring cleavage and mineralization (Pfaender and Alexander, 1972; Golovleva and Skryabin, 1981). However, anaerobic ring cleavage of several aromatic compounds was reviewed by Heider and Fuchs in 1997. Another concern was that anoxic conditions are better for dehalogenation of polyaromatic hydrocarbons (PAHs); however, Nadeau and colleagues (1997) demonstrated aerobic degradation of DDT. We believe that anaerobic conditions would not only lead to reductive dechlorination of the halogens on the aliphatic backbone of DDT but also to ring cleavage.

1.8 ANAEROBIC DEGRADATION PATHWAY

DDT degradation under anaerobic conditions is thought to follow a pathway proposed by Wedemeyer (1967), shown in Figure 1.1 (adapted from the University of Minnesota Biocatalysis/Biodegradation Database [UMBBD] website: http://umbbd.msi.umn.edu/ddt2/ddt2_image_map.html). The anaerobic degradation of DDT is proposed to involve reductive dechlorination to form DDD (Johnsen, 1976; Essac and Matsumura, 1980; Lal and Saxena, 1982; Rochkind-Dubinsky et al., 1987; Kuhn and Suflita, 1989).

One of the first reports demonstrating the reduction of DDT to DDD was a pure culture study with *Proteus vulgaris* isolated from a rodent intestine (Barker et al., 1965). Other pure culture studies also demonstrated DDT reduction under anoxic conditions (Table 1.1). Under anaerobic conditions, DDT may be reduced to DDD and successive reductions from DDD to DDMU (1,1-di(p-chlorophenyl)-2-chloroethylene), DDMS 1,1'-(2-chloroethane-1,1-diyl)bis(4-chlorobenzene), DDNU 1-chloro-4-[1-(4-chlorophenyl)ethenyl]benzene, DDOH 2,2,2-trichloro-1,1-bis(4-chlorophenyl)ethanol), DDA 2,2-bis(4-chlorophenyl)acetic acid, and DBP bis(4-chlorophenyl)methanone by pure cultures of *Escherichia coli* and *Enterobacter aerogenes* (Wedemeyer, 1967; Langlois et al., 1970; Pfaender and Alexander, 1972). DDT undergoes successive reductive dechlorinations, forming DDD, DDMU, DDMS, and DDNU. DDNU is oxidized to form DDOH, which is further oxidized to DDA. DDA undergoes decarboxylation, producing DDM, which may go to ring cleavage of one of the rings under aerobic conditions to form PCPA (4-chlorophenyl) acetic acid (Pfaender and Alexander, 1972). DBP accumulates under anaerobic conditions (Pfaender and Alexander, 1972).

It was thought that the major bacterial dechlorination was efficient under reducing conditions, whereas oxidative attacks were required for ring cleavage (Pfaender and Alexander, 1972; Golovleva and Skryabin, 1981). Pfaender and Alexander (1972) demonstrated that cell-free extracts of *Hydrogenomonas* species converted DDT to DDD, DDMS, DBP, and other products. However, no ring cleavage products were formed under anoxic conditions. Incorporating oxygen and fresh cells of *Hydrogenomonas* resulted in further metabolism of DDT to PCPA, implying that the enzyme system of a single organism could break down DDT to ring fission products. PCPA could, however, be further metabolized by

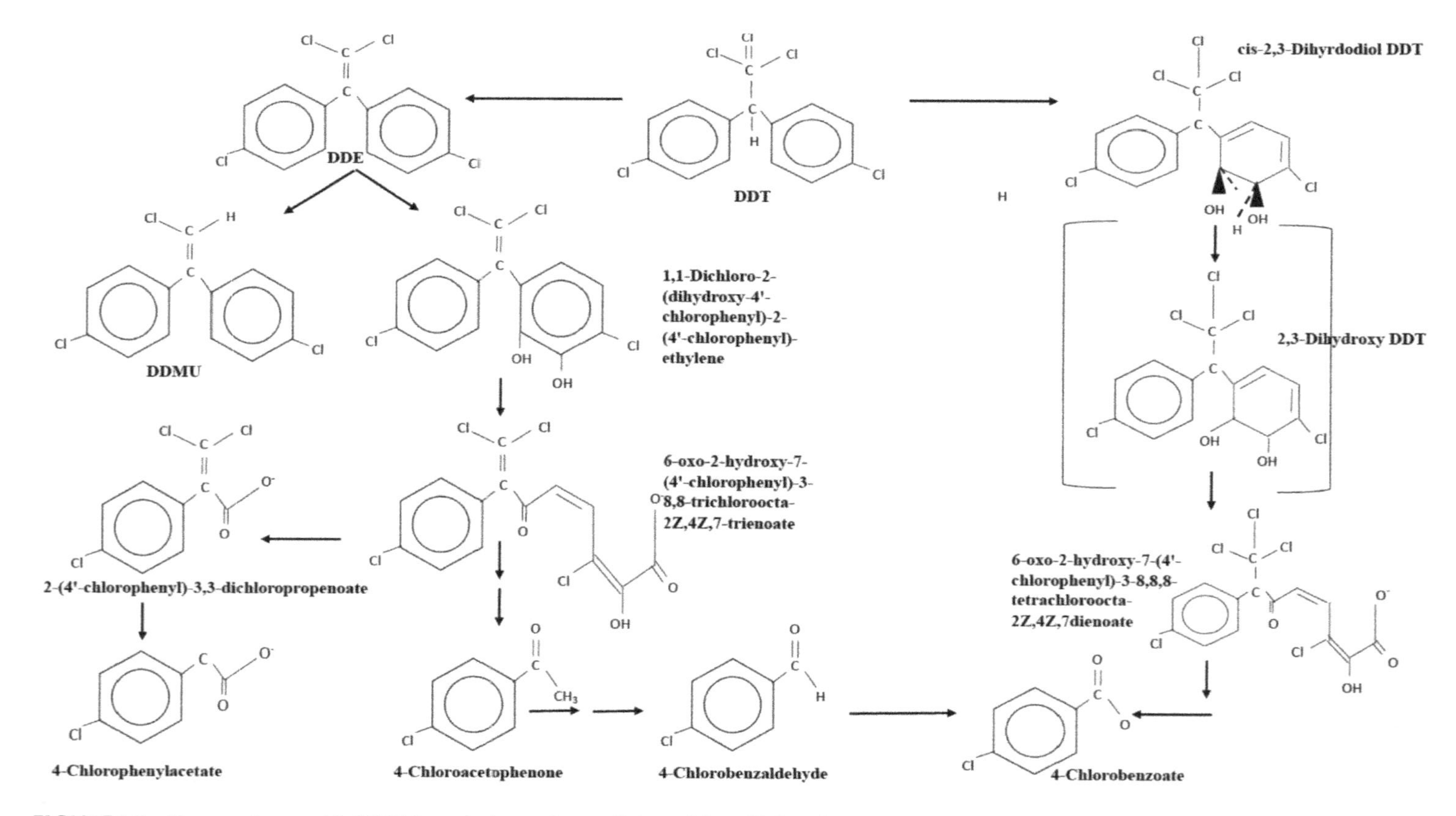

FIGURE 1.1 Proposed anaerobic DDT degradation pathway. (Adapted from University of Minnesota Biocatalysis/Biodegradation Database (UMBBD) website: http://umbbd.msi.umn.edu/ddt2/ddt2_image_map.html.)

Arthrobacter species. No PCPA was formed under anaerobic conditions, which suggested that aerobic conditions are required for ring cleavage (Pfaender and Alexander, 1972).

1.9 PROPOSED AEROBIC DDT DEGRADATION PATHWAY

The pathway for aerobic degradation of DDT is presented in Figure 1.2. In aerobic systems, DDT undergoes dehydrochlorination, which involves the replacement of HCl with a double bond between the carbon atoms on the alkyl chain of DDT. The first report of degradation of DDT under aerobic conditions was presented by Nadeau (Nadeau et al., 1994). The same group obtained the isolates by analog enrichment. Diphenylmethane was used as a primer to enrich and isolate responsible organisms. The first step in the pathway by *Ralstonia eutropha* A5 (previously named *Alcaligenes eutrophus*) is oxidation of DDT at the ortho and meta positions by a dioxygenase to form DDT-dihydrodiol. DDT-dihydrodiol is a transient metabolite that undergoes dehydration by dehydrogenase to form 2,3-dihydroxy-DDT. 2,3-dihydroxy-DDT undergoes meta cleavage, forming a yellow-colored metabolite, finally leading to formation of 4-chlorobenzoic acid. Similarly, using 4-chlorobiphenyl for analog enrichment, other organisms capable of cometabolizing DDT were isolated (Masse et al., 1989; Nadeau et al., 1994; Parsons et al., 1995). *R. eutropha* A5 also metabolizes DDD via meta-fission (Hay and Focht, 2000) and other aerobic bacteria namely *Terrabacter* species. DDE1 (Aislabie et al., 1999) and *Pseudomonas acidovorans M3GY* (Hay and Focht, 1998) metabolize DDE via meta-fission.

1.10 DEHALORESPIRATION

As mentioned previously, anaerobic reductive dehalogenation replaces one halogen atom with a hydrogen atom. Organisms that undertake such reactions may gain energy from reductive dechlorination via a process termed dehalorespiration (Suflita et al., 1982; Dolfing and Harrison, 1992; Mackiewicz and Wiegel, 1998). Shelton and Tiedje (1984) enriched a methanogenic consortium that they claimed was growing on 3-chlorobenzoate as the sole energy and carbon source. Later efforts to isolate the culture solely on 3-chlorobenzoate failed. These observations suggested that the consortium could be gaining energy from the substrate (Shelton and Tiedje, 1984). Dolfing (1990) demonstrated that energy was conserved via anaerobic reductive dechlorination of 3-chlorobenzoate. Further studies by Mohn and Tiedje (1990; 1991) substantiated this finding.

Biologically diverse environments harbor microbial communities with a broad range of enzymes with broad substrate specificities. This helps organisms use a variety of substrates, including xenobiotics, for anabolic metabolism or for energy generation. Generally, microbes could use organic or inorganic compounds as an electron donor or as the terminal electron acceptor (TEA). As electron donor, pollutants are oxidized and the electrons pass through a series of redox reactions, finally reducing the TEA.

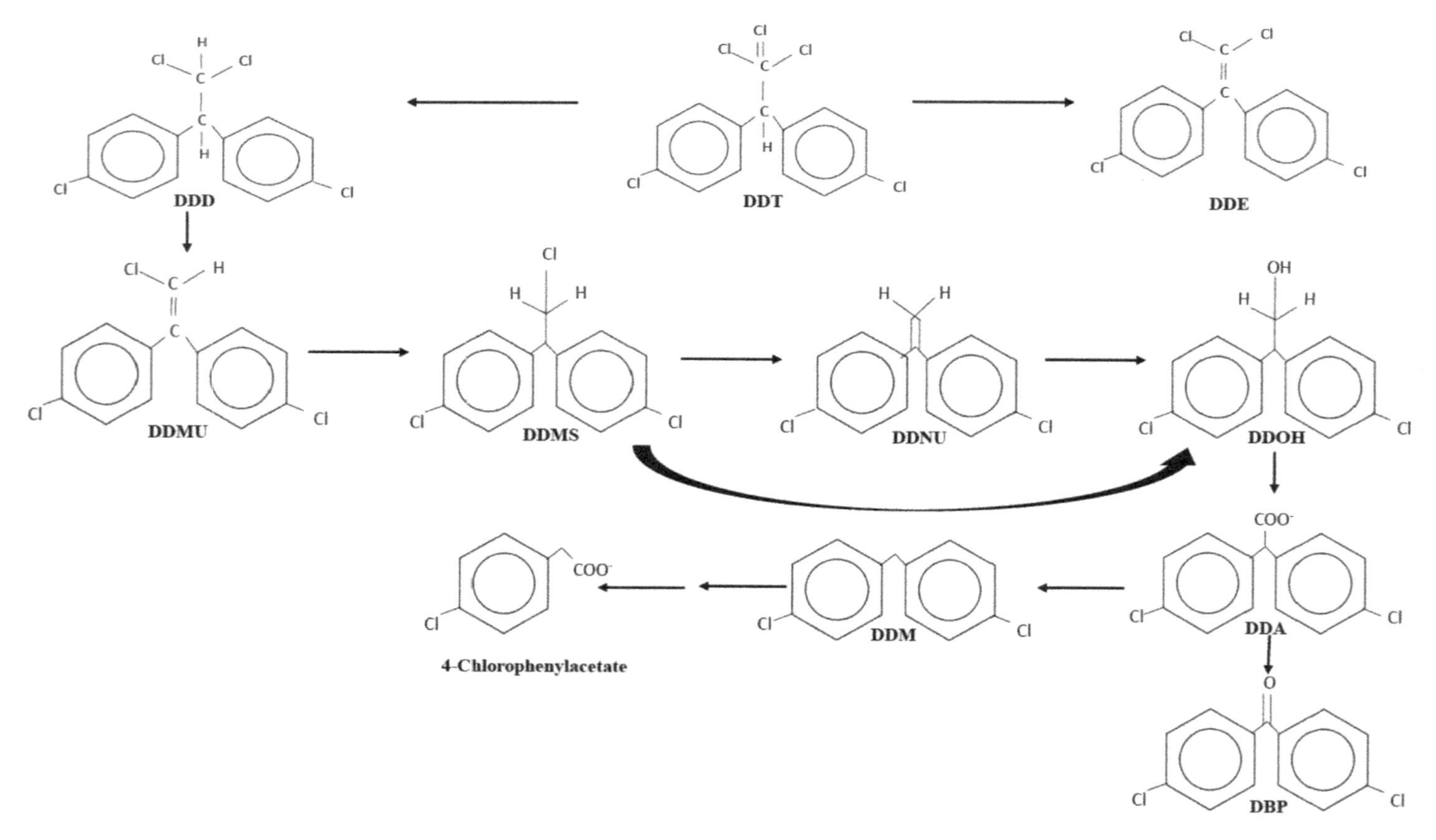

FIGURE 1.2 Proposed aerobic DDT degradation pathway. (Adapted from UMBBD website: http://umbbd.msi.umn.edu/ddt2/ddt2_image_map.html.)

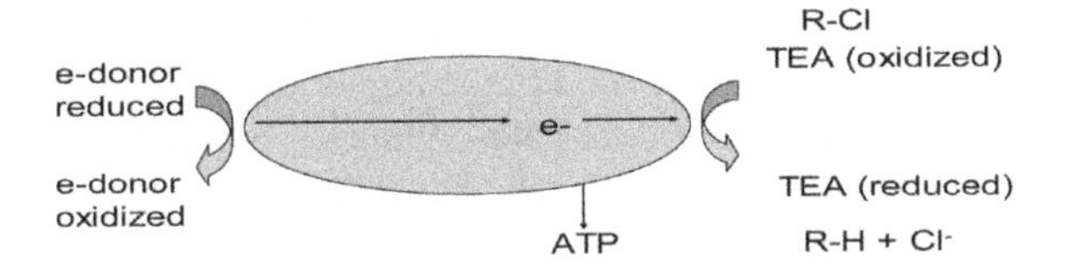

FIGURE 1.3 Reductive dehalogenation linked to ATP generation. (Adapted from Häggblom and Bossert, 2003.)

Conversely, while using such pollutants as TEA, energy is obtained by passing electrons to the pollutant, and hence reducing it (Figure 1.3). The halo-organic compound is used as a TEA and energy is conserved (Holliger et al., 1999; Smidt and de Vos, 2004). Energy is gained by a chemiosmotic gradient along the cell membrane, resulting in a proton motive force (PMF) which drives a membrane-bound ATPase to generate ATP (Mohn and Tiedje, 1991). Such a process can be stimulated at the field level, but requires selective stimulation of desirable organisms by the introduction of specific combinations of electron donors and acceptors (Suflita et al., 1988), as well as nutrients to meet the growth requirements of the enriched species. Hence, the choice of an electron donor and acceptor combination is crucial for the success of such processes.

1.11 THERMODYNAMIC CONSIDERATIONS AND PHYSIOLOGY FOR DEHALORESPIRATION

Thermodynamics of a reaction estimate the amount of energy obtained per reaction and control whether the reaction could sustain growth of the cell. One of the prerequisites for halo-organics to support growth is that the reaction they drive must be exergonic. A negative ΔG indicates that the reaction may be favorable for energy generation by microbes (Thauer et al., 1977).

Dolfing and Harrison (1992) showed that the potential of halogenated aromatics for use as TEA in reduced environments. ΔG values calculated by their group for various halo-organics ranged from −131.3 to −192.6 kJ/mol (Dolfing and Harrison, 1992). Reductive dehalogenation of halo-organics could yield a ΔG between −130 and −180 kJ/mol of halogen removed (Huang et al., 1996; Smidt and de Vos, 2004). This ensures that reductive dehalogenation is an exergonic and, hence, thermodynamically favorable process that enables the dehalorespiring bacteria to couple reductive dehalogenation to growth (Smidt and de Vos, 2004). Another example is ΔG value for dehalogenation of 1,2,3,4-tetrachlorodibenzo-p-dioxin with H_2 as electron donor: $\Delta G = -497$ kJ/mol (Huang et al., 1996; Dolfing, 2003). This suggests that hydrogenotrophic reductive dehalogenation of polychlorinated dioxins is sufficient to support the growth and survival of dehalorespiring organisms. These ΔG values were calculated for standard conditions, such that some reactions may not be exergonic or may not provide sufficient energy for growth under environmental conditions.

Redox potential (Eh) is related to the tendency of a substance to gain electrons compared to H_2. The more positive the Eh, the higher the affinity for electrons and the higher the energy generation. $E^{\circ\prime}$ for halo-organics is relatively higher than bicarbonate (HCO_3^-/CH_4) and SO_4^{2-}/H_2S, such that dehalorespiring organisms would be expected to outcompete methanogens or sulfate reducing bacteria (SRB) when reducing equivalents are limiting (Apajalahti and Salkinoja-Salonen, 1987).

Under anaerobic conditions, TEAs are frequently limiting and are competed for by a range of microbial processes. Hence, TEAs under anaerobic conditions will affect the community composition. Therefore, higher energy-generating TEAs would be preferred over lower energy-generating TEAs. Susaria and coworkers (1996) hypothesized that microbially mediated dehalorespiration sequentially uses higher to lower energy-yielding TEAs. To test this, they conducted anaerobic dechlorination experiments and demonstrated that dechlorination was a preferred reaction, with higher redox potential TEAs being used first. TEAs with highest reduction potentials were used preferentially, followed by the lower energy generating TEAs in the order $O_2 > NO_3^- >$ halo-organics $> SO_4^{2-} > HCO_3^-$. This indicates that dehalorespiration is feasible thermodynamically when compared to sulfate reduction, acetogenesis, or methanogenesis. One could predict from the redox potentials that such processes occur under anoxic environments (Dolfing, 2003) and that halo-organic compounds are important TEAs.

1.12 DIVERSITY OF BACTERIAL DEHALORESPIRERS

All of the dehalorespiring organisms known to date belong to the domain Bacteria and fall into three phylogenetic branches; the Firmicutes, Proteobacteria subgroups ε, α, or δ, and Chloroflexi (Holliger et al., 2003; Hiraishi, 2008). Electron donors used by dehalorespiring bacteria can range from electron-rich H_2 to organic acids, including formate, pyruvate, acetate, lactate, and butyrate (Table 1.2). Dehalorespiring populations are quite versatile in their ability to use electron donors and TEAs.

1.13 ELECTRON DONORS

There is competition for electron donors between dehalorespirers and other anaerobic microbes in the soil. Dehalorespiring organisms can use a broad range of electron donors (Table 1.2). H_2 is an important electron donor in dehalogenation and is used by a wide range of dehalorespiring organisms, as well as syntrophs (Schink, 1997). As stated before, dehalorespiring organisms can outcompete hydrogenotrophic methanogens, homoacetogens, and SRB based on the thermodynamic gains (Ballapragada et al., 1997; Fennell and Gossett, 1998).

1.14 FERMENTATIVE DEHALOGENATION OR SYNTROPHIC DEHALOGENATION

Syntrophy is a symbiotic relationship between two or more metabolically different bacteria that together may degrade a substance and satisfy the energy needs of the organisms (Atlas and Bartha, 1993). Organisms involved in this process can

TABLE 1.2
Phylogeny and Properties of Dehalorespiring Bacteria

Name	Dechlorinated Compound	Electron Donor	Phylogeny	References
Desulfomonile tiedjei	PCE, TCE, H_2, 3-chlorobenzoate, Pentachlorophenol	H_2, formate, pyruvate	Gram negative, SRB belong to δ-Proteobacteria	DeWeerd et al., 1991
Isolate 2CP-1	2-chlorophenol, 2,6-dichlorophenol	Acetate, formate, yeast extract	Facultative anaerobe, gram negative, δ-Proteobacteria	Cole et al., 1994
Desulfitobacterium chlororespirans	2,4,6-trichloro phenol, 3-Cl 4-OH phenylacetate	H_2, formate, pyruvate, lactate, butyrate, crotonate	Low G+C, gram positive	Sanford et al., 1996
Desulfitobacterium frappieri	2,4,6-trichloro phenol, 3-Cl 4-OH phenylacetate	Pyruvate	Low G+C, gram positive	Bouchard et al., 1996
Desulfitobacterium dehalogenans	PCE, 2,4, 6-trichlorophenol	H_2, formate, pyruvate	Low G+C, gram positive	Utkin et al., 1994
Desulfitobacterium strain *PCE 1*	3-Cl-4-OH-phenylacetate	Formate, pyruvate	Low G+C, gram positive	Gerritse et al., 1996
Dehalobacter restrictus	PCE, TCE	H_2	Gram positive, affiliated with desulfitobacterium, derive energy by dehalorespiration	Holliger et al., 1998
Isolate *TEA*	PCE, TCE	H_2	Affiliated with desulfitobacterium derive energy by dehalorespiration	Wild et al., 1996
Dehalospirillum multivorans	PCE, TCE	H_2, formate, pyruvate	Gram negative belong to δ-Proteobacteria	Scholz-Muramatsu et al., 1995
Desulfuromonas chloroethenica	PCE, TCE	Acetate, pyruvate	Gram negative, SRB	Krumholz et al., 1996
Dehalococcoides ethenogenes	PCE, TCE, DCE, chloroethenes	H_2	Eubacterium, only aerobe isolated so far with energy conservation though dehalorespiration and known to dechlorinate PCE all the way to ethene	Maymo-Gatell et al., 1997
Enterobacter strain *MS-1*	PCE, TCE	Formate, pyruvate, acetate	Facultative anaerobe, gram negative γ-Proteobacteria, energy conservation with dehalorespiration	Sharma and McCarty, 1996

(*Continued*)

TABLE 1.2 (*Continued*)
Phylogeny and Properties of Dehalorespiring Bacteria

Name	Dechlorinated Compound	Electron Donor	Phylogeny	References
Dehalococcoides ethenogenes strain *195*	PCE, TCE, cis-DCE,1,1-DCE, 1,2-DCA, VC	H_2	Green non-sulfur bacteria (*Chloroflexi*)	Maymo-Gatell et al., 1997; 1999
Dehalococcoides strain *CBDB1*	1,2,3-TCB, 1,2,4-TCB, 1,2,3,4-TeCB, 1,2,3,5-TeCB, 1,2,4,5-TeCB	H_2	Green non-sulfur bacteria (*Chloroflexi*)	Adrian et al., 2000

Note: *PCE = tetrachloroethene; TCE = trichloroethene; DCE = dichloroethene; TCB = trichlorobenzene; VC = vinyl chloride; DCB = dichlorobenzene; TeCB = tetrachlorobenzene; DCA = dichloroethane.*

together perform a reaction which neither can perform individually. Syntrophy plays an important role in reductive dehalogenation. Syntrophs have been previously shown to dehalogenate (Mohn and Tiedje, 1991; Yang and McCarty, 1998). Dehalorespiring populations participate in close syntrophic associations with hydrogen-producing organisms, which require low H_2 concentrations for energy gain (Drzyzga and Gottschal, 2002). Dehalorespirers could keep H_2 concentrations sufficiently low to allow growth of obligate syntrophs (Yang and McCarty, 1998; Dolfing, 2003; Sung et al., 2003).

1.15 PHYLOGENY OF DEHALORESPIRING POPULATIONS

Dehalorespiring organisms, because of their wide metabolic diversity, have been isolated from several different environments. As mentioned previously, all dehalorespirers described to date fall within the domain Bacteria and into three major phyla: Chloroflexi, Firmicutes, and Proteobacteria (Table 1.2) (Holliger et al., 2003; Hiraishi, 2008). *Desulfomonile tiedjei* DCB1 of the δ-Proteobacteria was the first organism described to conserve energy by dehalorespiration on 3-chlorobenzoate (Suflita et al., 1982).

Although dehalorespiration of haloaliphatics has been studied extensively, few reports are available regarding dehalorespiration on haloaromatic compounds. Reductive dehalogenation of polychlorinated biphenyl (PCB) has only been reported to date for a pure culture of *Dehalococcoides* (Yan et al., 2006; Baba et al., 2007) which belongs to the phylum Firmicutes. Based on their metabolic properties, dehalorespirers are further divided into two physiological types, the facultative and obligate dehalorespiring bacteria (Table 1.3).

TABLE 1.3
Phylogeny of Facultative and Obligate Dehalorespiring Bacteria

Genus	Species/Strain	Dechlorination Substrate or TEA	Microbiology and Physiology	Reference
Anaeromyxobacter	*Anaeromyxobacter dehalogens*	2-bromophenol, nitrate, fumarate, oxygen	Facultative anaerobic dehalorespiring myxobacteria, δ-Proteobacteria	Sanford et al., 2002
Desulfomonile	*Desulfomonile tiedje DCB-1*	3-chlorobenzoate, fumarate, sulfate, sulfite, thiosulfate, nitrate	Facultative dehalorespirers, δ-Proteobacteria	DeWeerd et al., 1990
	Desulfomonile limimaris	3-chlorobenzoate	Facultative dehalorespirer, δ-Proteobacteria	Sun et al., 2001
Desulfovibrio	*Desulfovibrio dechloroacetivorans*	2-chlorophenol	Facultative dehalorespirers belonging to group SRB, acetate oxidizers, δ-Proteobacteria	Sun et al., 2001
Desulfuromonas	*Desulfuromonas acetooxidans*	PCE, TCE	Facultative dehalorespirers belonging to anaerobic group of SRB and acetate oxidizing bacteria, δ-Proteobacteria	Pfennig and Biebl, 1976
	Desulfuromonas michiganesis strain BB1 and strain BRS1	PCE, TCE, fumarate, ferric iron	Facultative dehalorespirers, acetate oxidizers, δ-Proteobacteria	Sung et al., 2003
Geobacter	*Geobacter lovleyi strain SZ*	Metals, PCE	Facultative dehalorespirers metal reducers, δ-Proteobacteria	Sung et al., 2006
	Trichlorobacter thiogenes	Trichloroacetic acid	Facultative dehalorespirers, δ-Proteobacteria metal reducers	Nevin et al., 2007
Sulfurospirillum	*Sulfurospirillum halorespirans*	PCE to cis-DCE	Facultative dehalorespirers, ε-Proteobacteria	Luijten et al., 2003
Desulfitobacterium	*Desulfitobacterium dehalogenans*	2,4-dichlorophenol, hydroxylated PCBs and chloroalkenes	Facultative dehalorespirers, phylum Firmicutes	Utkin et al., 1994; Wiegel et al., 1999
	Desulfitobacterium sp. strain PCE1, PCE-S, Viet1	TCE (trichloro ethane) or cis-DCE (tetrachloro ethane)	Facultative dehalorespiring, low G+C, gram positive, phylum Firmicutes	Miller et al., 1997; Löffler et al., 1997; Gerritse et al., 1996; 1999
	Desulfitobacterium chlororespirans	3-chloro-4-hydroxybenzoate	Anaerobic lactate oxidizing organism, phylum Firmicutes	Sanford et al., 1996
	Desulfitobacterium hafniense	Pentachlorophenol, PCE, 2,4,6-trichlorophenol	Facultative dehalorespirer, phylum Firmicutes	Bouchard et al., 1996; Lanthier et al., 2000; Breitenstein et al., 2001.

1.16 FACULTATIVE DEHALORESPIRING ORGANISMS

Facultative dehalorespiring organisms are very versatile in their choice of electron donor and acceptor (Tables 1.3). Organisms falling under this physiological group belong to either the δ- or ε-Proteobacteria, or the phylum Firmicutes. *Desulfovibrio* and *Desulfuromonas* contain facultative dehalorespiring members. Representatives of the former dehalogenate 2-chlorphenol (Sun et al., 2002), whereas representatives of the latter utilize tetrachloroethene (PCE) and trichloroethene (TCE) (Krumholz, 1997; Sung et al., 2003). The genus *Geobacter* contains metal reducing members and belongs to the δ-Proteobacteria; *Geobacter lovleyi* is a metal reducer that is also capable of dehalorespiration (Sung et al., 2006). *Desulfitobacterium* belongs to the phylum Firmicutes, and most of the strains from this genus have been isolated with either PCE or TCE when grown with H_2 as an electron donor (Table 1.2). *Desulfitobacterium* isolates are so named as the first pure culture isolated grew on sulfite as TEA (Utkin et al., 1994) and most of them could repeat this as well (PCE1, PCE-S, Viet1). Being versatile with their choice of electron donor or TEA, some strains of *Desulfitobacterium* use chlorinated phenolic compounds and chloroethenes, whereas others could use either of the two. This is in accordance with other studies showing that different enzyme systems work with chloroaryl and chloroalkyl reduction, wherein respective substrates induce each enzyme system (Gerritse et al., 1999).

As presented in Table 1.2, the facultative dehalorespiring organisms have adapted very well with respect to the array of utilizable electron donors and acceptors. Little information is available regarding use of PCBs by dehalorespirers, with the exception of *Desulfitobacterium dehalogenans,* which has been reported to dehalogenate hydroxylated PCBs (Wiegel et al., 1999).

1.17 CONCLUSION

DDT and its major metabolites (DDE and DDD) accumulate in the environment, particularly in anoxic sediments, due to their low bioavailability and stable molecular structures. A variety of biochemical strategies to initiate attack of the molecules have been described, including oxygenation in aerobic systems and reductive dehalogenation of the trihalo backbone of the molecule in aerobic systems. If reductive dehalogenation precedes complete mineralization of DDT in anaerobic systems, stimulation of the appropriate microbial consortia would aid in the optimization of bioremediation of DDT in anoxic soils and sediments.

An understanding of the biochemistry and physiology of the microorganisms that participate in reductive dehalogenation, particularly those capable of respiratory dehalogenation (i.e., dehalorespiration), would be of value in designing remediation systems for anoxic systems. Coupling the appropriate electron donor with the appropriate acceptor is critical for anaerobic metabolism, and coupling the appropriate electron donor for dehalorespiration is critical for attacks on DDT using this strategy. It is also critical to identify potentially competing processes for those electron donors, such that a knowledge of the energy yield for DDT as a TEA compared with other processes, such as denitrification and sulfate reduction, is important.

Considering the relative energy yields from the different terminal electron accepting processes, it is likely that H_2 would be a valuable electron donor and that nitrate and sulfate would have to be maintained at low concentrations to suppress potential competition for that electron donor. These general hypotheses were tested in microcosms and larger mesocosm studies by Gohil et al. (2014), who found that H_2 and lactate were the optimum electron donors for stimulation of metabolism of DDT, DDD, and DDE in a muck soil. It is likely that lactate was fermented by syntrophs to H_2, which served as an electron donor for reductive dehalogenation and subsequent attack of the DDx.

More work is required to confidently identify the specific biochemical mechanisms at work in the biodegradation systems described by Gohil et al. (2014); however, dehalorespiration appears to have been an important step. A major limitation of biodegradation in many anoxic systems is availability of the chemical, which may be controlled by sorption to organic carbon. Much work remains to be done in this area.

REFERENCES

Adrian, L., Szewzyk U., Wecke J., Görisch H. 2000. Bacterial dehalorespiration with chlorinated benzenes. *Nature.* **408**:580–583.

Aislabie, J., Davidson A. D., Boul H. L., Franzmann P. D., Jardine D. R., Karuso P. 1999. Isolation of Terrabacter sp. strain DDE-1, which metabolizes 1,1,1-trichloro-2,2-bis-(4-chlorophenyl ethane) when induced with biphenyl. *Appl. Environ. Microbiol.* **65**:5607–5611.

Aislabie, J. M. 1997. Microbial degradation of DDT and its residues – a review. *New Zeal. J. Agr. Res.* **40**:269–282.

Alexander, J. M. 1995. How toxic are toxic chemicals in soil? *Environ. Sci. Technol.* **29**:2713–2717.

Alexander, M. 1997. Environmentally Acceptable Endpoints in Soil. In: Linz, D. G., Nakles, D. V. (eds.) American Academy of Environmental Engineers, Annapolis, MD.

Alexander, M. 2000. Aging, bioavailability, and overestimation of risk from environmental pollutants. *Environ. Sci. Technol.* **34**:4259–4265.

Apajalahti, J., Salkinoja-Salonen M. S. 1987. Complete dechlorination of tetrachlorohydroquinone by cell extracts of pentachlorophenol-induced Rhodococcus chlorophenolicus. *J. Bacteriol.* **169**:5125–5130.

Atlas, R. M., Bartha R. 1993. *Microbial Ecology: Fundamentals and application*, 3rd ed. Benjamin/Cummings Pub. Co., New York, NY.

Baba, D., Yasuta T., Yoshida N., Kimura Y., Miyake K., Inoue Y., Toyota K., Katayama A. 2007. Anaerobic biodegradation of polychlorinated biphenyls by a microbial consortium originated from uncontaminated paddy soil. *World J. Microbiol. Biotechnol.* **23**:1627–1636.

Ballapragada, B. S., Stensel H. D., Puhakka J. A., Ferguson J. F. 1997. Effect of hydrogen on reductive dechlorination of chlorinated ethenes. *Environ. Sci. Tech.* **31**:1728–1734.

Barker, P. S., Morrison F. O., Whitaker R. S. 1965. Conversion of DDT to DDD by Proteus vulgaris, a bacterium isolated from the intestinal flora of a mouse. *Nature.* **205**:621–622.

Baxter, R. M. 1990. Reductive dechlorination of certain chlorinated organic compounds by reduced hematin compared with their behavior in the environment. *Chemosphere.* **21**:451–458.

Beunink, J., Rehm H. J. 1988. Synchronous anaerobic and aerobic degradation of DDT by an immobilized mixed culture system. *Appl. Microbiol. Biotechnol.* **29**:72–80.

Bouchard, B., Beaudet R., Villemur R., McSween G., Lepine F., Bisaillon J. G. 1996. Isolation and characterization of Desulfitobacterium frappieri sp. nov., an anaerobic bacterium which reductively dechlorinates pentachlorophenol to 3-chlorophenol. *Int. J. Syst. Bacteriol.* **46**:1010–1015.

Breitenstein, A., Saano A., Salkinoja-Salone M., Andreesen J. R., Lechner U. 2001. Analysis of a 2,4,6-trichlorophenol-dehalogenating enrichment culture and isolation of the dehalogenating member Desulfitobacterium frappieri strain TCP-A. *Arch. Microbiol.* **175**:133–142.

Bumpus, J. A., Aust S. D. 1987. Biodegradation of DDT 1,1,1-trichloro-2,2-bis-(4-chlorophenyl ethane) by the white rot fungus Phanerochaete chrysosporium. *Appl. Environ. Microbiol.* **53**:2001–2008.

Burge, W. D. 1971. Anaerobic decomposition of DDT in soil. Acceleration by volatile components of alfalfa. *J. Agr. Food Chem.* **19**:375–378.

Beunink, J. and H. J. Rehm. 1988. Synchronous anaerobic and aerobic degradation of DDT by an immobilized mixed culture system. *Appl Microbial Biotech.* **29**:72–80.

Carson, R. 1962. *Silent Spring.* Penguin Books, London.

Castro, T. F., Yoshida T. 1974. Effect of organic matter on the biodegradation of some organochlorine insecticides in submerged soils. *Soil Sci. Plant Nutr.* **20**:363–70.

Chacko, C. J., Lockwood J. L., Zabik M. 1966. Chlorinated hydrocarbon pesticides: degradation by microbes. *Science.* **154**:893–895.

Chen, W., Mulchandani A. 1998. The use of 'live biocatalysts' for pesticide detoxification. *Trends Biotechnol.* **16**:71–76.

Cole, J. R., Cascarelli A. L., Mohn W. W., Tiedje J. M. 1994. Isolation and characterization of a novel bacterium growing via reductive dehalogenation of 2-chlorophenol. *Appl. Environ. Microbiol.* **60**:3536–3542.

Crooks, G. P., Copley S. D. 1993. A surprising effect of leaving group on the nucleophilic aromatic substitution reaction catalyzed by 4-chlorobenzoyl-CoA dehalogenase. *J. Am. Chem. Soc.* **115**:6422–6423.

Dagley, S. 1971. Catabolism of aromatic compounds by micro-organisms. *Adv. Microb. Physiol.* **6**:1–46.

Dagley, S. 1977. Microbial degradation of organic compounds in the biosphere. *Surv. Prog. Chem.* **8**:121–170.

DeWeerd, K. A., Concannon F., Suflita J. M. 1991. Relationship between hydrogen consumption, dehalogenation, and the reduction of sulfur oxyanions by Desulfomonile tiedjei. *Appl. Environ. Microbiol.* **57**:1929–1934.

DeWeerd, K. A., Mandelco L., Tanner R. S., Woese C. R., Suflita J. M. 1990. Desulfomonile tiedjei gen. nov. and sp. nov., a novel anaerobic, dehalogenating, sulfate-reducing bacterium. *Arch. Microbiol.* **154**:23–30.

Dimond, J. B., Owen R. B. 1996. Long-term residue of DDT compounds in forest soils in Maine. *Environ. Pollut.* **92**:227–230.

Dolfing, J. 1990. Reductive dechlorination of 3-chlorobenzoate is coupled to ATP production and growth in an anaerobic bacterium, strain DCB-1. *Arch. Microbiol.* **153**:264–266.

Dolfing, J. 2003. Thermodynamic consideration for dehalogenation. In: Häggblom, M. M., Bosser I. D. (eds.) *Dehalogenation: Microbial Processes and Environmental Applications.* Kluwer, Dordrecht, pp. 89–114.

Dolfing, J., Harrison B. K. 1992. The Gibbs free energy of formation of halogenated aromatic compounds and their potential role as electron acceptors in anaerobic environments. *Environ. Sci. Technol.* **26**:2213–2218.

Drzyzga, O., Gottschal J. C. 2002. Tetrachloroethene dehalorespiration and growth of Desulfitobacterium frappieri TCE1 in strict dependence on the activity of Desulfovibrio fructosivorans. *Appl. Environ. Microbiol.* **68**:642–649.

Environmental Protection Agency (EPA). 1975. *DDT: A review of scientific and economic aspects of the decision to ban its use as a pesticide.* U.S. Environmental Protection Agency, Washington, DC. EPA-540/1–75–022.

Environmental Protection Agency (EPA). 2002. Identification and listing of hazardous waste. U.S. Environmental Protection Agency. Code of Federal Regulations. 40 CFR 261.33(f).

Essac, E. G., Matsumura F. 1980. Metabolism of insecticides by reductive systems. *Pharmacol. Ther.* **9**:1–26.

Fennell, D. E., Gossett J. M. 1998. Modeling the production of and competition for hydrogen in a dechlorinating culture. *Environ. Sci. Technol.* **32**:2450–2460.

Focht, D. D., Alexander M. 1971. Aerobic cometabolism of DDT analogues by Hydrogenomonas sp. *J. Agric. Food Chem.* **19**:20–22.

Foght, J., April T., Biggar K., Aislabie J. 2001. Bioremediation of DDT-contaminated soils: a review. *Bioremed. J.* **5**:225–246.

Fuchs, G., Mohamed M. E., Altenschmidt U., Koch J., Lack A., Brackmann R., Lochmeyer C., Oswald B. 1994. Biochemistry of anaerobic biodegradation of aromatic compounds. In: Ratledge, C. (ed.) *Biochemistry of Microbial Degradation.* Kluwer Academic Publishers, Dordrecht, pp. 513–553.

Gerritse, J., Drzyzga O., Kloetstra G., Keijmel M., Wiersum L. P., Hutson R., Collins M. D., Gottschal J. C. 1999. Influence of different electron donors and acceptors on dehalorespiration of tetrachloroethene by Desulfitobacterium frappieri TCE1. *Appl. Environ. Microbiol.* **65**:5212–5221.

Gerritse, J., Renard, V., Pedro Gomes, T. M., Lawson, P. A., Collins, M. D. & Gottschal, J. C. 1996. Desulfitobacterium sp. strain PCE1, anaerobic bacterium that can grow by reductive dechlorination of tetrachloroethene or ortho-chlorinated phenols. *Arch. Microbiol.* **165**:132–140.

Gohil, H., Thomas J., Ogram A. 2014. Stimulation of anaerobic biodegradation of DDT and its metabolites in a muck soil: laboratory microcosm and mesocosm studies. *Biodegradation.* **25**:633–642. DOI 10.1007/s10532-014-9687-0.

Golovleva, L. A., Skryabin G. K. 1981. Microbial degradation of DDT. In: Leisinger, T., Cook A. M., Hutter R., Neusch J. (eds.) *Microbial Degradation of Xenobiotics and Recalcitrant Compounds.* Academic Press, London, pp. 287–292.

Guenzi, W. D., Beard W. E. 1967. Anaerobic biodegradation of DDT to DDD in soil. *Science.* **156**:1116–1117.

Guenzi, W. D., Beard W. E. 1968. Anaerobic conversion of DDT to DDD and aerobic stability of DDT in soil. *Soil Sci. Soc. Am. Proc.* **32**:522–527.

Häggblom, M. M., Bossert I. D. 2003. *Dehalogenation: Microbial Processes and Environmental Applications.* Kluwer Academic Publishers, Boston.

Harayama, S., Timmis K. N. 1989. Catabolism of aromatic hydrocarbons by Pseudomonas. In: Hopwood, D., Chater K. (eds.) *Genetics of Bacterial Diversity.* Academic Press, London, **15**, pp. 1–174.

Hay, A. G., Focht D. D. 1998. Cometabolism of 1,1,1-trichloro-2,2-bis-(4-chlorophenyl ethane) by Pseudomonas acidovorans M3GY grown on biphenyl. *Appl. Environ. Microbiol.* **64**:2141–2146.

Hay, A. G., Focht D. D. 2000. Transformation of 1,1-dichloro-2,2-(4-chlorophenyl)ethane (DDD) by Ralstonia eutropha strain A5. *FEMS Microbiol. Ecol.* **31**:249–253.

HazDat. 2002. Agency for Toxic Substances and Disease Registry (ATSDR) database, Atlanta, GA.

Heider, J., Fuchs G. 1997. Anaerobic metabolism of aromatic compounds. *Eur. J. Biochem.* **243**:577–596.

Hernandez, B. S. 1997. Terpene-utilizing isolates and their relevance to enhanced biotransformation of polychlorinated biphenyls in soil. *Biodegradation.* **8**:153–158.

Hiraishi, A. 2008. Biodiversity of dehalorespiring bacteria with special emphasis on polychlorinated biphenyl/dioxin dechlorinators. *Microbes Environ.* **23**:1–12.

Holliger, C., Hahn D., Harmsen H., Ludwig W., Schumacher W., Tindall B., Vazquez F., Weiss N., Zehnder A. J. B. 1998. Dehalobacter restrictus gen. nov. and sp. nov., a strictly anaerobic bacterium that reductively dechlorinates tetra- and trichloroethene in an anaerobic respiration. *Arch. Microbiol.* **169**:313–321.

Holliger, C., Regeard C., Diekert G. 2003. Dehalogenation by anaerobic bacteria. In: Häggblom, M. M., Bossert I. D. (eds.) *Dehalogenation: Microbial Processes and Environmental Applications.* Kluwer Academic Publishers, Norwell, pp. 115–157.

Holliger, C., Wohlfarth G., Diekert G. 1999. Reductive dechlorination in the energy metabolism of anaerobic bacteria. *FEMS Microbiol. Rev.* **22**:383–393.

Huang, C. L., Harrison B. K., Madura J., Dolfing J. 1996. Thermodynamic prediction of dehalogenation pathways for PCDDs. *Environ. Toxicol. Chem.* **15**:824–836.

Hunt, J. R., Sitar N. 1988. Non-aqueous phase liquid transport and clean-up analysis of mechanisms. *Water Resour. Res.* **24**:1247–1258.

Jacques, R., Okeke B., Bento F., Teixeira A., Peralba M., Camargo F. 2008. Microbial consortium bioaugmentation of a polycyclic aromatic hydrocarbons contaminated soil. *Bioresour. Technol.* **99**:2637–2643.

Jensen, S., Gothe R., Kindertedt M. O. 1972. Bis-(p-Chlorophenyl)-acetonitrile (DDN), a new DDT derivative formed in anaerobic digested sewage sludge and lake sediment. *Nature.* **240**:421–422.

Johnsen, R. E. 1976. DDT metabolism in microbial systems. *Residue Rev.* **61**:1–28.

Juhasz, A. L., Megharaj M., Naidu R. 1999. Bioavailability: the major challenge (constraint) to bioremediation of organically contaminated soils, in remediation of hazardous waste contaminated soils, 2nd ed. Vol **1**, *Engineering Considerations and Remediation Strategies, Section 1–1: Engineering Issues in Waste Remediation*, In: Wise, D. L., Tratolo D. J., Cichon E. J., Inyang H. I., Stottmeister U. (ed.) Marcel Dekker, New York, 217–241.

Juhasz, A. L., Naidu R. 2000. Bioremediation of high molecular weight polycyclic aromatic hydrocarbons: a review of the microbial degradation of benzo[a]pyrene. *Int. Biodeterior. Biodegradation* **45**:57–88.

Kantachote, D., Naidu R., Singleton I., McClure N., Harch B. D. 2001. Resistance of microbial population in DDT-contaminated and uncontaminated soils. *Appl. Soil Ecol.* **16**:85–90.

Kantachote, D., Singleton I., Naidu R., McClure N. C., Megharaj M. 2004. Sodium application enhances DDT transformation in a long-term contaminated soil. *Water Air Soil Pollut.* **154**:1–4.

Katayama, A., Fujimura Y. Kuwatsuka S. 1993. Microbioal degradation of DDT at extremely low concentrations. *J. Pesticide Sci.* **18**:353–359.

Keller, E., Rickabaugh J. 1992. Effects of surfactant structure on pesticide removal from a contaminated soil. *Hazard. Ind. Wastes.* **24**:652–661.

Krumholz, L. R. 1997. Desulfuromonas chloroethenica sp. nov. uses tetrachloroethylene and trichloroethylene as electron acceptors. *Int. J. Syst. Bacteriol.* **47**:1262–1263.

Krumholz, L. R., Sharp R., Fishbain S. S. 1996. A freshwater anaerobe coupling acetate oxidation to tetrachloroethylene dehalogenation. *Appl. Environ. Microbiol.* **62**:4108–4113.

Kuhn, E. P., Suflita J. M. 1989. Dehalogenation of pesticides by anaerobic microorganisms in soils and groundwater – a review. In: Sawhney, B. L., Brown K. (eds.) *Reactions and Movement of Organic Chemicals in Soils. Soil Sci. Soc. America Special Publication No. 22.* Soil Sci. Soc. America, Inc., Madison, WI, pp. 111–180.

Kunisue, T., Watanabe M., Subramanian A., Sethuraman A., Titenko A. M., Qui V., Prudente M., Tanabe S. 2003. Accumulation features of persistent organochlorines in resident and migratory birds from Asia. *Environ. Pollut.* **125**:157–172.

Lal, R., Saxena D. M. 1982. Accumulation, metabolism, and effects of organochlorine insecticides on microorganisms. *Microbiol. Rev.* **46**:95–127.

Langlois, B. E., Collins J. A., Sides K. G. 1970. Some factors affecting degradation of organochlorine pesticide by bacteria. *J. Dairy Sci.* **53**:1671–1675.

Lanthier, M., Villemur R., Lepine F., Bisaillon J. G., Beaudet R. 2000. Monitoring of Desulfitobacterium frappieri PCP-1 in pentachlorophenol-degrading anaerobic soil slurry reactors. *Environ. Microbiol.* **2**:703–708.

LeBlanc, G. 1995. Trophic-level differences in the bioconcentration of chemicals: implications in assessing environmental biomagnification. *Environ. Sci. Technol.* **29**:154–160.

Leštan, D., Leštan M., Chapelle J. A., Lamar R. T. 1996. Biological potential of fungal inocula for bioaugmentation of contaminated soils. *J. Ind. Microbiol.* **16**:286–294.

Lindqvist, R., Enfield, C. G. 1992. Biosorption of dichlorodephenyltrichloroethane and hexachlorobenzene in groundwater and its implication for facilitated transport. *Appl. Environ. Microbiol.* **58**:2211–2218.

Löffler, F. E., Ritalahti K. M., Tiedje J. M. 1997. Dechlorination of chloroethenes is inhibited by 2-bromoethanesulfonate in the absence of methanogens. *Appl. Environ. Microbiol.* **63**:4982–4985.

Luijten, M. L. G. C., De Weert J., Smidt H., Boschker H. T. S., De Vos W. M., Schraa G., Stams A. J. M. 2003. Description of sulfurospirillum halorespirans sp. nov., an anaerobic, tetrachloroethene-respiring bacterium, and transfer of dehalospirillum multivorans to the genus sulfurospirillum as sulfurospirillum multivorans comb. nov. *Int. J. Syst. Evol. Microbiol.* **53**:787–793.

Mackiewicz, M., Wiegel J. 1998. Comparison of energy and growth yields for Desulfitobacterium dehalogenans during utilization of chlorophenol and various traditional electron acceptors. *Appl. Environ. Microbiol.* **64**:352–355.

Masse, R., Lalnne D., Messier F., Sylvestre M. 1989. Characterization of new bacterial transformation products of 1,1,1-trichloro-2,2-bis-(4-chlorophenyl ethane) (DDT) by gas chromatography/mass spectrometry. *Biomed. Environ. Mass Spectrom.* **18**:741–752.

Maymo-Gatell, X., Anguish T., Zinder S. H. 1999. Reductive dechlorination of chlorinated ethenes and 1,2-dichloroethane by "Dehalococcoides ethenogenes" 195. *Appl. Environ. Microbiol.* **65**:3108–3113.

Maymo-Gatell, X., Chien Y.-T., Gossett J. M., Zinder S. H. 1997. Isolation of a bacterium that reductively dechlorinates tetrachloroethene to ethene. *Science.* **276**:1568–1571.

Megharaj, M., Kantachote D., Singleton I., Naidu R. 2000. Effects of long term contamination of DDT on soil microflora with special reference to soil algae and algal transformation of DDT. *Environ. Pollut.* **109**:35–42.

Mendel, J. L., Walton M. S. 1966. Conversion of p-p′-DDD by intestinal flora of the rat. *Science.* **151**:1527–1528.

Miller, E., Wohlfarth G., Diekert G. 1997. Studies on tetrachloroethene respiration in Dehalospirillum multivorans. *Arch. Microbiol.* **166**:379–387.

Mohn, W. W., Tiedje J. M. 1990. Strain DCB-1 conserves energy for growth from reductive dechlorination coupled to formate oxidation. *Arch. Microbiol.* **153**:267–271.

Mohn, W. W., Tiedje J. M. 1991. Evidence for chemiosmotic coupling of reductive dechlorination and ATP synthesis in Desulfomonile tiedjei. *Arch. Microbiol.* **157**:1–6.

Morgan, P., Watkinson R. J. 1989. The use of gel-stabilized model systems for the study of microbial processes in polluted sediments. *J. Gen. Microbiol.* **135**:549–555.

Nadeau, L. J., Menn F. M., Breen A., Sayler G. S. 1994. Aerobic degradation of 1,1,1-trichloro-2,2-bis-(4-chlorophenyl ethane) (DDT) by Alcaligens eutrophus A5. *Appl. Environ. Microbiol.* **60**:51–55.

Nadeau, L. J., Sayler G. S., Spain J. C. 1998. Oxidation of 1,1,1-trichloro-2,2-bis-(4-chlorophenyl ethane) (DDT) by Alcaligens eutrophus A5. *Arch. Microbiol.* **171**:44–49.

Nevin, K. P., Holmes D. E., Woodard T. L., Covalla S. F., Lovley D. R. 2007. Reclassification of trichlorobacter thiogenes as geobacter thiogenes comb. nov. *Int. J. Syst. Evol. Microbiol.* **57**:463–466.

Parsons, J. R., Goorissen H., Weiland A. R., de Bruijne J. A., Spraingael D., van der Lelie D., Mergeay M. 1995. Substrate range of the (chloro)biphenyl degradation pathway of Alcaligens sp. JB1. In: Hinchee, R. E., Vogel C. M., Brockman F. J. (eds.) *Microbial Processes for Bioremediation. Bioremediation.* Battelle Press, Columbus, OH, **8**, pp. 169–175.

Peterson, J. R., Adams R. S. Jr., Cutkomp L. K. 1971. Soil properties influencing DDT bioactivity. *Soil Sci. Soc. Am. Proc.* **35**:72–78.

Pfaender, F. K., Alexander M. 1972. Extensive degradation of DDT in vitro and DDT metabolism by natural communities. *J. Agric. Food Chem.* **20**:842–846.

Pfennig, N., Biebl H. 1976. Desulfuromonas acetoxidans gen. nov. and sp. nov., a new anaerobic, sulfur-reducing, acetate-oxidizing bacterium. *Arch. Microbiol.* **110**:3–12.

Quensen, J. F., III, Boyd S. A., Tiedje J. M. 1990. Dechlorination of four commercial polychlorinated biphenyl mixture (Aroclors) by anaerobic microorganisms from sediments. *App. Environ. Microbiol.* **56**:2360–2369.

Ratcliffe, D. A. 1967. Decrease in eggshell weight in certain birds of prey. *Nature.* **215**:208–210.

Rochkind-Dubinsky, M. L., Sayler G. S., Blackburn J. W. 1987. *Microbial Decomposition of Chlorinated Aromatic Compounds. Microbiology Series*, Vol. **18**, Marcel Dekker Inc., New York and Base.

Sanford, R. A., Cole J. R., Löffler F. E., Tiedje J. M. 1996. Characterization of Desulfitobacterium chlororespirans sp. nov., which grows by coupling the oxidation of lactate to the reductive dechlorination of 3-chloro-4-hydroxybenzoate. *Appl. Environ. Microbiol.* **62**:3800–3808.

Sanford, R. A., Cole J. R., Tiedje J. M. 2002. Characterization and description of Anaeromyxobacter dehalogenans gen. nov., sp. nov., an aryl-halorespiring facultative anaerobic myxobacterium. *Appl. Environ. Microbiol.* **68**:893–900.

Sayles, G. D., You G., Wang M., Kupferle M. J. 1997. DDT, DDD, and DDE dechlorination by zero-valent iron. *Environ. Sci. Technol.* **31**:3448–3454.

Schink, B. 1997. Energetics of syntrophic cooperation in methanogenic degradation. *Microbiol. Mol. Biol. Rev.* **61**:262–80.

Schink, B., Brune A., Schnell S. 1992. Anaerohic degradation of aromatic compounds. In: Winkelmann, G., (ed.) *Microbial Degradation of Rictrui-ul Products.* VCH, Weinheim, pp. 219–242.

Scholz-Muramatsu, H., Neumann A., MeMmer M., Moore E., Diekert G. 1995. Isolation and characterization of Dehalospirillum multivorans gen. nov. sp. nov., a tetrachloroethene-utilizing, strictly anaerobic bacterium. *Arch. Microbiol.* **163**:48–56.

Sharma, P. K., McCarty P. L. 1996. Isolation and characterization of a facultatively aerobic bacterium that reductively dehalogenates tetrachloroethylene to cis-1,2-dichloroethylene. *Appl. Environ. Microbiol.* **62**:761–765.

Sharma, S. K., Sadasivam K. V., Dave J. M. 1987. DDT degradation by bacteria from activated sludge. *Environ. Int.* **13**:183–190.

Shelton, D. R., Tiedje J. M. 1984. Isolation and partial characterization of bacteria in an anaerobic consortium that mineralizes 3-chlorobenzoic acid. *Appl. Environ. Microbiol.* **48**:840–848.

Smidt, H., de Vos W. M. 2004. Anaerobic microbial dehalogenation. *Annu. Rev. Microbiol.* **58**:43–73.

Smith, A. G. 1991. Chlorinated hydrocarbon insecticides. In: Hayes, W. J., Laws E. R. (eds.) *Handbook of Pesticides Toxicology.* Academic Press Inc., Sen Diego/New York, pp. 731–915.

Subba-Rao, R. V., Alexander M. 1985. Bacterial and fungal cometabolism of 1,1,1-trichloro-2,2-bis(4-chlorophenyl)ethane (DDT) and its breakdown products. *Appl. Environ. Microbiol.* **49**:509–516.

Suflita, J. M., Gibson S. A., Beeman R. E. 1988. Anaerobic biotransformations of pollutant chemicals in aquifers. *J. Ind. Microbiol.* **3**:179–194.

Suflita, J. M., Horowitz A., Shelton D. R., Tiedje J. M. 1982. Dehalogenation: a novel pathway for the anaerobic biodegradation of haloaromatic compounds. *Science.* **218**:1115–1117.

Sun, B., Cole J. R., Tiedje J. M. 2001. Desulfomonile limimaris sp. nov., an anaerobic dehalogenating bacterium from marine sediments. *Int. J. Syst. Evol. Microbiol.* **51**:365–371.

Sun, B., Griffin B. M., Ayala-del-Rio H. L., Hashsham S. A., Tiedje J. M. 2002. Microbial dehalorespiration with 1,1,1-trichloroethane. *Science.* **298**:1023–1025.

Sung, Y., Fletcher K. E., Ritalahti K. M., Apkarian R. P., Ramos-Hernandez N., Sanford R. A., Mesbah N. M., Löffler F. E. 2006. Geobacter lovleyi sp. nov. strain SZ, a novel metal-reducing and tetrachloroethene-dechlorinating bacterium. *Appl. Environ. Microbiol.* **72**:2775–2782.

Sung, Y., Ritalahti K. M., Sanford R. A., Urbance J. W., Flynn S. J., Tiedje J. M., Löffler F. E. 2003. Characterization of two tetrachloroethene-reducing, acetate-oxidizing anaerobic bacteria and their description as Desulfuromonas michiganensis sp. nov. *Appl. Environ. Microbiol.* **69**:2964–2974.

Susaria, S., Masunaga S., Yonezawa Y. 1996. Reductive dechlorination pathways of chloro organics under anaerobic conditions. *Water Sci. Technol.* **34**:489–494.

Sylvestre, M., Sandossi M. 1994. Selection of enhanced PCB-degrading bacterial strains for bioremediation: consideration of branching pathways. In: Chaudhry G. R. (ed.) *Biological Degradation and Remediation of Toxic Chemicals.* Chapman and Hall, New York.

Thauer, R. K., Jungermann K., Decker K. 1977. Energy conservation in chemotrophic anaerobic bacteria. *Bacteriol. Rev.* **41**:100–180.

Turusov, V., Rakitsky V., Tomatis L. 2002. Dichlorodiphenyltrichloroethane (DDT): ubiquity, persistence, and risks. *Environ. Health Perspect.* **110**:125–128.

United States Environmental Protection Agency (U.S. EPA). 2000. Test methods for evaluating solid waste. SW-846. 5th (ed.) *Office of Solid Waste and Emergency Response.* Washington, D.C.

Utkin, I., Woese C., Wiegel J. 1994. Isolation and characterization of Desulfitobacterium dehalogenans gen. nov., sp. nov., an anaerobic bacterium which reductively dechlorinates chlorophenolic compounds. *Int. J. Syst. Bacteriol.* **44**:612–619.

Vollner, L., Klotz D. 1994. Behaviour of DDT under laboratory and outdoor conditions in Germany. *J. Environ. Sci. Health.* **B29**:161–167.

Wedemeyer, G. 1967. Dechlorination of 1,1,1-trichloro-2,2-bis(p-chlorophenyl)ethane by Aerobacter aerogenes, I. Metabolic products. *Appl. Microbiol.* **15**:569–574.

Weissenfels, W. D., Klewer H. J., Langhoff J. 1992. Adsorption of polycyclic aromatic hydrocarbons (PAHs) by soil particles: influence on biodegradation and biotoxicity. *Appl. Micobiol. Biotechnol.* **26**:689–696.

Wiegel, J., Zhang X., Wu Q. 1999. Anaerobic dehalogenation of hydroxylated polychlorinated biphenyls by Desulfitobacterium dehalogenans. *Appl. Environ. Microbiol.* **65**:2217–2221.

Wild, A., Hermann R., Leisinger T. 1996. Isolation of an anaerobic bacterium which reductively dechlorinates tetrachloroethene and trichloroethene. *Biodegradation.* **7**:507–511.

World Health Organization (WHO). 1979. DDT and its derivatives. Environmental health criteria 9. Geneva: World Health Organization, United Nations Environment Programme, 1979.

Yan, T., LaPara T. M., Novak P. J. 2006. The reductive dechlorination of 2,3,4,5-tetrachloro-biphenyl in three different sediment cultures: evidence for the involvement of phylogenetically similar dehalococcoides-like bacterial populations. *FEMS Microbiol. Ecol.* **55**:248–261.

Yang, Y., McCarty P. L. 1998. Competition for hydrogen within a chlorinated solvent dehalogenating anaerobic mixed culture. *Environ. Sci. Technol.* **32**:3591–3597.

You, G., Sayles G. D., Kupferle M. J., Bisshop P. L. 1995. Anaerobic bioremediation of DDT-contaminated soil with nonionic surfactants. In: Hinchee, R. E., Hoeppel R. E., Anderson D. B. (eds.) *Bioremediation of Recalcitrant Organics.* Battelle Press, Columbus, U.S.A., pp. 137–144.

Zoro, J. A., Hunter J. M., Eglinton G., Ware G. C. 1974. Degradation of p,p′-DDT in reducing environments. *Nature.* **247**:235–237.

2 Biological Degradation and Detoxification of Toxic Contaminants in Leachate

An Environmentally Friendly Perspective

Indu Shekhar Thakur, Juhi Gupta, and Pooja Ghosh*

CONTENTS

* Corresponding author: E- mail: isthakur@hotmail.com

2.1 INTRODUCTION

The shifting lifestyle of the urbanized population is leading to an irrevocable upsurge in the production of municipal solid waste (MSW). Currently, Delhi's contribution to MSW generation is more than 11000 tons/day, which is projected to increase to three times this figure by the year 2021 (Talyan et al., 2008). The research world, with its intellectual minds, is innovating new methods to deal with the rising waste; however, landfills are still the most feasible and common dumping alternative. The major drawback of an economically feasible landfill is the generation of contaminated harmful leachate. Leachate pollution can be challenged by careful environmental surveillance of landfill operation to further prevent surface and groundwater pollution (Mor et al., 2006).

2.2 PRESENT LAND-FILLING STATUS

The main purpose of a landfill is to concentrate the generated waste at a separate location in order to cut its contact with the existing ecological life (Narayana, 2009). Conversely, the Indian landfill sites are not efficiently managed and engineered, which exposes the waste and leads to groundwater infiltration (Kaushal et al., 2012). The poor segregation and dumping activity is responsible for the generation of concentrated leachate, which is produced once the rainwater passes through the layers of waste (Sahu, 2007). The combined physical and chemical activity, along with the microbial flora, is responsible for causing changes in the waste line and transferring further metabolic products to the generated leachate, which trickles down to pollute the groundwater (Christensen and Kjeldsen, 1989; Saarela, 2003). The annual leachate generation from the multinational city Delhi is reported to be approximately 1000 m^3 which is continuously rising with the urbanized population (Figure 2.1) (Kumar et al., 2002). Due to poor collection systems, the landfills pose significant issues of leachate generation and greenhouse gas emissions, causing deleterious changes in the environment (Talyan et al., 2008). The surmounting burden has led to shutdown of many landfill sites, including 17–19 sites in Delhi. Currently, Delhi has three large existing functional landfill sites at Bhalswa, Okhla, and Narela, while the Ghazipur landfill site has been recently closed due to overburden and exploitation. An engineered dump-yard involves a tough lining at the bottom, which would prevent the leachate percolation and avoids further contamination of nearby ecological spots (Mor et al., 2006). The dumping sites at Delhi have a nonsystematic design and posean environmental threat due to the continuous production of leachate even after closure (Kurniawan et al., 2006).

2.3 LEACHATE GENERATION: A CONCERNING CAUSE

The composition of the leachate generated from a landfill is highly influenced by the waste type and age of the landfill (Slack et al., 2005). The water-based leachate is a combination of mainly four types of contaminants: inorganic macro-nutrients (anions and cations like ammonia, chloride, and sulphate), soluble organic matter (acid, alcohol, aldehydes, etc.), heavy metals, and recalcitrant xenobiotic compounds

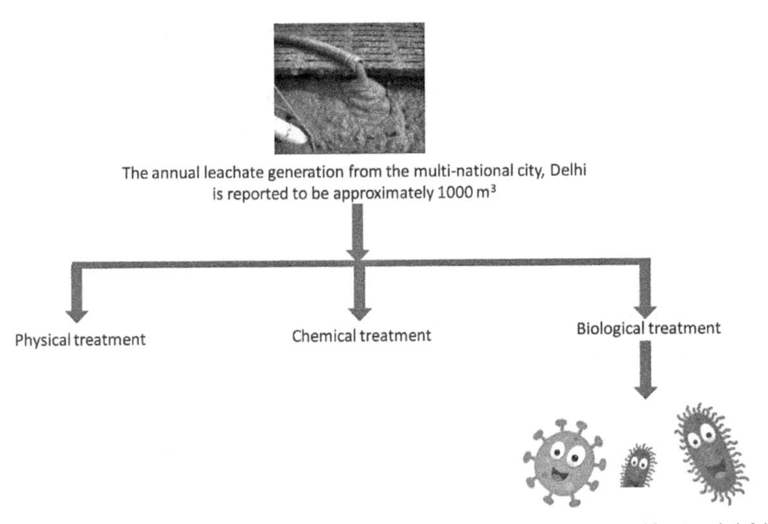

FIGURE 2.1 Treatment technologies for leachate generation.

like dioxinsand polychlorinated biphenyls (PCBs) (Pivato and Gaspari, 2005). Due to such a harmful composition, leachate presents many issues to the nearby water sources, spreading toxicity and carcinogenicity of the involved chemicals to both aquatic and terrestrial life (Matejczyk et al., 2011). The dose of the toxic chemical also plays a significant role, and the negative effects are highly biomagnified while passing through a food chain synergistically, ultimately affecting all life (Tillitt et al., 2010). With a shifting and highly civilized lifestyle, there is a remarkable transition in the type and quality of the products used like batteries, electrical appliances, paint products, medicines, and cosmetics. Nowadays, these commodities have harmful additives which are ultimately transformed to emerging contaminants and further drained to form the leachate resulting in deleterious impacts on human and ecological settings. The former conclusion has been drawn through a study conducted by Eggen et al. (2010), which showed the prevalence of emerging contaminants in leachate drawn out of the daily products of usage. The most common chemicals reported were metabolic products of insecticides, personal care cosmetic products, anti-inflammatory drugs, and polycyclic compounds. The presence of such harmful compounds, including polycyclic aromatic hydrocarbons (PAHs), xenobiotic aromatic compounds, esters of phthalate, etc., has also been reported by Ghosh et al. (2015) in the leachate of landfill sites (Ghazipur, Okhla, Bhalswa) in Delhi. The same study also indicated the toxic effects of leachate on a human liver cell line, with an EC_{50} value between 11% and 20% occurring as a result of noticeable DNA damage. This concludes that the potential leachate generation from landfill sites is highly harmful for the environment.

The overloaded development is leading to nonmanageable production of solid waste, ultimately increasing the amount of leachate generated, which is harming the entire planet. There is a wide scope to develop a specific database in order to understand the characteristics of leachate efficiently with respect to toxicity and composition. India like other developing nations immediately require a sustainable alternative to better manage and effectively treat the increasing leachate.

2.4 LEACHATE TREATMENT

2.4.1 Physicochemical Methods and Limitations

Leachate treatment is conducted using a combination of both physical and chemical methods, such as reagent precipitation, ultra-reverse osmosis, membrane filtration, ion exchange, and adsorption (Beszedits, 1983). However, these methods present the limitation of being cost-demanding and restricted in terms of versatility due to which the biological waste treatment is gaining attention and increasingly investigated (Kapoor and Viraraghavan, 1995).

2.4.2 Biological Treatment

Biological microorganisms are the most potential nature recyclers, primarily fungi and bacteria. They utilize and transform such natural and artificial chemicals, including the harmful components, into their sources of survival required during metabolic growth. Due to this, biological waste treatment is more feasible, cost-effective, and environmentally friendly in comparison to physical or chemical remediation. The unexplored and tapped microbial resources still need to be investigated properly for their employment in effective bioremediation (Kumar et al., 2011).

2.4.2.1 Algae and Biological Treatment

As discussed in Section 2.4.2, biological treatment is a very attractive and feasible alternative to the limited activity of physical and chemical processes. Among the active biological life, usage of algae is highly appreciated in the treatment of wastewater as concluded by many reports (Zimmo et al., 2004; Lin et al., 2007). The algae-based waste treatment is gaining attention because of providing two-way benefits; it is environmentally friendly, and on the other hand, it results in the generation of useful value-added products like biogas and other algal metabolites (Oswald, 2003). As reported by Tchobanoglous et al. (2003), the synergistic relationship between bacteria and a micro-algae proves to be more economically beneficial during the remediation of contaminants. There are different parts of an aerobic waste treatment process, each one of which is energy intensive or extensive—for example, aeration accounts for more than half the percentage of the whole energy requirement. So employing algae will ensure a better supply of oxygen, which will ultimately aid Biological Oxygen Demand (BOD) removal by heterotrophic aerobic microbes. Such synergistic associations are safer with respect to release of aerosols and/or pollutants, as detailed by a study conducted by Safonova et al. (1999, 2004) showed in which a bacteria-algae consortium was utilized to degrade black oil and detoxify industrial wastewater in Russia (Hamoda, 2006).

An integrated algal system involves large-scale production of algae, which itself comes with many challenges, like the transfer and exchange of gases, nutrient recycling, availability of water and land, intensity of photosynthetic active radiation (PAR), and culture maintenance. Each of these factors is responsible for various limitations at increased culture densities (Zijffers et al., 2008). With progressive research, some innovations have been proposed to overcome these limitations, like the selection of flocculating strains, application of immobilized biomass, or employing better flocculants, which will help in remediating toxic organic contaminants. Microorganisms have their own limitations, which under high pollutant loads don't survive; as with algae, their growth is hampered due to increased nitrogen and ammonia conditions. Similarly, the growth of *Pseudokircheneriella subcapitata* was inhibited (EC_{50} 0.004–0.013 mg L^{-1}) by pentachlorophenol (PCP) when operated in an air-closed setting (Chen and Lin, 2006). Algae are sensitive to both NH_3 and pH values as NH_3 disturbs the photosystem II by hindering the electron transport system and tries to replace H_2O during the oxidation reactions (Azov and Goldman, 1982). When compared to algae, bacteria have high resistance to the pollutant load and resist the stress efficiently. For example, *Chlorella sorokiniana* was completely hindered by 10 mg phenanthrene per liter while a combined consortium of *Chlorella* with *Pseudomonas* easily bioremediated phenanthrene (25 mg/L) (Borde et al., 2003).

Due to a larger size, heterotrophic bacteria have a faster growth pattern than microalgae (Fenchel, 1974). This happens even with the fastest-growing algae. For example, a specific growth rate of 0.4–0.8 per hour is seen by a toluene-metabolizing *Pseudomonas* sp. (Reardon et al., 2000), while the maximum growth rate achieved by the fastest-growing *Chlorella* is equal to or less than 0.2 h^{-1} (Lee, 2001). This concludes that growth pattern of algae is a significant limiting factor during pollutant remediation.

Non-hazardous effluent has a low concentration of suspended solids and is efficiently generated in the case of biomass harvesting. Neither the physical nor the chemical industrial approaches, like ultracentrifugation, filtration, and micro-straining, are economic enough to remove large-scale microalgae (Hoffman, 1998). As discussed earlier, total suspended solids (TSS) removal is a tedious task due to which wastewater effluents are therefore often characterized by high TSS values. This hinders the treatment process because a microbial biomass might contain harmful hydrophobic organic compounds and/or toxic heavy metals.

2.4.2.2 Bacteria and Biological Treatment

The versatile compounds ranging from simple phenols, to aromatic compounds, to highly complex lignin are degraded by a wide variety of pathways followed by bacteria (Chen et al., 2012). Every microorganism serves a different advantage; for example, bacteria have a faster growth pattern when compared with fungi, allow overexpression of required proteins, and are also resistant to genetic manipulation (Bugg et al., 2011). Indigenous microbes are exposed to harmful contaminants at a polluted site, while resistant microbial strains are preferably obtained during acclimation to toxic pollutants, genetic manipulation, or harnessing the contaminated sites (Malik, 2004). Significant reports have been published addressing the importance of bacteria for degrading lignin, PAHs, dioxins, PCBs, and other harmful aromatic compounds concluding its action versatility (Bugg et al., 2011) like the genera *Serratia*, *Aneurinibacillus*, *Klebsiella*, *Pseudomonas* and *Bacillus* (Furukawa, 2000; Kanaly and Harayama, 2000; Chandra et al., 2007, 2008; Field and Sierra-Alvarez, 2008).

Unlike fungi, the hindering disadvantage faced by bacteria during bioremediation is the required pre-exposure to a pollutant. This means an induction is necessary to activate the degradation enzymes in the presence of a significant concentration of pollutant. As with fungi, lignin degradation is not chemical dependent and follows nonspecific free-radical reaction sequence for remediation. However, bacterial metabolism comprises enzymatic transformations and follows Michaelis–Menton kinetics for contaminant degradation, due to which the Km values need to be considered. Km values are solubility dependent, and therefore the solubility of a pollutant also governs the degradation process (Barr and Aust, 1994). Fungal degradation follows nonspecific pathways to metabolize nonsoluble complex pollutants like Arochlor (Yadav et al., 1995).

2.4.2.3 Fungi and Biological Treatment

Considering the large amount of agricultural waste generated in India, lignin is a very significant contributor to it. Lignin is a recalcitrant compound, and its degradation is important for a cleaner society. Fungi are efficient lignin degraders; for example, white rot fungi, a basidiomycetes, mineralizes lignin and leads to white rot decay of wood (Buswell and Odier, 1987). There are three important lignin-modifying enzymes (LME) of white rot fungi comprising manganese peroxidase (MnP), lignin peroxidase (LiP), and laccase enzyme. Former enzymes are nonspecific and broadly degrade a wide array of ecological contaminants such as PCBs, dioxins, nitro-phenols, PAHs and pesticides (Bumpus et al., 1985). These fungi do not utilize organic pollutants easily as a carbon source for their growth, while soft inexpensive lignin sources such as nut hulls, straw, and saw dust are feasible for metabolism to further promote biomass and fungi colonization (Reddy and Mathew, 2001). Filamentous fungi are competitively advanced over the single cells of bacteria and yeasts, as they prefer hyphal extension growth and colonize easily using the insoluble substrates (Baldrian, 2008). There are fungi, like *Trametes hirsute* IFO4917, which have degraded a range of harmful and toxic endocrine disruptors found in dump-yard leachate like bisphenol A and estradiol (Ike et al., 2005). As discussed earlier, a mutual combination of bacteria and fungi proves more beneficial while treating polluted water sites and a higher color reduction, COD and BOD maintenance, and removal of toxic compounds have been reported (Kaushik et al., 2010). Until today, only few reports have utilized the step-wise treatment of combined bacteria-fungi for pollutant detoxification of a contaminated site (Ghosh, Swati, et al., 2014). In spite of the greater scope presented by the magic potential of the bacteria-fungi combination, they are not extensively utilized in leachate treatment. Therefore, this area of bioaugmentation should be enhanced to promote a better removal of heavy metals and recalcitrant pollutants from leachate.

2.4.2.4 Advantage of Using Both Fungal and Bacterial Enzymes Sequentially

The combined action of bacterial and fungal enzymes has been investigated, which showed their advantage of coming together. Discussing high-molecular-weight PAHs, it was found that fungal coenzymes effectively attack them due to their diffusive approach to insoluble elements and compounds. On the other hand, bacterial intracellular enzymes are slow to operate in insoluble conditions. The bacterial enzymes like PAH-dioxygenases are significantly cell bound due to their requirement of NADH (Nicotinamide Adenine Dinucleotide-hydrogen [reduced]) as a cofactor. Once the initial attack is initiated, degraded oxidation products are formed, which are more soluble and available to the microbial flora. Therefore, a combined action should be always promoted during bioremediation.

2.5 IN VITRO BIOASSAYS FOR MONITORING ENVIRONMENTAL CONTAMINANTS AND BIOREMEDIATION

With advancing technologies, the use of bio-detectors for analyzing ecological contaminants is continuously growing and highly appreciated, like enzyme immune and bioassays, biomarkers, etc. (Behnisch et al., 2001). A bioassay quantifies the amount of a substance which can be easily exposed to an organism without expecting any adverse effects. Unlike the chemical analysis, which can only quantify the targeted known chemical, bioassays have an additional advantage of detecting the combination of both known and unknown compounds using the same action mode. Bioassays provide us a better picture of the effects induced by such compounds in comparison to the conventional chemical approach (Thomas et al., 2008).

Bioassays have been performed using fish, invertebrates, plants, algae, bacteria, and mammalian cells. In comparison to in vivo systems, in vitrobioassays are more sensitive, quick, and cost-effective. A fair share of reproducibility is seen with cancerous mammalian cell lines, which imitate the toxic effects to in vivo studies (Das et al., 2012). In contrast to primary cell lines, toxicity evaluation is better performed with cancerous cell lines as with secondary cell lines, the inherency of tissue and organ such as human liver cell line HepG2 is retained. It is a well-characterized model line to perform toxicity studies (Westerink and Schoonen, 2007). These cell lines evidently show quite a few similar phenotypic and genotypic characteristics of regular liver cells, different liver-specific responses to variety of drugs, and a well-defined glutathione framework and plays an important role in detoxifying genotoxic carcinogens (Jondeau et al., 2006). Thesc cell lines with various metabolic enzymes (cytochrome P450 monooxygenase) are better predictors to define the biological effect of chemicals on humans (Natarajan and Darroudi, 1991). In comparison to other cell lines, like Hek and HeLa, CHO K1: HepG2 is a much more sensitive cell line to toxic compounds (Schoonen et al., 2005). Additionally, the liver is the main site where toxicity effects of contaminants are first observed, and there are two phases of metabolism followed in HepG2 cells which reduce the possible false-positive responses in genotoxic experiments (Kirkland et al., 2007). Toxicity of various environmental samples is tested with cytotoxicity, receptor (AhR)-mediated toxicity (dioxin-like toxicity), and genotoxicity.

2.5.1 Cytotoxicity

Cytotoxicity is an assay preformed to measure cell viability, which also evaluates a compound's efficacy leading to cell death. There are assays that measure cytotoxicity which also involves MTT (3-[4,5-dimethylthiazol-2-yl]-2,5- diphenyl tetrazolium bromide) a colorimetric assay that quantifies the percentage of cell viability and measures cell death, occurring as a result of different cell damage or alterations (Gupta et al., 2019). It was developed by Mosmann (1983) and as of today is still the most feasible and preliminary screening performed. The test is quantified by the generation of insoluble purple formazan crystals by the action of mitochondrial dehydrogenase enzyme (Gupta et al., 2019). It is a very sensitive and rapid colorimetric test which ends up with reproducible results and efficiently measures cell viability. Due to its feasible behavior, this test has successfully predicted the cytotoxic nature of wastewater and drinking water samples (Zegura et al., 2009). Ghosh, Das, et al. (2014) also highlighted importance of MTT assay for cytotoxicity in evaluating the efficiency of bacteria for the treatment of leachate.

2.5.2 Genotoxicity

Genotoxicity is measured to analyze any breakage in DNA strands, excision and repair of DNA strands, or any formation of alkali labile salts which may occur due to exposure to various contaminants. The comet assay has been frequently employed to evaluate the genotoxicity of environmental samples. Unlike other genotoxic assays, the comet assay is more advantageous and has a high sensitivity to even detect low levels of DNA damage, less cell requirement per sample, short time requirement, and low cost. Additionally, this assay is quite flexible to detect different DNA damages and modifies enough to adapt to different experimental settings (Tice et al., 2000). It is found that genotoxic effects are occurring due to the generation of free radicals and oxidative damage reported in the affected organs like the heart, spleen, liver, and brain (Li et al., 2006a, b). Such tests have been applied to observe the toxicity of landfill leachate in erythrocytes from the gill cells of goldfish (*Carassius auratus*) (Deguchi et al., 2007). Thus, the comet assay is a promising test to evaluate the gene-based effects of contaminants on DNA and other related organs.

2.6 FUTURE PERSPECTIVES

With the increasing load on the present landfills, the amount of leachate generation is out of control. Developing countries like India are putting forth their best efforts to properly characterize this leachate both chemically and toxicologically. And it is important to combine both chemical analyses for the identification of organic and inorganic contaminants present in leachate, as well as toxicological evaluation to assess the toxic potency of leachate. This will help in the assessment of risk factors and in implementation of regulatory actions and designing an efficient treatment method.

2.7 CONCLUSIONS

Toxic leachate generation is very harmful for both humans and their ecological surroundings and affects our natural sources. It is highly necessary to design a sustainable alternative to treat this effectively before discharging it into the environment. Compared to the conventional physical and chemical methods of treatment, biological methods are more efficient as well as cost-effective. Also, monitoring the course of bioremediation using a suite of bioassays is of utmost importance to evaluate the efficacy of the treatment method.

REFERENCES

Azov, Y., Goldman, J. C. (1982). Free ammonia inhibition of algal photosynthesis in intensive cultures. Appl Environ Microbiol. 43:735–739.

Baldrian, P. (2008). Wood-inhabiting lignolytic basidiomycetes in soils: ecology and constraints for applicability in bioremediation. Fungal Ecol. 1:4–12.

Barr, D. P., Aust, S. D. (1994). Pollutant degradation by white rot fungi. Rev Environ Contam Toxicol. 138:49–72.

Behnisch, P. A., Hosoe, K., Sakai, S. (2001). Bioanalytical screening methods for dioxins and dioxin-like compounds—a review of bioassay/biomarker technology. Environ Int. 27:413–439.

Beszedits, S. (1983). Heavy metals removal from waste waters. Eng Dig. 18–25.

Borde, X., Guieysse, B., Delgado, O., Munoz, R., Hatti-Kaul, R., Nugier-Chauvin, C., Patin, H., Mattiasson, B. (2003). Synergistic relationships in algal-bacterial microcosms for the treatment of aromatic pollutants. Bioresour Technol. 86:293–300.

Bugg, T. D. H., Ahmad, M., Hardiman, E. M., Rahmanpour, R. (2011). Pathways for degradation of lignin in bacteria and fungi. Nat Prod Rep. 28:1883–1896.

Bugg, T. D. H., Ahmad, M., Hardiman, E. M., Singh, R. (2011). The emerging role for bacteria in lignin degradation and bio-product formation. Curr Opin Biotechnol. 22:394–400.

Bumpus, J. A., Tien, M., Wright, D., Aust, S. D. (1985). Oxidation of persistent environmental pollutants by a white rot fungus. Science. 228:1434–1436.

Buswell, J. A., Odier, E. (1987). Lignin biodegradation. Crit Rev Microbiol. 15:141–168.

Chandra, R., Raj, A., Purohit, H. J., Kapley, A. (2007). Characterisation and optimisation of three potential aerobic bacterial strains for kraft lignin degradation from pulp paper waste. Chemosphere. 67:839–846.

Chandra, R., Singh, S., Reddy, M. M. K., Patel, D. K., Purohit, H. J., Kapley, A. (2008). Isolation and characterization of bacterial strains *Paenibacillus* sp. and *Bacillus* sp. for kraft lignin decolorization from pulp paper mill waste. J Gen Appl Microbiol. 54:399–407.

Chen, C. Y., Lin, J. H. (2006). Toxicity of chlorophenols to *Pseudokirchneriellasubcapitata* under air-tight test environment. Chemosphere. 62:503–509.

Chen, Y. H., Chai, L. Y., Zhu, Y. H., Yang, Z. H., Zheng, Y., Zhang, H. (2012). Biodegradation of kraft lignin by a bacterial strain *Comamonas* sp. B-9 isolated from eroded bamboo slips. J Appl Microbiol. 112:900–906.

Christensen, T. H., Kjeldsen, P. (1989). Basic biochemical processes in landfills. In: Christensen, T. (ed.), Sanitary Landfilling: Process, Technology and Environmental Impact. Academic Press, London, UK, pp. 29–49.

Das, M. T., Budhraja, V., Mishra, M., Thakur, I. S. (2012). Toxicological evaluation of paper mill sewage sediment treated by indigenous dibenzofuran-degrading *Pseudomonas* sp. Bioresour Technol. 110:71–78.

Deguchi, Y., Toyoizumi, T., Masuda, S., Yasuhara, A., Mohri, S., Yamada, M., Inoue, Y., Kinae, N. (2007). Evaluation of mutagenic activities of leachates in landfill sites by micronucleus test and comet assay using goldfish. Mut Res. 5:178–185.

Eggen, T., Moeder, M., Arukwe, A. (2010). Municipal landfill leachates: a significant source for new and emerging pollutants. Sci Total Environ. 408:5147–5157.

Fenchel, T. (1974). Intrinsic rate of natural increase: the relationship with body size. Oecologia. 14:317–326.

Field, J. A., Sierra-Alvarez, R. (2008). Microbial degradation of chlorinated dioxins. Chemosphere. 71:1005–1018.

Furukawa, K. (2000). Biochemical and genetic bases of microbial degradation of polychlorinated biphenyls (PCBs). J Gen Appl Microbiol. 46:283–296.

Ghosh, P., Das, M. T., Thakur, I. S. (2014). Mammalian cell line-based bioassays for toxicological evaluation of landfill leachate treated by *Pseudomonas* sp. ISTDF1. Environ Sci Pollut Res. 21(13):8084–8094.

Ghosh, P., Gupta, A., Thakur, I. S. (2015). Combined chemical and toxicological evaluation of leachate from municipal solid waste landfill sites of Delhi, India. Environ Sci Pollut Res. 22:9148–9158.

Ghosh, P., Swati, Thakur, I. S. (2014). Enhanced removal of COD and color from landfill leachate in a sequential bioreactor. Bioresour Technol. 170:10–19.

Gupta, J., Rathour, R., Singh, R., Thakur, I. S. (2019). Production and characterization of extracellular polymeric substances (EPS) generated by a carbofuran degrading strain *Cupriavidus* sp. ISTL7. Bioresour Technol. 282:417–424.

Hamoda, M. F. (2006). Air pollutants emissions from waste treatment and disposal facilities. J Environ Sci Health ATox Hazard Subst Environ Eng. 41:77–85.

Hoffman, J. P. (1998). Wastewater treatment with suspended and nonsuspended algae. J Phycol. 34:757–763.

Ike, M., Kusunoki, K., Ueno, T., Soda, S., Fujita, M. (2005). Fungal bioreactor with ultra-membrane separation for degradation of colored and endocrine disrupting substances. Annual Report of FY 2004. The Core University Program between Japan Society for the Promotion of Science (JSPS) and Vietnamese Academy of Science and Technology (VAST), pp. 155–158.

Johnsen, A. R., Wick, L. Y., Harms, H. (2005). Principles of microbial PAH-degradation in soil. Environ Pollut. 133:71–84.

Jondeau, A., Dahbi, L., Bani-Estivals, M. H., Chagnon, M. C. (2006). Evaluation of the sensitivity of three sublethal cytotoxicity assays in human HepG2 cell line using water contaminants. Toxicology. 226:218–228.

Kanaly, R. A., Harayama, S. (2000). Biodegradation of high molecular weight polycyclic aromatic hydrocarbons by bacteria. J Bacteriol. 182:2059–2067.

Kapoor, A., Viraraghavan, T. (1995). Fungal biosorption—an alternative treatment option for heavy metal bearing wastewaters: a review. Bioresour Technol. 53:195–206.

Kaushal, R. K., Varghese, G. K., Chabukdhara, M. (2012). Municipal solid waste management in India—current state and future challenges: areview. Int J Eng Sci Technol. 4:1473–1489.

Kaushik, G., Gopal, M., Thakur, I. S. (2010). Evaluation of performance and community dynamics of microorganisms during treatment of distillery spent wash in a three stage bioreactor. Bioresour Technol. 101:4296–4305.

Kirkland, D. J., Aardema, M., Banduhn, N., Carmichael, P., Fautz, R., Meunier, J. R., Pfuhler, S. (2007). In vitro approaches to develop weight of evidence (WoE) and mode of action (MoA) discussions with positive in vitro genotoxicity results. Mutagenesis. 22:161–175.

Kumar, A., Bisht, B. S., Joshi, V. D., Dhewa, T. (2011). Review on bioremediation of polluted environment: amanagement tool. Int J Environ Sci. 1:1079–1093.

Kumar, D., Khare, M., Alappat, B. J. (2002). Threat to the Groundwater from the municipal landfill sites in Delhi, India. 28th WEDC Conference, Calcutta, India, 377–380.
Kurniawan, T. A., Lo, W. H., Chan, G. Y. (2006). Physico-chemical treatments for removal of recalcitrant contaminants from landfill leachate. J Hazard Mater. 129:80–100.
Lee, Y. K. (2001). Microalgal mass culture systems and methods: their limitation and potential. J Appl Phycol. 13:307–315.
Li, G., Sang, N., Guo, D. (2006a). Oxidative damage induced in hearts, kidneys and spleens of mice by landfill leachate. Chemosphere. 65:1058–1063.
Li, G., Sang, N., Wang, Q. (2006b). Oxidative damage induced in brains and livers of mice by landfill leachate. Ecotox Environ Safe. 65:134–139.
Lin, L., Chan, G. Y. Jiang, B. L., Lan, C. Y. (2007). Use of ammoniacal nitrogen tolerant microalgae in landfill leachate treatment. Waste Manag. 27:1376–1382.
Malik, A. (2004). Metal bioremediation through growing cells. Environ Int. 30:261–278.
Matejczyk, M., Płaza, G. A., Jawecki, G. N., Ulfig, K., Markowska-Szczupak, A. (2011). Estimation of the environmental risk posed by landfills using chemical, microbiological and ecotoxicological testing of leachates. Chemosphere. 82:1017–1023.
Mor, S., Ravindra, K., Dahiya, R. P., Chandra, A. (2006). Leachate characterization and assessment of groundwater pollution near municipal solid waste landfill site. Environ Monit Assess. 118:435–456.
Mosmann, T. (1983). Rapid colorimetric assay for cellular growth and survival: application to proliferation and cytotoxicity assays. J Immunol Methods. 65:55–63.
Narayana, T. (2009). Municipal solid waste management in India: From waste disposal to recovery of resources? Waste Manag. 29:1163–1166.
Natarajan, A. T., Darroudi, F. (1991). Use of human hepatoma cells for in vitro metabolic activation of chemical mutagens/carcinogens. Mutagenesis. 6:399–403.
Oswald, W. J. (2003). My sixty years in applied algology. J Appl Phycol. 15:99–106.
Pivato, A., Gaspari, L. (2005). Acute toxicity test of leachates from traditional and sustainable landfills using luminescent bacteria. Waste Manag. 26:1148–1155.
Reardon, K. F., Mosteller, D. C., Rogers, J. D. B. (2000). Biodegradation kinetics of benzene, toluene, and phenol as single and mixed substrates for *Pseudomonas putida* F1. Biotechnol Bioeng. 69:385–400.
Reddy, C. A., Mathew, Z. (2001). Bioremediation potential of white rot fungi. In: Gadd. G. M. (ed.), Fungi in Bioremediation. Cambridge University Press, London, pp. 52–78.
Saarela, J. (2003). Pilot investigations of surface parts of three closed landfills and factors affecting them. Environ Monit Assess. 84:183–192.
Safonova, E., Dmitrieva, I. A., Kvitko, K. V. (1999). The interaction of algae with alcanotrophic bacteria in black oil decomposition. Resourc Conserv Recycl. 27:193–201.
Safonova, E., Kvitko, K. V., Iankevitch, M. I., Surgko, L. F., Afri, I. A., Reisser, W. (2004). Biotreatment of industrial wastewater by selected algal-bacterial consortia. Eng Life Sci. 4:347–353.
Sahu, A. K. (2007). Present scenario of municipal solid waste (MSW) dumping grounds in India. Proceedings of the International Conference on Sustainable Solid Waste Management, Chennai, India, pp. 327–333.
Schoonen, W. G., De Roos, J. A., Westerink, W. M., Débiton, E. (2005). Cytotoxic effects of 110 reference compounds on HepG2 cells and for 60 compounds on HeLa, ECC-1 and CHO cells: II Mechanistic assays on NAD (P) H, ATP and DNA contents. Toxicology in vitro, 19(4):491–503.
Slack, R. J., Gronow, J. R., Voulvoulis, N. (2005). Household hazardous waste in municipal landfills: contaminants in leachate. Sci Total Environ. 337:119–137.
Talyan, V., Dahiya, R. P., Sreekrishnan, T. R. (2008). State of municipal solid waste management in Delhi, the capital of India. Waste Manag. 28:1276–1287.

Tchobanoglous, G., Burton, F. L., Stensel, H. D. (2003) Wastewater Engineering: Treatment and Reuse. McGraw-Hill, New York.

Thomas, D. J. L., Tyrrel, S. F., Smith, R., Farrow, S. (2008). Bioassays for the evaluation of landfill leachate toxicity. J Toxicol Environ Health B Crit Rev. 12:83–105.

Tice, R. R., Agurell, E., Anderson, D., Burlinson, B., Hartmann, A., Kobayashi, H., Miyamae, Y., Rojas, E., Ryu, J. C., Sasaki, Y. F. (2000). Single cell gel/comet assay: guidelines for in vitro and in vivo genetic toxicology testing. Environ Mol Mutagen. 35:206–221.

Tillitt, D. E., Papoulias, D. M., Whyte, J. J., Richter, C. A. (2010). Atrazine reduces reproduction in fathead minnow (*Pimephalespromelas*). Aquat Toxicol. 99:149–159.

Westerink, W. M., Schoonen, W. G. (2007). Cytochrome P450 enzyme levels in HepG2 cells and cryopreserved primary human hepatocytes and their induction in HepG2 cells. Toxicol In Vitro. 21:1581–1591.

Yadav, J. S., Quensen J. F. 3rd, Tiedje, J. M., Reddy, C. A. (1995). Degradation of polychlorinated biphenyl mixtures (Aroclors 1242, 1254, and 1260) by the white rot fungus *Phanerochaetechrysosporium* as evidenced by congener-specific analysis. Appl Environ Microbiol. 61:2560–2565.

Zegura, B., Heath, E., Cernosa, A., Filipic, M. (2009). Combination of in vitro bioassays for the determination of cytotoxic and genotoxic potential of wastewater, surface water and drinking water samples. Chemosphere. 75:1453–1460.

Zijffers, J. F., Janssen, M., Tramper, J., Wijffels, R. H. (2008). Design process of an area-efficient photobioreactor. Mar Biotechnol. 10:404–415.

Zimmo, O. R., van der Steen, N. P., Gijzen, H. J. (2004). Nitrogen mass balance across pilot-scale algae and duckweed-based wastewater stabilisation ponds. Water Res. 38:913–920.

3 Microbial Strategies for the Decolorization and Degradation of Distillery Spent Wash Containing Melanoidins

*P. Mahla and N. Bhatt**

CONTENTS

* Corresponding author: E-mail: bhattnikhil2114@gmail.com; and bhatt@gujaratvidyapith.org

3.1 INTRODUCTION

Industrialization and urbanization are the signs of progress of any country, but along with this progress, a huge amount of waste is also generated from the various industries. The increase in in distillery industries is due to the high demand, and the necessity of distillery products is continuously growing for molasses-based alcohol industries. Most sugar industries extend the distillery plant for managing molasses waste and make a profit by the production of alcohols, as sugarcane molasses is used as a raw material in alcohol production (Satyawali and Balakrishnan, 2008). It was reported in the last decade that 319 distillery industries in India alone produced more than 40 billion L of spent wash compared to 2.7 billion L of alcohol produced (Pant and Adholeya, 2007). Alcohol-producing distillery industries produce a huge amount of dark brown and highly acidic wastewater known as DSW. History has shown that in the distillation process, alcohol ranges between 5% and 12% by volume, and in this process, it produces the huge amount of spent wash: 88–95% by volume. Each sugar-based distillery industry generates 8–15 L spent wash per 1 L alcohol production (Wagh and Nemade, 2015).

TABLE 3.1
Characteristics of Raw Distillery Spent Wash and Anaerobically Treated Distillery Spent Wash

Sr. No.	Parameters	Distillery Spent Wash	Anaerobically Treated Distillery Spent Wash
1	pH	3.0–4.5	7.5–8
2	Total Solids (mg/L)	110000–190000	70000–75000
3	Total Volatile Solids (mg/L)	80000–120000	68000–70000
4	Total Suspended Solids (mg/L)	13000–15000	38000–42000
5	Total Dissolved Solids (mg/L)	90000–150000	30000–32000
6	Phenolics (mg/L)	8000–10000	7000–8000
7	Sulfate (mg/L)	7500–9000	3000–5000
8	Phosphate (mg/L)	2500–2700	1500–1700
9	Chloride (mg/L)	5000–8000	7000–9000
10	Total Nitrogen (mg/L)	5000–7500	4000–4200
11	BOD (mg/L)	50000–60000	8000–10000
12	COD (mg/L)	110000–190000	45000–52000

Among all types of wastewater, spent wash is one of the most complex hazardous materials to dump in environmental bodies because of high dissolved chemical moieties with dark brown pigmentation, acidic pH, high temperature and ash content (Pant and Adholeya, 2007). The spent wash is a hydrophilic viscous liquid amended with high organic and inorganic content is only reason for its high COD and high biological oxygen demand (BOD) (Tiwari et al., 2012). As per Chowdhary et al. (2018), COD ranges between 80000 and 100000 mg/L and BOD between 40000 and 50000 mg/L. Chemical content and melanoidin content are responsible for the complexity of spent wash (Chowdhary et al., 2018). According to Chavan et al. (2006), spent wash is considered generated liquid waste that has a strong objectionable odor.

Table 3.1 shows a comparison between the values of general DSW and anaerobically digested DSW characteristics (Mohana et al., 2009).

3.1.1 Major Pollutant: Melanoidin

The key pollutant of DSW is the recalcitrant compound termed melanoidin. DSW contains 2% melanoidin, which is responsible for dark brown color of spent wash (Arimi et al., 2014). Reducing sugars and amino acids is carried out by nonenzymatic reaction and yields melanoidins. This reaction is called the Maillard reaction (Martin et al., 2009). pH and temperature play a key role in this reaction as intensifiers for the process, e.g. 4–7 and ≥70–80°C, respectively (Mohana et al., 2009).

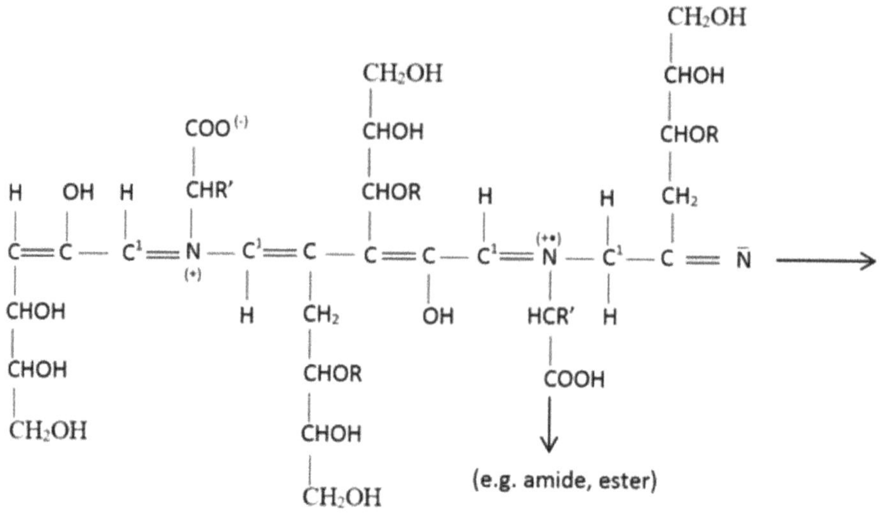

FIGURE 3.1 Melanoidin polymer's general structure.

3.1.2 Structure of Melanoidin Polymer

The complexity of the Maillard reaction is responsible for the complicated chemical structure of the melanoidin compound. Because of the fluctuating conditions of alcohol production, melanoidin composition cannot be assured, so its fundamental structure may change every time. As a result, melanoidin never produces repeating units of its composition molecules, and the general structure of synthetic melanoidin is prepared from monosaccharides and glycine (Jiranuntipon et al., 2009). The fundamental chemical structure is shown in Figure 3.1.

Also, Munde and Bhattacharjee (2015) had suggested a basic assembly of composition of the melanoidin structure containing an amino acid and carbohydrate compound, which is shown in Figure 3.2.

3.1.3 Nature of Melanoidin Polymer

The distillery and fermentation industries primarily release melanoidins, but some other food and drinking product industries also release melanoidins (Santal and Singh, 2013). When sugarcane molasses is distilled, not only are melanoidins produced by the Maillard reaction but also a dark brown-colored caramel formation materialized. Caramel is also responsible for the dark brown pigmentation of spent wash. Melanoidins can act as anionic hydrophilic polymers and create complexes with metal cations. Due to that, melanoidins are very recalcitrant in nature and difficult for microorganisms to degrade (Bharagava and Chandra, 2010). Melanoidins consist of acidic, long chains of polymers and extremely colloidal

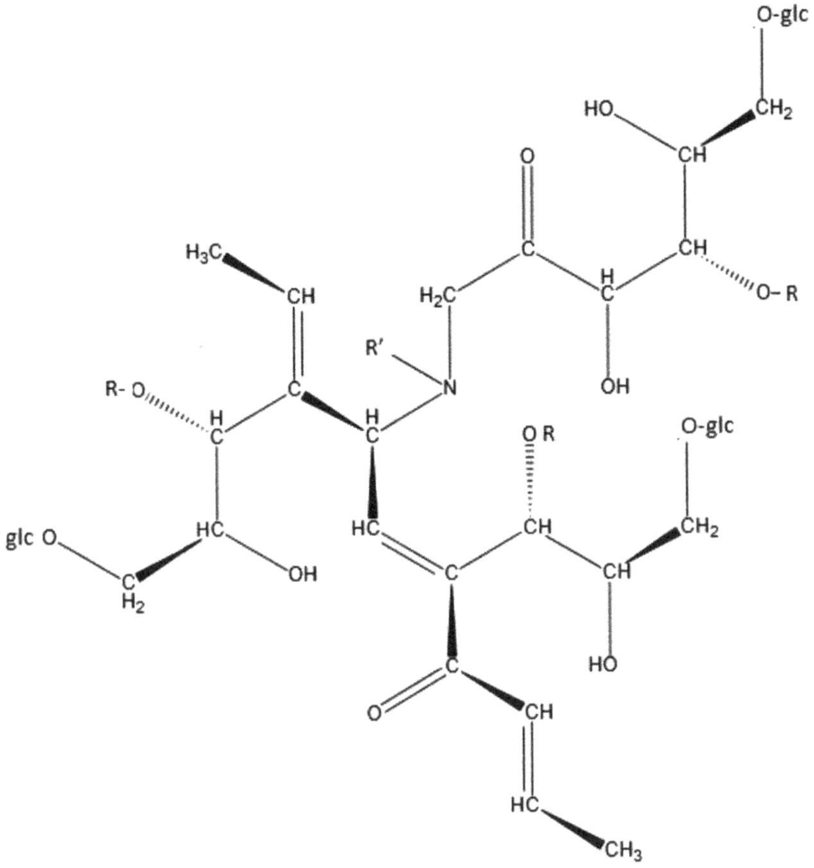

FIGURE 3.2 Defined structure of melanoidin.

negatively charged particles due to the breakdown of phenolic groups and carboxylic acids (Naik et al., 2008).

3.1.4 Chemistry and Formation of Melanoidins

At the beginning, Hayase et al. (1984) suggested the preparation of synthetic melanoidin from glucose and glycine and hydrogen peroxide, and achieved some facts about the functional groups and structure of melanoidins. It was verified as CH-COR molecules and C-terminal organizations of melanoidin from glucose, and the functional H or OH groups present in the melanoidin as are subsequently arranged. Through that acquaintance, Cammerer et al. (2002) synthesized melanoidin with several mixtures of sugars and amino acids, like glucose with glycine, maltose with glycine and lactose with glycine.

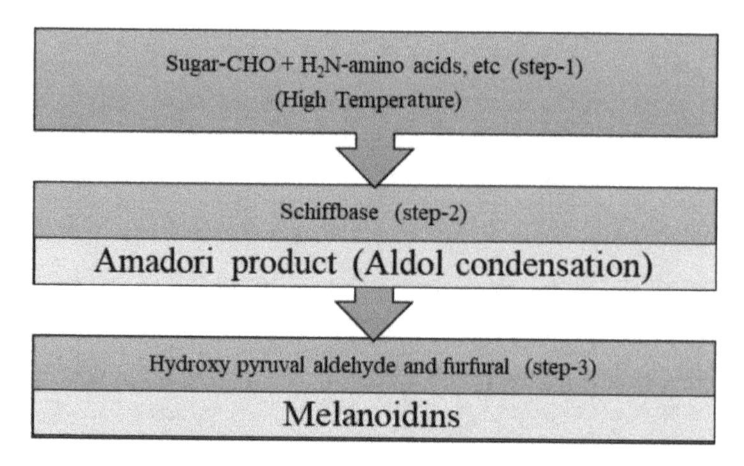

FIGURE 3.3 Melanoidin formation.

At high temperature (>60°C), sugar and amino acids react and produce melanoidins. As a result, Arimi et al. (2014) tried to define the melanoidin formation pathway in three steps, which is shown in Figure 3.3.

Another study involved synthetic melanoidins isolated at three different steps of polymers (Bharagava et al., 2009).

Step 1: One of the polymers had a nitrogen compound in their structure, e.g. $C_7H_{11}NO_4$.

Step 2: Another two polymers did not have nitrogen in their structure, e.g. $(CH_2O)_n$.

Step 3: The basic structure of glucose-derived melanoidins has the ability of metal chelating and acts as a complex material with anionic, which is hydrophilic (Cammerer et al., 2002).

After establishing an understanding of the melanoidin formation pathway, many successful attempts were carried out for synthetic melanoidin preparation and trends for investigation of degradation and decolorization. The synthetic melanoidin was prepared by heating an amino acid, sugar and Na_2CO_3 at above 100°C for a few hours (Bharagava et al., 2009).

Recently, Tiwari and Gaur (2019) reported a scheme of melanoidin generation, which is shown in Figure 3.4; the convincing Maillard reaction pathway with its derivatives and melanoidins as a final product were properly demonstrated. It was proved that melanoidins are acidic in nature. The high temperature and long reaction period result in an increase in the unsaturation of molecules, and due to that, the total carbon content rises. The polymerization degree increases with the color intensity. The amount of colorization commonly measured by absorbance at 420 nm is usually adapted to measure the extent of the Maillard reaction and its products.

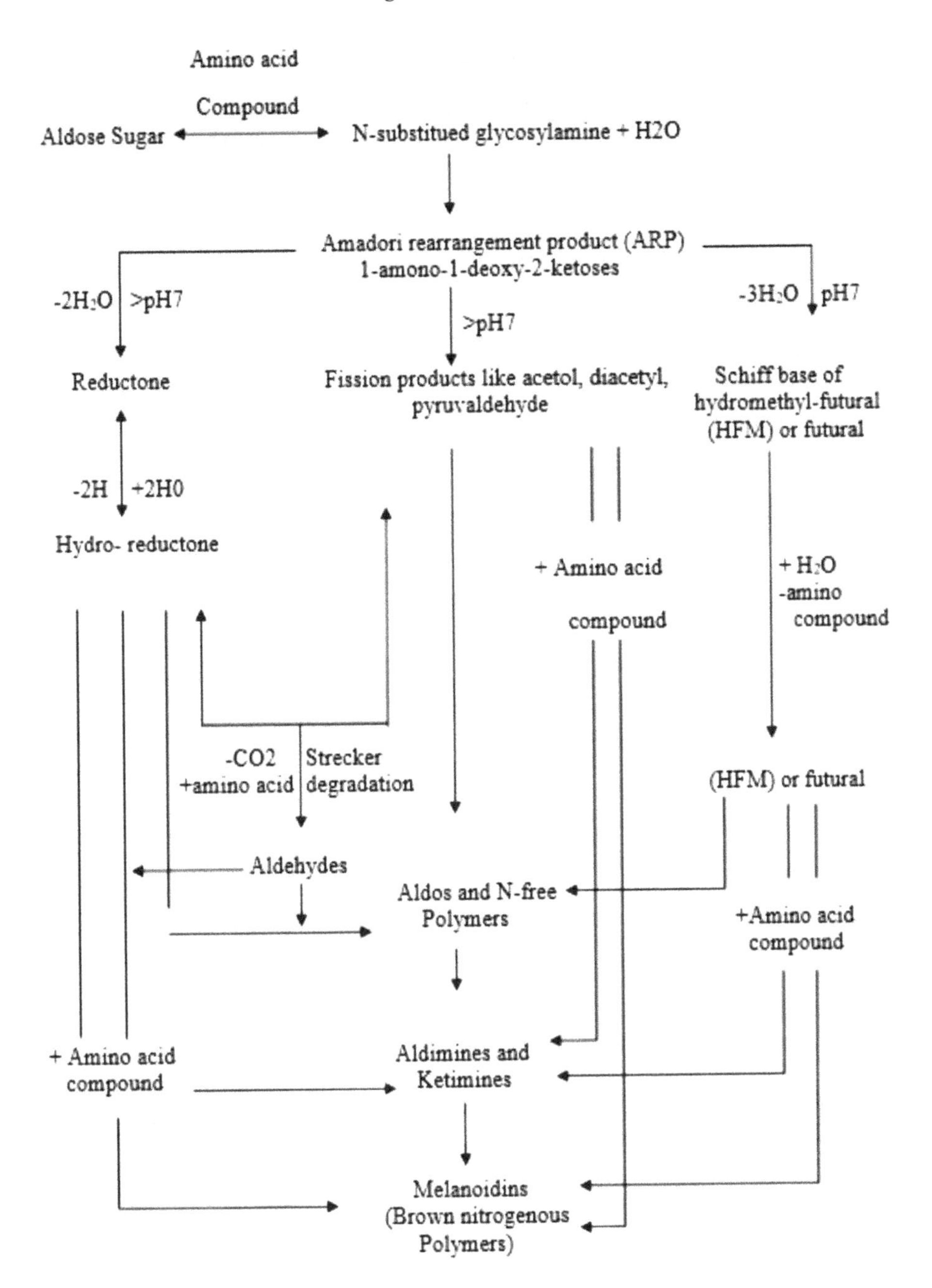

FIGURE 3.4 Pattern of melanoidins formation.

3.1.5 Other Components of Spent Wash

Spent wash contains not only melanoidin as pollutants but also other colorants and pollutants that create more complexity. Due to so many organic and complex materials, spent wash is more critical to treat and reuse.

3.1.5.1 Caramel

Reports have shown that the caramels derived from sucrose have been used commercially in food industries as a pigment agent (Myers and Howell, 1992). The reaction between proteins and sugars at high temperature yields a small amount of caramel. It is a brownish black and viscous fluid with a high molecular weight that may be greater than 10 kDa. Dehydration of monosaccharides to reactive groups creates a complex polymerization between them and produces a number of reaction series. While sugars are boiled at high temperature as well as acids and bases catalyzing that reaction, the caramelization reaction arises. In the sugar production mode, the caramels are shaped when the sucrose treacle is elevated above 210°C, predominantly in the crystallization stage, when crystals of sugars come in close contact with hot planes, vessels or containers (Arimi et al., 2014).

3.1.5.2 Polyphenol

Phenol groups containing polyphenols are antioxidants in nature, and their study is significant because of their distinct characteristics like antioxidancity, antimicrobial properties and anticarcinogenic characteristics (Jimoh et al., 2008; Dai and Mumper, 2010). Polyphenols look initially colorless, but after interaction with air, they can be oxidized and become a brownish-yellow colored compound. So many phenolic acids have been determined from DSW, such as benzoic acid and its derivatives gallic acid, cinnamic acid coumaric acid, caffeic acid, chlorogenic acid and ferulic acid. Also, spent wash was classified in three wide classes, e.g. phenolic acids, flavonoids and tannins (Payet et al., 2005, 2006; Incedayi et al., 2010).

3.1.5.3 Another Notable Component

The alkaline degradation products of hexoses (ADPH) with a short carbon chain (fewer than six carbons) are usually pigment-less with negligible antimicrobial properties, whereas a long chain with more than six carbons has diverse configurations according to the physical and chemical contributions in the method. Those extended APDH products that undergo polymerization yield colored compounds (Coca et al., 2008). Bacterial consortia had the ability to remove 75% of the color of ADPH color compounds from molasses distillery wastewater (Yadav and Chandra, 2012). Other heavy metals and sulfides have also been reported as components of spent wash (Yadav and Chandra, 2019).

3.2 HAZARDOUS EFFECTS OF SPENT WASH TOWARD ENVIRONMENT

The untreated DSW dumped into water bodies and on the land is lethal and contains critical pollutants. A high amount of COD and other components of spent wash can result in the eutrophication of water bodies such as river, ponds and oceans. Because the spent wash is a highly colored compound, it reduces sunlight diffusion in water bodies, which turns decreases both photosynthesis and dissolved oxygen concentration, upsetting the marine life.

3.2.1 Antimicrobial Activity of Melanoidin

The melanoidins have antioxidant characteristics first, which are responsible for their toxicity to microbes and a second recalcitrance characteristic to all biological treatment (Dandi et al., 2010). *Bacillus* spp., *Pseudomonas* spp., *Aeromonas* spp., *Fumigatus* and *Neurospora* intermediate have been used for the decolorization degradation of DSW (Ghosh et al., 2004). The comparison between the raw spent wash and anaerobically digested spent wash is recycled by irrigation. The raw spent wash resulted in a reduced count of bacteria (nitrogen fixers *Rhizobium* and *Azotobacter*) and actinomycetes, whereas the fungal count was increased due to the acidic condition. Anaerobically digested spent wash also showed similar results but not as much as the raw spent wash (Jiranuntipon et al., 2009).

3.2.2 Spent Wash Disposal Affects the Environment

Untreated or partially treated spent wash contains different inorganic and organic chemical content, and dumping of that in natural bodies of water affects the oxygen ingesting capability of a fish, *Labeo rohita*, during respiration. The dissolved oxygen consumption also is reduced because of coagulation and resulting asphyxiation. Different concentrations of spent wash had different toxic effects, so there are different toxicities for the freshwater crab, *Barytelphusa guerini* (Mohana et al., 2009). Discarding DSW directly on the land can affect the ecosystem: it is lethal to the plants and reduces soil alkalinity and manganese availability from the soil, and hence seed germination is affected (Santal and Singh, 2013). A study on the phytotoxicity of *Vigna radiata* seeds with 5% spent wash showed an effect on growth and germination because it had a low concentration of spent wash (Kannan and Upreti, 2008). Throwing DSW into the environment without appropriate treatment can also be extremely harmful for the quality of groundwater. Its physicochemical characteristics are altered, such as pigmentation, pH, conductivity, viscosity and others (Jain et al., 2005).

3.3 TREATMENTS BASED ON PHYSICOCHEMICAL METHODS

3.3.1 Adsorption

The activated charcoal method is the traditional method for colorant waste. Among the physicochemical treatment methods, adsorption on activated carbon is extensively used for the removal of color and specific organic pollutants. Activated carbon is considered a perfect adsorbent because its characteristics are favorable for the absorption process, such as extended surface area, microporous, immense adsorption capacity and large degree of surface reactivity. Activated carbon is able to decolorize spent wash by the adsorption method, which is profitable, as well as any indigenously synthetic or any prepared material (Satyawali and Balakrishanan, 2008). Activated carbon is helpful to remove melanoidin, but the disadvantages of modified plant materials give a low biodegradability to spent wash (Zhang et al., 2010).

3.3.2 Oxidation Processes

The oxidation process needs a great oxidizer, and ozone is a potent oxidant for water and wastewater treatment. Ozone and hydroxyl radicals have been reported as potent oxidant compounds and are able to oxidize various moieties (Pala and Erden, 2005). Biologically treated spent wash was treated by ozone oxidation and could achieve 80% decolorization as well as 15.25% COD reduction. Oxidation also resulted in improved biodegradability of the effluent (Pena et al., 2003). The combination of ozone and ultraviolet (UV) radiation boosted spent wash degradation by immediate removal of COD. By exposing this to ultrasound for 2 hours, it displayed 44% COD removal (Arimi et al., 2014).

3.3.3 Coagulation and Flocculation

For the removal of chemical content such as chromophore groups, total solids (TS), COD and BOD coagulation and flocculation methods are more effective. Coagulation is the method by which deterioration occurs in colloidal particles through neutralization and creates the forces that keep particles apart until the process is complete (Wagh and Nemade, 2015). The total dissolved solids (TDS) of wastewater can form small particles, which can be detached in the liquid stage. The most commonly used coagulant agents are Fe (iron), Al (aluminum) and Ca (calcium) salts. Some experimental data was reported that the low-molecular-weight products are more effectively removed by $FeCl_3$ (Liang et al., 2009). As per Dwyer et al. (2009), the coagulant aluminum had very low molecular weight, e.g. less than 10 kDa.

3.3.4 Membrane Treatment

Nataraj et al. (2006) reported a prime trial on hybrid nanofiltration (NF) as well as the reverse osmosis (RO) process for treatment. NF is mainly efficient at removing the color and colloidal material by a maximum of 80% and 45%, respectively. A cation-anion charged membrane is used for electrodialysis for desalting the spent wash and achieved 50–60% removal of potassium (Fan et al., 2011).

3.3.5 Evaporation and Combustion

An evaporation system with thermal vapor recompression is used as one of the treatments of spent wash, which comprises solids around 4% and could be concentrated at a high point of 40% TS in a quintuple-effect evaporator (Bhandari et al., 2004).

3.3.6 Other Treatments

Earlier, Pikaev (2001) treated DSW by radiation technology. That study included a combination of electron rays and coagulant $Fe_2(SO)_3$, which yielded an optical absorption reduction noted in the UV region with a maximum 65–70% treated DSW. Reports have also been produced on ultrasound technology used for the treatment of DSW. Experiments were carried out to check the effectiveness of the ultrasonic rays as a pretreatment phase, and it was noted that ultrasound treatment improved the biodegradability of the DSW (Sangave and Pandit, 2004). The catalytic thermolysis method was used for retreatment or a catalyzed thermolysis process and also recovered energy with COD and BOD removal. This process resulted in the configuration of solid remains that settled at the bottom, and the slurry obtained after the thermolysis exhibited very good filtration (Chaudhari et al., 2008).

3.4 TREATMENTS BASED ON BIOLOGICAL METHODS

Earlier several microorganisms were isolated and established as degraders of melanoidins. There were several aerobic and anaerobic methods used for decolorization and degradation of spent wash containing melanoidins. With the help of microbes, various bioreactors were developed for the treatment of DSW.

3.4.1 Sequential Strategies for the Treatment of DSW

The microbial treatments are eco-friendly and cheap compared to the chemical and physical methods. Living systems used for the treatment of spent wash can be aerobic, anaerobic or microaerophilic, and so many methods could be combined or sequential. Using different aerobic microorganisms for anaerobically treated DSW has also been widespread.

DSW was treated in a two-stage anaerobic digestion format. That two-stage method successfully achieved total solids reduction (66%), COD removal (81.7%) and an organic loading rate (OLR) of 7.02 kg TS m^3/day; total production of methane was 0.21 m^3 CH_4/kg with total solids supplemented (Wang and Banks, 2003).

Decolorization using *Cladosporium cladosporioides* of the anaerobically treated DSW was successfully investigated and showed a maximum decolorization of 52.6% and COD removal of 62.5% (Ravikumar et al., 2011).

Emericella nidulans var. lata, *Neurospora intermedia* and *Bacillus* spp. were used for the treatment of DSW in a three-stage bioreactor. Optimized parameters were conducted in the shaking condition. The entire procedure was first carried out with the fungi followed by *Bacillus* spp. The treated DSW presented decolorization of 82% and COD removal of 93% after 30 hours (Kaushik et al., 2010).

Anaerobically treated spent wash has also been treated with the physical-biological method and achieved maximum decolorization of 97.2% with the combination of coagulant and potent fungus, *Aspergillus niger* ATCC No. 26550 (Shukla et al., 2014). The potent fungi *Cladosporium cladosporioides* and cyanobacteria *Phormidium valdernium* was used for the two-stage sequential system for the treatment of DSW. The *Cladosporium cladosporioides* employed in the first-stage bioreactor achieved maximum decolorization of 68.5% and 81.37% COD removal with the fungus. The second stage of treatment was conducted with cyanobacteria, and COD reduction was 89.5% and color removal was 92.7% (Ravikumar and Karthik, 2015).

3.4.2 Anaerobic Bioreactors for the Treatment of DSW

As earlier reported, DSW is highly acidic and contains a high amount of COD, so the primary recommended treatments are anaerobic with a bioreactor.

3.4.2.1 Anaerobic Fixed Film Reactors

Perez-Garcia et al. (2005) deliberated the effect of the pH of the influents on the execution of the biodegradation process. Acharya et al. (2008) performed treatment of spent wash using anaerobic fixed film reactors with a comparative study of low-cost packing materials. Coconut coir was established to be the greatest behind material (that is, material on which biofilm developed), as the system supports the treatment at very elevated OLR of 31 kg COD m^3/day with 50% COD reduction. In fixed film reactors, the reactor has a biofilm support structure for biomass attachment. The representation of an anaerobic fixed film reactor offers the benefits of plainness of construction, removal of mechanical mixing, better stability even at high loading rates and capability to endure toxic shock loads. The reactors can recover very quickly after a period of starvation (Acharya et al., 2010).

3.4.2.2 Upflow Anaerobic Sludge Blanket (UASB) Reactors

The UASB is one of the most basic processes that has been successfully used for not only spent wash but also the treatment of various types of wastewaters. The UASB method is included as a highly anaerobic condition for the treatment of wastewater, and therefore it is preferable worldwide and is used in the design of reactors for spent wash treatment. The formation of active granules that can settle is responsible for the success of UASB. Granules form due to the aggregation of anaerobic bacteria, and they would self-immobilize and create compact arrangements. The development of this system enhances the satiability of the biomass of the reactor and leads to a valuable retention of bacteria. The designs of that type of reactor were only attractive because of its characteristics that included automatic mixing, recycling biomass of sludge and ability to make up when it goes to high loading rates (Sharma and Singh, 2002). Whiskey production has also produced spent wash and developed UASB reactors for primary treatment. The one- and two-stage UASB reactors were also carried out at psychrophilic temperature, e.g. 4–10°C for the treatment of DSW (Mohana et al., 2009).

3.4.2.3 Anaerobic Fluidized Bed Reactors

Highly motile organisms could also reduce impurities of wastewater. An anaerobic fluidized bed (AFB) reactor was also reported for DSW treatment. AFB contains a liquid or semiliquid medium which supports bacteria for attachment and helps in the floating state. A medium could be used such as sand, which contains small particles and activated carbon. In that fluid state, each medium makes available a huge surface area for biofilm development and the growth of microorganisms. It could maintain a steady state and enables the achievement of high bioreactor biofilm development (Jiranuntipon et al., 2009).

3.4.2.4 Anaerobic Lagoon

Anaerobic lagoons are the simplest choice for the anaerobic treatment of molasses wastewater. That was a traditional method for research in the field of distillery waste management by employing two anaerobic lagoons sequentially, resulting in BOD removal ranging from 82% to 92%. However, the lagoon systems are often operational, with being a repeated phenomenon (Singh et al., 2004; Naik et al., 2008).

3.4.3 Aerobic Approaches for the Treatment of DSW

Different pure cultures of fungi, bacteria and algae have been investigated particularly for their capability to decolorize the sugarcane spent wash. However, the presentation of fungal decolorization was limited by an extensive growth cycle and reasonable decolorization rate. In the spent wash or post-methanated spent wash, microbial treatments employing a pure bacterial culture have been reported frequently in the literature. A detailed list of several potent microorganisms have been used by different researchers for the decolorization of distillery effluent, as shown in Table 3.2. Fungi, bacteria and yeast were used for treatment.

3.4.3.1 Bacterial Strains for DSW Treatment

In the decolorization of spent wash, three bacterial strains—*Xanthomonas fragariae*, *Bacillus megaterium* and *Bacillus cereus*—were used and were immobilized on calcium alginate beads for the treatment of 33% predigested distillery effluent (Jain et al., 2002). The ability to decolorize distillery wastewater has been studied with lactic acid bacteria (Kaletunc et al., 2004), acetogenic bacteria, *Pseudomonas* spp. (Ghosh et al., 2004), *Paracoccus* spp. (Santal et al., 2016), *Bacillus* spp. (Tiwari et al., 2012), mixed cultures of bacteria of the genus *Bacillus* (Krzywonos et al., 2008) and consortium comprising *Bacillus licheniformis*, *Bacillus* spp. and *Alcaligenes* spp. (Bharagava and Chandra, 2010). The treated wastewater of the distillery industry leads to better growth of rice crops due to adequate nitrogen content and therefore can be used as a low-cost fertilizer (Satyawali and Balakrishnan, 2008). The bacteria *Bacillus* spp., *Pseudomonas* spp., *Aeromonas* spp., *Fumigatus* and *Neurospora* intermediates have been reported for the decolonization of diluted DSW (Dandi et al., 2013), but scanty efforts have been attempted to improve its remediation strategy.

3.4.3.2 Fungal Strains for DSW Treatment

Several fungi have been investigated for their ability to decolorized melanoidins, and spent wash *Coriolus* spp. in the Class Basidiomycetes was the first strain used due to its ability to remove melanoidins from spent wash; it obtained 80% decolorization in darkness under optimum conditions (Satyawali and Balakrishnan, 2008). Previously, the major work was done with fungus that had the potency to degrade and decolorize DSW: *Aspergillus,* e.g. *Aspergillus fumigatus* G-2-6, *Aspergillus niger, Aspergillus niveus, Aspergillus oryzae* JSA-1 and *Aspergillus fumigatus* UB260 carried out an

TABLE 3.2
Various Potent Microorganisms Performing a Key Role in the Treatment of Distillery Spent Wash

Culture	COD Removal (%)	Color Removal (%)
Fungi		
Marine basidiomycetes NIOCC	-	100
Mycelia sterilia	-	93
Rhizoctonia spp. D-90	-	90
Pleurotus spp.	-	90
Pleurotus flurida Eger EM1303	-	86.3
Coriolus no. 20	-	85
Phanerochaete chrysosporium	-	85 (Free),59 (Immobilized)
Phanerochaete chrysosporium NCIM 1106	-	82
Phanerochaete chrysosporium NCIM	-	82
Trametes versicolor	75	80
Coriolus hirsutus	-	80
Flavodon flavus	-	80
Flavadofiavus	-	80
Phanerochaete chrysosporium JAG-40	-	80
Aspergillus niger UM2	-	80
Phanerochaete chrysosporium NCIM 1197	-	76
Phanerochaete chrysosporium 1557	-	75
Aspergillus oryzae Y-2-32	-	75
Aspergillus UB2	-	75
Coriolus versicolor	70	71.5
Geotrichum candidum	-	70
Citeromyces spp. WR-43	-	68.91
Aspergillus niger	-	63
Coriolus versicolor	49	63
Pycnoporus coccineus	-	60
Bacteria		
Phanerochaete chrysosporium	73	53.5
Bacillus subtilis	-	85
Mixed culture of Bacillus subtilis and Pseudomonas aeruginosa	-	84.45
Pediococcus acidilactici B-25	85	79
Acetogenic bacteria BP103	-	76
Pseudomonas fluorescens	-	76
Xanthomonas fragariae	-	76
Bacillus cereus	81	75
Yeast		
Pseudomonas putida U	44.4	60
Candida tropicalis RG-9	-	75
Citeromyces spp. WR-43-6	99.38	68.91
Candida spp.	60	60
Cyanobacteria		
Oscillatoria boryana BDU 92181 (marine cyanobacteria)	-	75

average of 69–75% decolorization and 70–90% COD reduction (Mohammad et al., 2006; Agnihotri and Agnihotri, 2015). Pant and Adholeya (2007) isolated three fungi for the treatment of DSW, and the cultures were identified as *Penicillium pinophilum* TERI DB1, *Alternaria gaisen* TERI DB6 and *Pleurotus florida* EM 1303.

3.4.3.3 Yeast Strains for DSW Treatment

Some experiments on the semi-pilot and pilot scales were carried out with *Citeromyces* for checking the capability of *Citeromyces* spp. that achieved decolorization successfully (Sirianuntapiboon et al., 2004). Gupta et al. (2011) isolated and identified a yeast strain from the paper and pulp effluent, which was *Candida* spp. It could decolorize DSW almost 60% within 4 days incubation period at 38±1°C and pH 5.6. In addition that strain could reduce BOD and COD of the spent wash almost 78% and 60%, respectively. Tiwari et al. (2012) isolated potent yeasts which were thermotolerant in nature. Among 24 yeasts, Y-9 was able to achieve maximum decolorization, and that one was identified as *Candida tropicalis.*

3.4.3.4 Mixed Consortium Treatment for DSW Treatment

For treatment of spent wash, so many methods and strategies have been adapted. Mixed consortium treatment is the one of them. For example, Adikane et al. (2006) reported decolorization by 69% was obtained using 10% (w/v) soil and 12.5% (v/v) spent wash within a 7-day incubation period using mixed consortium. Tiwari et al. (2014) developed a consortium of a mixture of bacterium and yeast which resulted in the highest (82±1.5%) decolorization under static conditions for 24 hours and at 45°C. Those cultures were identified by 16S rDNA analysis as *Pediococcus acidilactici* and *Candida tropicalis.* Hereafter, it was proved that a consortium could degrade the spent wash potently. The bacterial consortia are more potent and significant than pure culture due to the preservation of the microorganism along with co-metabolism to boost the proficiency of melanoidin decolorization. It was also found that some vital supplementary nutrients and dilution of the spent wash obtained the best microbial activity as well as pollutant reduction (Arimi et al., 2014). Bioremediation of DSW by mixed consortia containing *Pseudomonas aeruginosa* and *Aspergillus niger* was reported. These studies found that the mixed consortia had the potency to degrade concentrated spent wash in a range of 75–80% (Nikam et al., 2014).

3.4.3.5 Immobilization of Microbial Cells for DSW Treatment

Various investigations have investigated the possibility of using vital cell immobilization in aerobic wastewater treatment. Early trials using certain pure cultures were immobilized on a solid surface for the degradation of specific toxic moieties. Later, there were developed immobilized consortia of two or more specific strains (Pant and Adholeya, 2007). Previous research in the last decade suggests that microorganisms such as *Pseudomonas, Bacillus* and *Flavodon flavus* had notable ability for decolorization and degradation by immobilization methods. Dahiya et al. (2001); Raghukumar et al. (2004); and Tiwari et al. (2014) experimented with the consortium immobilized technique by using *Pediococcus acidilactici* B-25 and *Candida tropicalis* RG-09, which exhibited maximum decolorization. Immobilized consortium showed the highest (85%) decolorization at optimized conditions like 2% (w/v) sodium alginate and 2%

(w/v) calcium chloride for a 16-hour preservation period; the beads formed were 2 mm in diameter from 5 g alginate beads. That study showed that the beads can be reused for almost 18 cycles without any decrease in their activity. Hence, the immobilized cells of the consortium within alginate beads are more effective for the spent wash treatment.

3.4.3.6 Algae and Cyanobacteria for DSW Treatment

Algal growth of a potential bioassay is a standard method to determine the potential of water bodies, natural waters and wastewaters bodies to support or inhibit the growth of microalgae. Algal growth potency was determined by anaerobically digested DSW with a combination of an anaerobic-aerobic system (Valderrama et al., 2002; Pant and Adholeya, 2007). These strategies involved in the treatment were mostly microalgae or algae with higher plant combinations (Mata et al., 2011) and also cyanobacteria and heterotrophic bacterial strain combinations (Satyawali and Balakrishnan, 2008). Most of the currently tested methods of anaerobically digested DSW remediation are methods for cultivating green microalgae (*Chlorophyta*), specifically the genus *Chlorella* was used the most because of its extraordinary stress tolerance and ability for mixotrophic growth (Perez-Garcia et al., 2005; Alcantara et al., 2015).

3.4.3.7 Phytoremediation for DSW Treatment

Phytoremediation of wastewater is still developing, but it appears to be a low-cost system for the removal of pollutants, including metals, from industrial wastewater. Kumar and Chandra (2006) successfully treated distillery effluent in a two-stage process involving the transformation of recalcitrant coloring components of the effluent by a bacterium, *Bacillus thuringiensis*, followed by subsequent reduction of the remaining load of pollutants by a macrophyte, *Spirodela polyrrhiza*. A similar biphasic treatment of the effluent was carried out in a constructed wetland with *Bacillus thuringiensis* and *Typha angustata* by Chandra et al. (2008), which resulted in 98–99% BOD, COD and color reduction after 7 days. The phytoremediation study of the accumulation of heavy metal and ultra-structural deviations used *Phragmites communis*, *Typha angustifolia* and *Cyperus esculentus* under both in situ and ex situ environments (Chandra and Sangeeta, 2011).

3.5 MECHANISM OF THE MICROBIAL DEGRADATION WITH RESPONSIBLE ENZYMES

The current trend for environmental pollution control is biodegradation processes with the help of various microorganisms. Every pollutant required specific enzymes to degrade or alter its configuration and convert the pollutant into nonhazardous waste. Melanoidins could also be decolorized and degraded by several enzymes, for example, peroxidases, oxidoreductases, cellulolytic enzymes, cyanidase, proteases and amylases (Santal and Singh, 2013). Among them significant ligninolytic enzymes are manganese peroxidase (MnP), lignin peroxidase (LiP) and laccase, reported by many authors for the degradation and detoxification of melanoidins (Sangeeta et al., 2011; Sangeeta and Chandra, 2012; Sangeeta and Chandra, 2013). Those enzymes have been reported for the decolorization and degradation of many recalcitrant compounds, as well as dyes. They also included as oxidoreductases two categories of peroxidases, e.g. lignin peroxidase and manganese peroxidase.

The following valuable information regarding these enzymes was reported by Tiwari and Gaur (2019):

3.5.1 Laccases

Laccase has been included in a multi-copper enzyme oxidase due to its ability to oxidize phenolic and aromatic complexes. That one is a type of benzenediol: oxygen oxidoreductase enzyme. It does not contain a heme group and does not require a co-substrate. Laccases act as a catalyzing agent in the process of oxidation of various aromatic hydrogen donors, along with reduction of O_2 to H_2O. Laccase frequently supports a glycosylation at a high degree, which helps it increase its resistance capacity when attacked by proteases.

3.5.2 Manganese Peroxidases (MnP)

Mostly MnP are produced by all white-rot fungi, decomposing fungi and some bacteria. These enzymes contain glycosylated glycoproteins bounded with an iron protoporphyrin IX prosthetic group. Their molecular weight is reported between 32 and 62.5 kDa. MnP is oxidized and stabilized by chelates such as oxalic acid and converts Mn^{2+} into Mn^{3+}. Chelated Mn^{3+} is extremely reactive with H_2O. Hence, MnP are capable of oxidation and depolymerization of recalcitrant substrates.

3.5.3 Lignin Peroxidases (LiP)

LiP act as catalysis in the process of the oxidation of non-phenolic aromatic lignin containing compounds and similar compounds. LiP has been reported for the mineralization of several recalcitrant and aromatic moieties, for example, polycyclic aromatic hydrocarbons (PAHs), some polychlorinated biphenyls and many dyes. Most LiP are extracellular, and the molecular weight ranges between 38 and 47 kDa, including a heme group on the active site, and react similar to the classical peroxidase mechanism. LiP needs H_2O_2 as a substrate, as well as mediator, like veratryl alcohol, to degrade lignin and other phenolic compounds. In this process H_2O_2 reduces to H_2O by gaining an electron from LiP, and oxidized LiP returns to its original reduced state by accepting an electron from veratryl alcohol and oxidizing it to veratryl aldehyde. That veratryl aldehyde gains an electron from lignin or analogous structures such as xenobiotic pollutants and is reduced to veratryl alcohol.

3.5.4 Versatile Peroxidases (VP)

The hybrid product of MnP and LiP is categorized as third group of peroxidases, e.g. versatile peroxidases. VP is capable not only of oxidizing Mn^{2+} but also phenolic and non-phenolic aromatic compounds, including dyes. VP has been reported

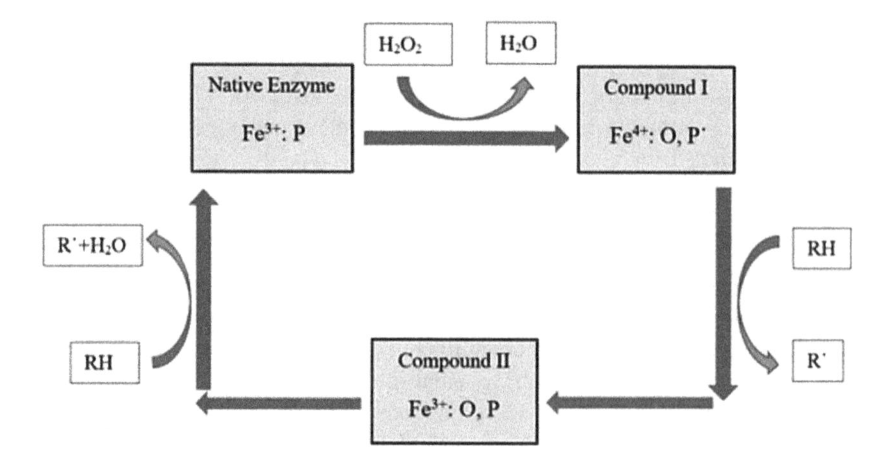

FIGURE 3.5 Cyclic pathway of catalytic peroxidases.

for species of *Pleurotus* and *Bjerkandera*. It shows combined characteristics for substrate as well as oxidation processes. It has a homologous Mn binding site similar to MnP and a tryptophan residue similar to the veratryl alcohol oxidation by LiP. The catalytic properties of that new peroxidase are due to a hybrid product arrangement, which is a combination of a diversity of substrate binding locations and oxidation sites.

Figure 3.5, modified from Yadav and Chandra (2019), shows the pathway of the catalytic cycle of common peroxidases. MnP, horseradish peroxidase (HRP) and LiP can resemble the cyclic pathway of heme peroxidases—they also were included as the native ferric enzyme. They had the reactive intermediate compounds (I and II). Among all peroxidases, MnP needs Mn^{2+} as the foremost substrate as an electron donor. Peroxidases' catalytic cycle could start with the reaction between native ferric enzyme Fe^{3+}: P, e.g. porphyrin and H_2O_2 producing compound I which is considered for reducing a substrate (RH) and producing a radical cation (R). The complex yielded is a porphyrin cation radical with a high-valent oxo-iron, e.g. Fe^{4+}: O, P. Compound I oxidized the enzyme intermediate compound II with one electron, e.g. Fe^{4+}: O, P. Subsequently, the oxidation of substrate particle by one electron yielded returning of enzyme to the native ferric state, concluding the catalytic cycle. As with MnP, activity is started by binding H_2O_2 or organic peroxide to the native ferric enzyme and establishing an iron-peroxide compound. Two electrons were necessary for the peroxide oxygen-oxygen bond and heme, resulting in MnP compound I formation, which was a Fe^{4+}-oxo-porphyrin-radical complex. Subsequently, the dioxygen bond cleaves heterolytically and one water mole is ejected. Later, reduction continues and yields the MnP compound II, e.g. Fe^{4+}-oxo-porphyrin complex. The Mn^{2+} ion is monochelated and serves as one electron donor for that porphyrin intermediate and converts Mn^{3+} via oxidation. The reduction process of compound II takes place in a similar way as reported earlier, and another Mn^{3+} is formed from Mn^{2+} by releasing the second molecule of water via transfer to the native enzyme (Yadav and Chandra, 2019).

Chandra et al. (2008) reported that the melanoidins decolorizing activity (MDA) of *Coriolus versicolor* was mainly due to intracellular enzymes that are induced by melanoidins. The induced enzyme consisted of two components, namely a sugar-dependent enzyme forming two-thirds and the other sugar-independent part constituting one-third of the system. Research has demonstrated the purified melanoidins decolorizing enzyme (MDE) isolated from *C. versicolor* and reported that the MDE of this strain was an intracellular enzyme consisting of a major P-fraction and a minor E-fraction. The P-fraction has at least five enzymes, which were of two types and required sugar or no sugar for the decolorizing activity.

3.6 BIOMETHANATED DSW AS A SUBSTRATE FOR VARIOUS PRODUCTS

DSW has various hazardous effects, but with some applicable angle, it is significant for many applied sectors and research areas. Either DSW or anaerobically treated DSW is commonly used currently for the purpose of producing valuable products such as biohythane, citric acid, enzymes, biohydrogen, biocompost, bioelectricity, biogas and methane, etc.

Biocomposting is another trend which is reported by Kannan and Upreti (2008). It is an aerobic method activated by a bioconversion process. Heterotrophies could access the carbonaceous materials and also the existence of inorganic resources significant for their growth. Biocompost was further used in the agriculture sector so it was significant trial. Anaerobically treated DSW and lignocellulosic biomass were used for laccase production by *Aspergillus heteromorphus*. The lignocellulosic biomass was rice straw, wheat straw and sugarcane bagasse. Several mediums were tested for enzyme production with and without lignocellulosic biomass supplements. The results showed the other media with anaerobically treated DSW induced more laccase production than the mineral medium. The addition of a lignocellulosic biomass enhances laccase production, and the highest laccase activity was obtained in 5% anaerobically treated DSW medium with rice straw (Singh et al., 2010). *Aspergillus ellipticus* was used for the production of cellulases by utilizing anaerobically treated DSW. The entire process was conducted under solid-state fermentation using wheat straw as a substrate. The physicochemical characteristics were included and modeled using a response surface methodology, as well as the Box–Behnken design (BBD). At optimized conditions, cellulose assayed with three substrates like the filter paper, b-glucosidase and endo-b-1,4-glucanase. The activities achieved 13.38, 26.68 and 130.92 U/g, respectively. Also the partial purification and characterization of endo-b-1,4-glucanase and b-glucosidase were achieved through the ammonium sulfate partial purification method, followed by desalting. The partially purified enzymes showed maximum activity at 60°C (Acharya et al., 2010).

The UASB reactor developed for the treatment of spent wash gained COD 93 (±3%) removal with 6.87 (±3.93) kg COD/m^3/day COD removal rate. When comparing nonlinear regression models, the projected multiple inputs and multiple outputs (MIMO) fuzzy-logic-based model exhibited superior production of both biogas and methane. The production rate was achieved with suitable fortitude coefficients above

0.98 (Aydinol and Yetilmezsoy, 2010). Roy et al. (2012) experimented with biohydrogen production with thermophilic mixed culture from spent wash using an anaerobic digester. Maximum hydrogen production was 3985 mL/L and yielded glucose 2.7 mol/mol. Mishra and Das (2014) investigated biohydrogen production from DSW by *Enterobacter cloacae* IIT-BT 08. It was estimated that an optimum hydrogen yield of 7.4 mol H_2/kg COD was reduced, and after supplementation, the hydrogen production rate was 80 mL/L/h average. A single-stage bioreactor with 34 L capacity was developed for biohythane production from spent wash treatment. After five cycles, the maximum biohythane production was 147.5±2.4 L along with a high production rate of 4.7±0.1 L/h. Also, there was 60% COD reduction during biohythane production. The composition of biohythane depended on the volatile fatty acids (VFA) production and consumption of a hybrid biosystem (Pasupulet and Mohan, 2015). Deval et al. (2016) reported bioelectricity production from distillery wastewater using a bioelectrochemical system with simultaneous waste remediation. All conditions were optimized and adapted a surface response methodology with BBD parameters such as the concentration of antifoam, pH and resistance. 8.3 and 1000 U resistance were selected as the optimum with no antifoam in the process, and 31.49 Wm-3 bioelectricity was generated from distillery spent wash. Recently, Aghera and Bhatt (2019) produced citric acid by *Aspergillus fumigatus* utilizing biomethanated DSW. At all optimized conditions 10.2 g/L production of citric acid was achieved.

3.7 SIGNIFICANT VISION FOR THE FUTURE

- Demand for alcohol and alcohol-containing products from distillery industries is widespread on a global level. But the generation of spent wash still needs to be addressed to achieve harmless dumping, and pollution needs to be removed.
- So many aerobic and anaerobic methods can be developed for the treatment of DSW.
- Potent microbes have the ability to decolorize and degrade under favorable conditions. Those microorganisms are bacteria, fungi, algae and cyanobacteria, and they can treat spent wash by enzymatic treatment, e.g. MnP, LiP and laccase.
- After treatment, detoxified spent wash can be recycled so it can be utilized in biocomposting and other agricultural applications.
- Biomethanated spent wash can be utilized in several industrial products like citric acid, hydrogen and many enzymes.

3.8 CONCLUSION

Raw DSW disposal is a critical issue for environmental health. But after primary anaerobic digestion, it becomes less harmful. Still, all impurities cannot be removed from biomethanated DSW, like the dark brown color of melanoidins and high amount of organic and inorganic TDS. All these complications can be treated by microorganisms with various strategies. Aerobic, microaerophilic and anaerobic bioreactors and enzymatic treatments play a key role. However, it is still a challenge for researchers and industries alike. The structure and chemistry of melanoidins are still not completely understood. In addition, due to the biodegradation of melanoidins and other compounds, valuable products can be achieved and may also play a role in the industrial sector.

REFERENCES

Acharya, B. K., Mohana, S., Jog. R., Divecha, J., Madamwar, D. 2010. Utilization of anaerobically treated distillery spent wash for production of cellulases under solid-state fermentation. *Journal of Environmental Management.* 91(10):2019–2027.

Acharya, B. K., Mohana, S., Madamwar, D. 2008. Anaerobic treatment of distillery spent wash: a study on up-flow anaerobic fixed film bioreactor. *Bioresource Technology.* 99:4621–4626.

Adikane, H. V., Dange, M. N., Selvakumari, K. 2006. Optimization of anaerobically digested distillery molasses spent wash decolourization using soil as inoculum in the absence of additional carbon and nitrogen source. *Bioresource Technology.* 97:2131–2135.

Aghera, P., Bhatt, N. 2019. Biosynthesis of citric acid using distillery spent wash as a novel substrate. *Journal of Pure and Applied Microbiology.* 13(1):599–607.

Agnihotri, S., Agnihotri, S. 2015. Study of bioremediation of biomethanated distillery effluent by *Aspergillus oryzae* JSA-1, using electron micrography and column chromatography techniques. *International Journal of Current Microbiology and Applied Science.* 4:308–314.

Alcantara, C., Fernandez, C., Garcia-Encina, P. A., Munoz, R. 2015. Mixotrophic metabolism of Chlorella sorokiniana and algal-bacterial consortia under extended dark-light periods and nutrient starvation. *Applied microbiology and biotechnology.* 99(5):2393–2404.

Arimi, M. M., Zhang, Y., Gotz, G., Kirimi, K., Geiben, S. U. 2014. Antimicrobial colorants in molasses distillery waste water and their removal technologies. *International Biodeterioration and Biodegradation.* 87:34–43.

Aydinol, F. I. T., Yetilmezsoy, K. 2010. A fuzzy-logic-based model to predict biogas and methane production rates in a pilot-scale mesophilic UASB reactor treating molasses wastewater. *Journal of Hazardous Materials.* 182:460–471.

Bhandari, H. C., Mitra, A. K., Kumar, S. 2004. Crest's integrated system: reduction and recycling of effluents in distilleries. In: Tewari, P. K. (ed.), Liquid Asset, Proceedings of Indo-EU Workshop on Promoting Efficient Water Use in Agro-based Industries. 167–169.

Bharagava, R. N., Chandra, R. 2010. Biodegradation of the major colour containing compounds in distillery wastewater by an aerobic bacterial culture and characterization of their metabolites. *Biodegradation.* 21:703–711.

Bharagava, R. N., Chandra, R., Rai, V. 2009. Isolation and characterization of aerobic bacteria capable of the degradation of synthetic and natural melanoidins from distillery wastewater. *World Journal of Microbiology and Biotechnology.* 25:737–744.

Cammerer, B., Jaluschkov, V., Kroh, L. W. 2002. Carbohydrates structures as part of the melanoidins skeleton. *International Congress Series.* 1245:269–273.

Chandra, R., Bharagava, R. N., Rai, V. 2008. Melanoidins as major colorant in sugarcane molasses based distillery wastewater and its degradation. *Bioresource Technology.* 99:4648–4660.

Chandra, R., Sangeeta, Y. 2011. Phytoremediation of Cd, Cr, Cu, Mn, Fe, Ni, Pb and Zn from aqueous solution using Phragmites communis, Typha angustifolia and Cyperus esculentus. *International Journal of Phytoremediation.* 13:580–591.

Chaudhari, P. K., Mishra, I. M., Chand, S. 2008. Effluent treatment for alcohol distillery: catalytic thermal pre-treatment (Catalytic thermolysis) with energy recovery. *Chemical Engineering Journal.* 136:14–24.

Chavan, M. N., Kulkarani, M. V., Zope, V. P., Mahulikar, P. P. 2006. Microbial degradation of melanoidins in distillery spent wash by an indigenous isolate. *Indian Journal Biotechnology*. 5:416–421.

Chowdhary, P., Raj, A., Bharagava, R. N. 2018. Environmental pollution and health hazards from distillery wastewater and treatment approaches to combat the environmental threats: a review. *Chemosphere*. 194:229–246.

Coca, M., Garcıa, M. T., Mato, S., Cart, A., Gonalez, G. 2008. Evolution of colorants in sugarbeet juices during decolorization using styrenic resins. *Journal of Food Engineering*. 89:429–434.

Dahiya, J., Singh, D., Nigam, P. 2001. Decolourisation of molasses wastewater by cells of *Pseudomonas fluorescens* immobilized on porous cellulose carrier. *Bioresource Technology*. 78:111–114.

Dai, J., Mumper, R. J. 2010. Plant Phenolics: extraction, analysis and their antioxidant and anticancer properties. *Molecules*. 15:7313–7352.

Dandi, B. N., Dandi, N. D., Shelar, R. D., Chaudhari, A. B., Chincholkar, S. B. 2010. Biological decolorization of high concentration distillery spent wash using newly isolated microbial strains. *Journals of Advances in Science and Technology*. 13(2):73–78.

Dandi, N. D., Dandi, B. N., Chaudhari, A. B. 2013. Bioprospecting of thermo- and osmotolerant fungi from mango pulp-peel compost for bioethanol production. *Antonie van Leeuwenhoek*. 103(4):723–736.

Deval, A. S., Parikh, H. A., Kadier, A., Chandrasekhar, K., Bhagwat, A. M., Dikshit, A. K. 2016. Sequential microbial activities mediated bioelectricity production from distillery wastewater using bio-electrochemical system with simultaneous waste remediation. *International Journal of Hydrogen Energy*. 1–12.

Dwyer, J., Griffiths, P., Lant, P. 2009. Simultaneous colour and don removal from sewage treatment plant effluent: alum coagulation of melanoidin. *Water Resource*. 43:553–561.

Fan, L., Nguyen, T., Roddick, F. A. 2011. Characterisation of the impact of coagulation and anaerobic bio-treatment on the removal of chromophores from molasses wastewater. *Water Resource*. 45:3933–3940.

Ghosh, M., Verma, S. C., Mengoni, A. 2004. Enrichment and identification of bacteria capable of reducing chemical oxygen demand of anaerobically treated molasses spentwash. *Journal of Applied Microbiology*. 96:1278–1286.

Gupta, M., Mishra, P. K., and Kumar, A. 2011. Decolorization of molasses Melanoidin by *Candida Sp. Indian Journal of Applied Pure Biology*. 26:199–204.

Hayase, F., Kim, S. B., Kato, H. 1984. Decolourisation and degradation production of the melanoidin by hydrogen peroxide. *Agriculture and Biological Chemistry*. 48:2711–2717.

Incedayi, B., Tamer, C. E., Copur, U. C. 2010. A research on the composition of pomegranate molasses. *Journal of Food Agriculture and Environment*. 24:37–47.

Jain, N., Bhatia, A., Kausik, R. 2005. Impact of post methanation distillery effluent irrigation on ground water quality. *Environmental Monitoring and Assessment*. 110:243–255.

Jain, N., Minocha, A. K., Verma, C. L. 2002. Degradation of predigested distillery effluent by isolated bacterial strains. *Indian Journal of Experimental Biology*. 40:101–105.

Jimoh. F., Adedapo, A., Aliero, A., Afolayan, A. 2008. Polyphenolic contents and biological activities of Rumex ecklonianus. *Pharmaceutical Biology*. 46:333–340.

Jiranuntipon, S., Delia, M. L., Albasi, C., Damronglerd, S., Chareonpornwattana, S. 2009. Decolorization of molasses based distillery wastewater using a bacterial consortium. *Science Asia*. 35:332–339.

Kaletunc, G., Lee, J., Alpas, H., Bozoglu, F. 2004. Evaluation of structural changes induced by high hydrostatic pressure in *Leuconostoc mesenteroides*. *Applied and Environmental Microbiology*. 70(2):1116–1122.

Kannan, A., Upreti, R. K. 2008. Influence of distillery effluent on germination and growth of mung bean (Vignaradiata) seeds. *Journal of Hazardous Materials*. 153:609–615.

Kaushik, G., Gopal, M., Thakur, I. S. 2010. Evaluation of performance and community dynamics microorganisms during treatment of distillery spent wash in three stage bioreactor. *Bioresource Technology.* 101:4296–4305.

Krzywonos, M., Cibis, E., Miskiewicz, T., Kent, C. A. 2008. Effect of temperature on the efficiency of the thermo- and mesophilic aerobic batch biodegradation of high-strength distillery wastewater (potato stillage). *Bioresource Technology.* 99:7816–7824.

Kumar, P., Chandra, R. 2006. Decolourisation and detoxification of synthetic molasses melanoidins by individual and mixed cultures of *Bacillus* spp. *Bioresource Technology.* 97:2096–2102.

Liang, Z., Wang, Y., Zhou, Y., Liu, H. 2009. Coagulation removal of melanoidins from biologically treated molasses waste water using ferric chloride. *Chemical Engineering Journal.* 152:88–94.

Martin, M. A., Ramos, S., Maotes, E., Ruffian-Henares, J. A., Morales, F. J., Bravo, L., Goya, L. B. 2009. Biscuit melanoidins of different molecular masses protect human HepG2 cells against oxidative stress. *Journal of Agriculture Food Chemistry.* 57:7250–7258.

Mata, T. M., Martins, A. A., Sikdar, S. 2011. Sustainability considerations of biodiesel based on supply chain analysis. *Clean Technologies and Environmental Policy.* 13:655–671.

Mishra, P., Das, D. 2014. Biohydrgen production from *Enterobacter cloacae* IIT-BT 08 using distillery effluent. *International Journal of Hydrogen Energy.* 39:7496–7507.

Mohammad, P., Azarmidokht, H., and Fatollah, M. 2006 Application of response surface methodology for optimization of important parameters in decolorizing treated distillery wastewater using *Aspergillus fumigatus* UB2.60. *International Biodeterioration and Biodegradation.* 57:195–199.

Mohana, S., Acharya, B. K., Madamwar, D. 2009. Distillery spent wash: treatment technologies and potential applications. *Journal of Hazardous Materials.* 163:12–25.

Munde, S., Bhattacharji, T. 2015. Distillery spent wash treatment: model degradation of synthetic melanoidin by in-situ chemical oxidation. *International Journal of Chemical and Physical Sciences.* 4:128–136.

Myers, D. V., Howell, J. C. 1992. Characterization and specification of caramel colours: an overview. *Food and Chemical Toxicology.* 30:359–363.

Naik, N. M., Jagadeesh, K. S., Alagawadi, A. R. 2008. Microbial decolorization of spentwash: a review. *Indian Journal of Microbiology.* 48(1):41–48.

Nataraj, S. K., Hosamani, K. M., Aminabhavi, T. M. 2006. Distillery wastewater treatment by the membrane-based nano-filtration and reverse osmosis processes. *Water Research Journal.* 40(12):2349–2356.

Nikam, S. B., Saler, R. S., Bholay, A. D. 2014. Bioremediation of distillery spent wash using *Pseudomonas aeruginosa*, *Aspergillus niger* and mixed consortia. *Journal of Environmental Research and Development.* 9(1):129.

Pala, A., Erden, G. 2005. Decolorization of a beker's yeast industry effluent by fenton's oxidation. *Journal of Hazardous Materials.* 127:141–148.

Pant, D., Adholeya, A. 2007. A biological approaches for treatment of distillery wastewater: a review. *Bioresource Technology.* 98:2321–2334.

Pasupulet, S. B., Mohana, V. S. 2015. Single-stage fermentation process for high value biohythane production with the treatment of distillery spent wash. *Bioresource Technology.* 189:177–185.

Payet, B., Shum Cheong Sing, A., Smadja, J., 2005. Assessment of antioxidant activity of cane brown sugars by ABTS and DPPH radical scavenging assays: determination of their polyphenolic and volatile constituents. *Journal of Agricultural and Food Chemistry.* 53:10074–10079.

Payet, B., Shum Cheong Sing, A., Smadja, J., 2006. Comaparison of the concentrations of phenolic constituents in cane sugar manufacturing products with their antioxidant activities. *Journal of Agricultural and Food Chemistry.* 54:7270–7276.

Pena, M., Coca, M., Gonzalez, R., Rioja, R., Garcia, M. T. 2003. Chemical oxidation of wastewater from molasses fermentation with ozone. *Chemosphere.* 51:893–900.

Perez-Garcia, M., Romero-García, L. I., Rodríguez-Cano, R., Sales-Márquez, D. 2005. Effect of the pH influent conditions in fixed film reactors for anaerobic thermophilic treatment of wince-distillery wastewater. *Water Science and Technology.* 51:183–189.

Pikaev, A. K. 2001. New environmental applications of radiation technology. *High Energy Chemistry.* 35:148–160.

Raghukumar, C., Mohandass, C., Kamat, S., Shailaja, M. S. 2004. Simultaneous detoxification and decolorization of molasses spent wash by the immobilized white-rot fungus *Flavodon flavus* isolated from a marine habitat. *Enzyme and Microbial Technology.* 35(2–3):197–202.

Ravikumar, R., Karthik, V. 2015. Effective utilization and conversion of spent distillery liquid to valuable products using an intensified technology of two-stage biological sequestration. *Chemical and Biochemical Engineering Quarterly.* 29(4):599–608.

Ravikumar, R., Vasanthi, N. S., Saravanan, K. 2011. Single factorial experimental design for decolorizing anaerobically treated distillery spent wash using cladosporium cladosporioides. *International Journal of Environmental Science and Technology.* 8:97–106.

Roy, S., Ghosh, S., Das, D. 2012. Improvement of hydrogen production with thermophilic mixed culture from rice spent wash of distillery industry. *International journal of hydrogen energy.* 37(21):15867–15874.

Sangave, P. C., Pandit, A. B. 2004. Ultrasound pretreatment for enhanced biodegradability of the distillery wastewater. *Ultrasonics Sonochemistry.* 11:197–203.

Sangeeta, Y., Chandra, R. 2012. Biodegradation of organic compounds of molasses melanoidin (MM) from biomethanated distillery spentwash (BMDS) during the decolourisation by a potential bacterial consortium. *Biodegradation.* 23(4):609–620.

Sangeeta, Y., Chandra, R. 2013. Effect of pH on melanoidin extraction from post methanated distillery effluent (PMDE) and its decolorization by potential bacterial consortium. *International Journal of Recent Scientific Research.* 4(10):1492–1496.

Sangeeta, Y., Chandra, R., Vibhuti, R. 2011. Characterization of potential MnP producing bacteria and its metabolic products during decolourisation of synthetic melanoidins due to biostimulatory effect of d-xylose at stationary phase. *Process Biochemistry.* 46:1774–1784.

Santal, A. R., Singh, N. 2013. Biodegradation of melanoidin from distillery effluent: role of microbes and their potential enzymes. *Biodegradation of Hazardous and Special Products.* 71–100.

Santal, A. R., Singh, N. P., Saharan, B. S. 2016. A novel application of *Paracoccus pantotrophus* for the decolorization of melanoidins from distillery effluent under static conditions. *Journal of Environmental Management.* 169:78–83.

Satyawali, Y., Balakrishnan, M. 2008. Wastewater treatment in molasses-based alcohol distilleries for COD and colour removal: a review. *Journal of Environmental Management.* 86(3):481–497.

Sharma, J., Singh, R. 2002. Effect of nutrient supplementation anaerobic sludge development and activity for treating distillery effluent. *Bioresource Technology.* 79:203–206.

Shukla, S. K., Tripathi, A., Mishra, P. K. 2014. Fungal decolorization of anaerobically biodegradation distillery effluent (ABDE) following coagulant pre-treatment. *International Journal of Science, Environment.* 3(2):723–734.

Singh, A., Bajar, S., Bishnoi, N. R., Singh, N. 2010. Laccase production by *Aspergillus heteromorphus* using distillery spent wash and lignocellulosic biomass. *Journal of Hazardous Materials.* 176:1079–1082.

Singh, P. N., Robinson, T., Singh, D. 2004. In treatment of industrial effluents distillery effluent. In: Pandey, A. (ed.), *Concise Encyclopedia of Bioresource Technology.* Food Products Press, New York, pp. 135–141.

Sirianuntapiboon, S., Phothilangka, P., Ohmomo, S. 2004. Decolourisation of molasses wastewater by a strain no. BP 103 of acetogenic bacteria. *Bioresource Technology.* 92:31–39.

Tiwari, S., Gaur, R. 2019. In treatment and recycling of wastewater from distillery. *Advances in Biological Treatment of Industrial Waste Water and their Recycling for a Sustainable Future.* Springer, Singapore, pp. 117–166.

Tiwari, S., Gaur, R., Rai, P., Tripathi, A. 2012. Decolorization of distillery effluent by thermotolerant *Bacillus subtilis. American Journal of Applied Sciences.* 9(6):798–806.

Tiwari, S., Gaur, R., Singh, A. 2014. Distillery spentwash decolorization by a novel consortium of *Pedicoccus acidilactici* and *Candia tropicalis* under static condition. *Pakistan Journal of Biological Sciences.* 17:780–791.

Valderrama, L. T., Del Campo, C. M., Rodriguez, C. M. 2002. Treatment of recalcitrant wastewater from ethanol and citric acid using the microalgae *Chlorella vulgaris* and the macro *Phytema minuscule. Water Research.* 36:4185–4192.

Wagh, M. P., Nemade, P. D. 2015. Treatment processes and technologies for decolorization and COD removal of distillery spent wash. *International Journal of Innovative Research in Advanced Engineering.* 7:2163–2349.

Wang, Z., Banks, C. J., 2003. Evaluation of two stage anaerobic digester for the treatment of mixed abattoir wastes. *Process Biochemistry.* 38:1267–1273.

Yadav, S., Chandra, R. 2012. Biodegradation of organic compounds of molasses melanoidin (MM) from biomethanated distillery spent wash (BMDS) during the decolorization by a potential bacterial consortium. *International Biodeteroation and Biodegradation,* 23:609–620.

Yadav, S., Chandra, R. 2019. In environmental health hazardous of post-methanated distillery effluent and its biodegradation and decolorization. *Environmental Biotechnology: For Sustainable Future.* Springer, Singapore, pp. 73–101.

Zhang, X., Li, Z., Lei, L. 2010. Biodegradation of reactive blue 13 in a two stageanaerobic/aerobic fluidized beds system with a *Pseudomonas spp.* Isolation. *Bioresource Technology.* 101:34–40.

4 Bioremediation of Polycyclic Aromatic Hydrocarbons (PAHs): An Overview

*Shalini Gupta and Bhawana Pathak**

CONTENTS

* Corresponding author: Email: bhawana.pathak@cug.ac.in

4.1 INTRODUCTION

Petroleum industries provide energy and other petro products used in daily life. This industrial sector has given rise to various serious environmentally threatening organic chemical compounds. There are clear environmental concerns related to the exploitation of oil reserves and the release of hydrocarbons in the ecosystem. International Energy Outlook (2014) highlights the universal consumption of petrogenic products and fuel escalation to 0.098 billion barrels/day in 2020 and 0.119 billion barrels/day in 2040 (Kumari et al., 2016). Polycyclic aromatic hydrocarbon (PAH) compounds are hazardous organic chemical compounds of petroleum-related products. Oil refineries are the chief source of PAH compounds and various toxic heavy metal pollutants due to meager control and execution, defective tools, and lack of conservation control measures (Tiwari et al., 2011). PAH compounds are particularly significant organic pollutants due to their carcinogenic and mutagenic characteristics—these are categorized under highly chained aromatic carbon compounds, including benzene rings with delocalizing double bonds. The US EPA listed PAH compounds as persistent organic pollutants consisting of fused aromatic rings in various structural configurations. Industry discharges effluent containing PAHs due to accidental spills that cause a hazardous effect on the terrestrial and aquatic environments. The remediation of PAH-contaminated environments is still a challenge today.

Presently, bioremediation is an evolving area of interest for the removal and complete mineralization of these complex compounds. Bioremediation is a developing and sustainable strategy, which includes the use of biological agents—specifically microorganisms—for the elimination of persistent pollutants. PAHs of low or high molecular weight may be biologically degraded by indigenous microorganisms able to synthesize enzymes that disrupt the strong carbon bond linkages of PAH compounds. Various bacteria and fungi have an inherent ability to utilize a single PAH compound as a sole carbon source—bacteria- and fungi-mediated degradation has been studied thoroughly (Radzi et al., 2015). Henceforth, it is necessary to focus on the issue of microbes distributed within soil, water, and sediment communities, containing PAHs at contaminated sites; future research on the utilization of potential microbes for the rapid mineralization of PAH via metabolic processes; and enzymatic activity with changes in environmental factors.

4.2 PHYSICOCHEMICAL PROPERTIES OF PAHs

PAHs are complex organic compounds including several fused benzene rings with resonance. PAH compounds with more aromatic ring structures are poorly soluble in organic solvents as well as in lipids. The PAH compounds generally possess high melting and boiling points and low vapor pressure (Masih and Taneja, 2006). The physical and chemical properties of 16 PAH compounds are listed in Table 4.1 and in Figure 4.1.

TABLE 4.1
Physicochemical Properties of PAHs

Compounds	Number of Rings (5 or 6)	Melting Points (°C)	Boiling Points (°C)	Vapor Pressure at 25°C (Pa)	Density	N-Octanol Water Partition Coefficient (log K_{ow})	Solubility in Water at 25°C (μg/L)	Henry Law Constant at 25°C (kPa)
Acenaphthylene	2(1)	92–93	–	8.9×10^{-1}	0.899	4.07	–	1.14×10^{-3}
Acenaphthene	2(1)	95	279	2.9×10^{-1}	1.024	3.92	3.93×10^{-3}	1.48×10^{-4}
Fluorene	2(1)	115–116	295	8.0×10^{-2}	1.203	4.18	1.98×10^{-3}	1.01×10^{-2}
Anthracene	3	216.4	342	8.0×10^{-4}	1.283	4.5	73	7.3×10^{-2}
Phenanthrene	3	100.5	340	1.6×10^{-2}	0.98	4.6	1.29×10^{-3}	3.98×10^{-3}
Fluoranthene	3(1)	108.8	375	1.2×10^{-3}	1.252	5.22	260	6.5×10^{-4}
Pyrene	4	150.4	393	6×10^{-4}	1.271	5.18	135	1.1×10^{-3}
Benzo[α] anthracene	4	160.7	400	2.8×10^{-5}	1.226	5.61	14	–
Chrysene	4	253.8	448	8.4×10^{-5}	1.274	5.91	2.0	–
Benzo[β]fluoranthene	4(1)	168.3	481	6.7×10^{-5}	–	6.12	1.2(20°C)	5.1×10^{-5}
Benzo[*k*]fluoranthene	4(1)	215.7	480	1.3×10^{-8}	–	6.84	0.76	4.4×10^{-5}
Benzo[α]pyrene	5	178.1	496	7.3×10^{-7}	1.351	6.5	3.8	3.54×10^{-5}
Benzo[*g,h,i*] perylene	6	278.3	545	1.4×10^{-8}	1.329	7.1	0.26	2.7×10^{-5}
Indeno[1,2,3-*c,d*]pyrene	5(1)	163.6	536	1.3×10^{-8}	–	6.58	62	2.9×10^{-5}(20°C)
Dibenzo[*a,h*]anthracene	5	266.6	524	1.3×10^{-8}	1.282	6.5	0.5 (27°C)	7×10^{-6}
Benzo[*e*]pyrene	–	–	–	–	–	–	–	–

Note: (1) here stands for 2 benzene ring with one central carbon 9 in structural arrangement.

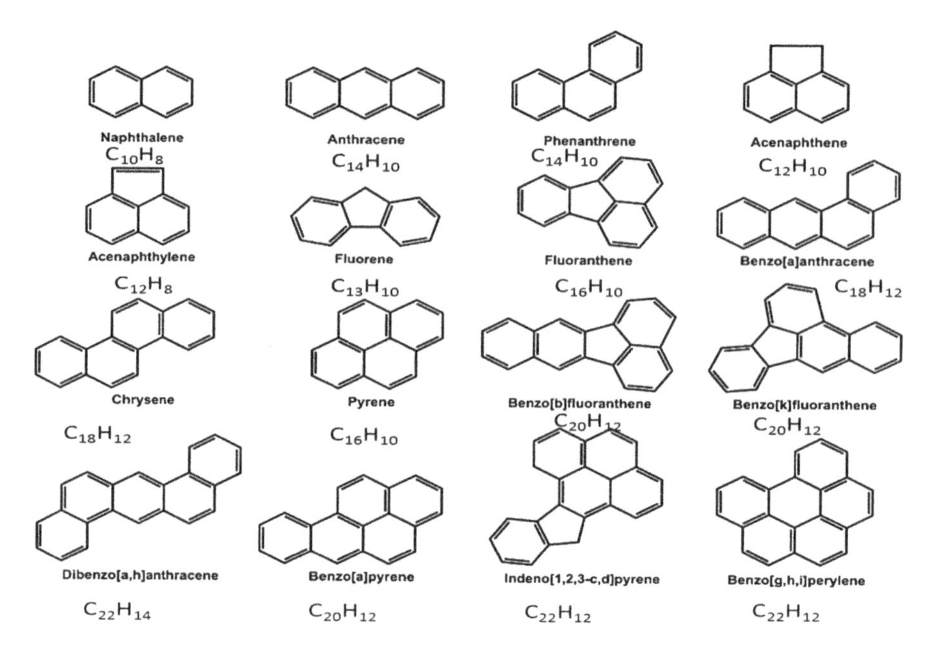

FIGURE 4.1 Structure of the 16 PAH compounds enlisted as priority pollutants by the US Environmental Protection Agency (USEPA).

4.3 INDUSTRIAL APPLICATION OF PAH COMPOUNDS

PAH compounds are synthesized chemically for industrial application and utilized as intermediates in different industrial sectors (Table 4.2). Asphalt utilized for the construction of roads and roofing tar contains different PAH compounds. Specific PAHs and other refined chemicals are used in the field of microelectronics and serviceable plastics.

TABLE 4.2
Application of Different PAH Compounds

PAH Compounds	Applications
Naphthalene	Production of azo dye, mothballs, agrichemicals, and insecticide
Acenaphthene	Dyes, plastics, and pharmaceuticals
Anthracene	Wood preservatives and pigments
Phenanthrene	Dyes, explosives, and drugs
Fluoranthene	Agrochemicals, dyes, and pharmaceuticals
Fluorene	Pharmaceuticals, pigments, dyes, pesticides, and thermoset plastic
Phenanthrene	Resins and pesticides
Pyrene	Pigments

4.4 SOURCES OF PAHs

Significant concentrations of PAH compounds that exist in the environment could be classified into different groups depending on the source, *viz.* pyrogenic and petrogenic. Pyrogenic PAH compounds are produced by partial ignition of organic matter exposed to high temperatures in the absence of oxygen. Emission sources of PAHs are mainly categorized into four classes: industrial emission, mobile sources, agricultural activities, and natural sources. The disparaging condensation of coal into coke and thermal cracking of petroleum give simpler hydrocarbon compounds at temperature ranges from 350°C to 1200°C or higher. The petrogenic PAHs are found in oil and oil-related stuffs. Pyrogenic PAHs can be distinguished from petrogenic PAHs on the basis of the five-membered benzene ring structure in the PAHs. Five-membered benzene ring hydrocarbons are present enormously in fuel-originated hydrocarbons as compared to pyrogenic hydrocarbons, because widespread hydrocarbon production favors the structural arrangement of benzene rings.

4.5 ENVIRONMENTAL FATE

4.5.1 PAHs in the Atmosphere

PAHs enter the atmosphere mainly from the partial ignition of organic substances. PAH compounds are found in higher concentrations in urban areas than in rural areas due to vehicle exhaust and industrial activity. PAH compounds exist in distinct phases, such as gaseous and solid phases, in the atmosphere where they sorb to suspended particles present in the atmosphere (Ravindra et al., 2008; Wang et al., 2013). Photooxidation is an important pathway of elimination of PAH compounds from the air. A correlation has been found among the different concentrations of dust in the air and PAH concentrations in the particulate matter (Kuo et al., 2013).

4.5.2 PAHs in the Land

PAH compounds present in the atmosphere are constantly deposited to the terrestrial ecosystem through automobile exhaust from nearby roadway sources (Figure 4.2). Major factors influencing PAH mobilization of suspended particles in the sub-surface would be the size of the sorbent particles or the pore size of the soil. Such soil pores could be defined as tiny pores among distinct particles of soil (Hussain et al., 2018). The tendency of PAHs to be sorbet to soil depends on both the physico-chemical characteristic of PAH compounds and the soil. The correlation between the partition coefficient of soil's physical and chemical properties depends on the percentage of organic carbon (EPRI, 2000).

4.5.3 PAHs in Water and Sediment

Coastal areas, lacustrine, and river environments receive PAHs from petroleum oil or accidental leaks of products, or by natural seepages; treated or untreated industrial and municipal wastewater, domestic stormwater; and recreational and commercial

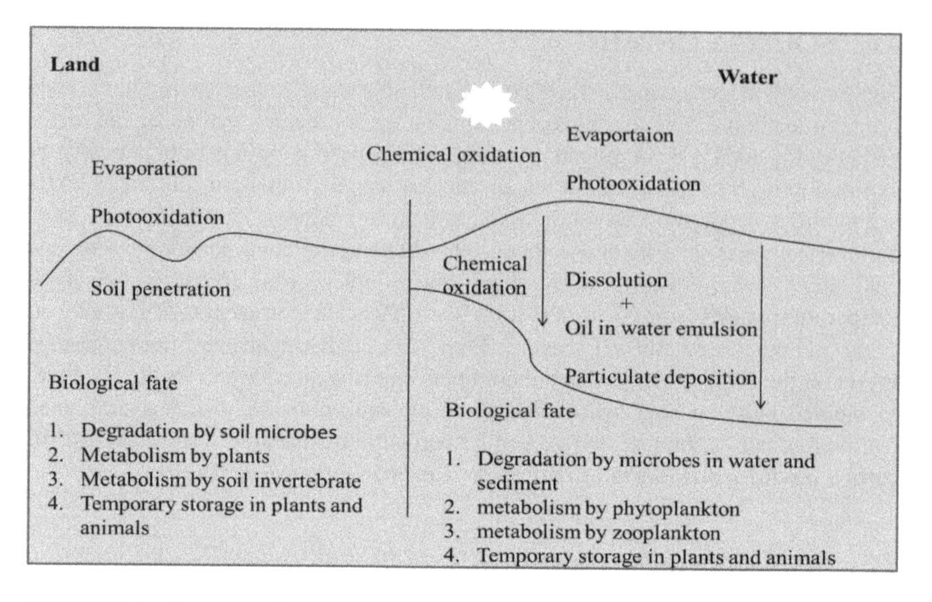

FIGURE 4.2 Representation of environmental fate of PAHs.

boats. Low-molecular-weight PAH compounds (one- and two-ring) are readily soluble in water. Seven-ring aromatics have low vapor pressure and are nearly insoluble in the aqueous medium. PAH compounds in contact with ultraviolet (UV) light in the atmosphere or the aqueous phase could be transformed into polar oxidized reactive chemical species. PAH compounds absorbed to suspended particulate matter present in air may settle down on the hydraulic surface, river, and ocean through desiccation and rain in rural areas. Carbon-based colloids can increase the concentration of PAH compounds due to the sorption capacity of PAHs on organic colloids, which results in transport of these compounds through the pore spaces present in the sediments and thus stimulates the bioavailability (Eisler, 2000; Abdel-Shafy and Mansour, 2016).

4.6 PAHs: EFFECT ON THE ECOSYSTEM

Ecotoxicology focuses on adverse effects at different levels of biological organization of anthropogenic compounds in ecosystems. Photooxidation of PAHs in the presence of sunlight may lead to the adsorbtion of the dust particles in the atmosphere. The biological availability of toxic pollutants is the essential parameter for uptake of concentration at the target sites in organisms and subsequent level of biotic ecosystems. The half-life of different PAH compounds in the presence of diverse oxidizers in the environment is listed in Table 4.3.

4.6.1 PAHs' Toxicity to Plants

The effects of PAHs on plant cell membranes were studied among various species and different hydrocarbons. PAHs do not eradicate plants, but they do inhibit plant development such as reduction in biomass or length of root and shoot. PAH noxiousness

TABLE 4.3
Half-Life of PAH Compounds by Different Environmental Oxidizers

PAHs	RO_2	Singlet Oxygen	O_3 (Water)	O_3 (Air)	Cl_2	HO
Anthracene	1,38,000	1,5				
Phenanthrene	2×10^{-4}, 2×10^{-8}					
Naphthalene					All PAHs	All PAHs
Benzopyrene		1,5	1, 1.05	1, 870	have $t_{1/2}$ <	have $t_{1/2}$
Dimethylanthracene		100, 0.5			0.5 hours	10 hours
Diphenylanthracene		8, 0.6				
Pyrene	0.12, 2.10^{-5}		1.5, 0.68	1.5, 560		
Pyrelene	1,38,000					
Benzanthracene		2, 10	2.4, 0.45			

Source: Radding et al. (1976).
Note: Half-life in hours.

could be employed in plant parts by damaging cell membranes, and hence it prevents the nutrients and metabolite transport in plants. PAHs are commonly recognized to cause genetic mutations, restriction in developmental processes, and may alter the physiological behavior of the plant to different kinds of environmental stress factors (Maliszewska-Kordybach and Smreczak, 2000). Alkino et al. (2005) reported that phenanthrene adversely affected the growth of *Arabidopsis thaliana*, indicated by a decrease in shoot and root elongation, deformation in trichomes, and chlorosis. Phenanthrene also affected the phenology of the experimental plant. Another research study on toxic effects of PAHs on plants showed that phenanthrene, pyrene, fluoranthene, and fluorine reduced the fresh and dry biomass of *Lolium perenne*, *Trifolium pratense*, and *Salix alba* (Sverdrup et al., 2003). Chouychai et al. (2007) found the lowest seed germination rate of edible crops such as corn, groundnut, cow pea, and mung bean in acidic soil treated with phenanthrene and pyrene.

4.6.2 PAHs' Toxicity to Terrestrial and Aquatic Invertebrate and Vertebrate Life

PAH compounds are discreetly recalcitrant in the ecosystem due to their persistence and hydrophobicity, hence leading to their bioaccumulation. Biological accumulation may also be observed in land-inhabiting invertebrates (Tudoran and Putz, 2012; Abdel-Shafy and Mansour, 2016). PAHs adversely affect organisms by their toxic action; the toxicity mechanism is believed to be interference with the function of cellular metabolic and enzymatic reactions that are linked to the cell membrane. PAH compounds may reflect toxic effects on animals by the formation of reactive metabolites, epoxides, and dihydrodiols, and may bind to DNA that results in DNA adduction. The biochemical turbulences and cellular damage may lead to mutation and abnormal cell growth (Santodonato et al., 1981; Neff, 1985; Varanasi, 1989;

Eisler, 2000; Albers, 2003). High-molecular-weight (more than three rings) PAH compounds have massive carcinogenic potential as compared to low-molecular-weight (two or three rings) PAH compounds (Varanasi, 1989; Albers, 2003). Van Brummelen et al. (1996) reported the ecological toxicity of complex PAH compounds in different isopod species exposed to contaminated food with fluorene, phenanthrene, and fluoranthene at specific concentrations.

4.7 EFFECT ON HUMAN HEALTH AND ENVIRONMENTAL REGULATION

PAH compounds have been known as a key concern due to the exposure and adverse effect on human health. PAHs may cause metabolic disruptions and cellular damage, which results in mutations, tumors, and cancer. Chronic exposure to PAH compound may lead to reduced immune function and respiration illness, and its regular skin contact could cause inflammation (Khairy et al., 2009). Naphthalene exposure may lead to the breakdown of RBCs if inhaled or ingested at levels above a permissible limit.

Mixtures of PAH compounds are carcinogenic and mutagenic to the human body (Table 4.4)—mainly workers exposed to occupational sources, and the chronic exposure of PAH results in increased risk of dermis inflammation, disturbance in respiratory organs, and gastrointestinal cancer (Lippman and Hawk, 2009). Benzo[*a*]pyrene

TABLE 4.4
Potential Carcinogenicity and Bioactivity of PAH

PAHs	Carcinogenicity Potential	Bioactivity	Exposure Limits
Naphthalene	0	0	10 ppm (OSHA)
Anthracene	0	0	0.2 mg/m^3 (OSHA)
Phenanthrene	0	0	0.2 mg/m^3 (OSHA)
Fluoranthene	0	CC	NDA
Pyrene	0	TI	NDA
Benzo[*a*]anthracene	+	TI	NDA
Chrysene	+	TI	0.2 mg/m^3 (OSHA)
Benzo[*b*]fluoranthene	++	C, TI	NDA
Benzo[*k*]fluoranthene	0	0	NDA
Benzo[*a*]pyrene	+++	C, TI	0.2 mg/m^3 (OSHA)
Dibenzo[*a,h*]anthracene	+++	C, TI	NDA
Benzo[*g,h,i*]pyrene	0	CC	0.1 and 1.3 ng/m^3 (WHO)
Indeno [1,2,3-*c,d*] pyrene	+	TI	NDA

Source: WHO (1971), IARC (1973), NIOSH (1977), OSHA (1983).
Note: + to +++active CC = carcinogen with BaP, TI = tumor initiator, C = complete carcinogen, 0 = inactive, NDA = no data available.

is the most common PAH to cause cancer in animals, and it is identified as the first carcinogenic chemical (IARC, 1987; Rengarajan et al., 2015). To control the PAH concentrations in different environment matrixes, several government bodies and the EPA have drawn a permissible range of concentrations acceptable in the environment. Various environmental regulation bodies have laid down standards that are applicable to PAH compound exposure under the Occupational Safety and Health Administration's (OSHA) Air Contaminants (mainly volatiles) Standard for fossil fuel emissions. It is important for industry workers to use engineering controls and work practices to avoid chronic PAH exposure at the workplace (Abdel-Shafy and Mansour, 2016).

4.8 PHYSICAL AND CHEMICAL TREATMENT TECHNOLOGY

PAH pollutants can be screened out, separated, and removed by using physical and chemical treatment methods. In situ treatment for a PAH-affected site is mainly carried out by organic solvents or distilled water, but due to their toxicity and high cost, this method is ineffective in the removal of higher-molecular-weight PAH compounds. Various solvents, e.g., ethanol, 2-propanol, acetone, and ethyl acetate, are known to remove PAH compounds, but more complex PAH compounds still remain in the soil after extraction (Khodadoust et al., 2000; Lee et al., 2001). Song et al. (2012) reported that the extraction process with distilled water and ultrasonic power could stimulate the desorption of phenanthrene from soil, improving the performance of the soil washing with distilled water separately. This study provides a basis for future prospects of research to adapt different extraction agents in place of distilled water in the combination of washing the soil with an ultrasonic treatment approach for PAH-contaminated sites (Lau et al., 2014).

Oxidation techniques are also used for PAH removal using hydrogen peroxide, ozone, and permanganate (Crimi et al., 2009; Gitipour et al., 2018). In an advanced oxidation technique, the Fenton oxidation method is used to increase the removal rate of PAH components (Gan and Ng, 2012a, 2012b; Veignie et al., 2009; Gitipour et al., 2018; Gitipour et al., 2018). Reclamation of soil by ozone injection into liquid or air inside wells is performed to clean up the low-molecular-weight PAH compounds (Rivas, 2006; Gan et al., 2009; Gitipour et al., 2018).

4.9 BIOREMEDIATION TECHNOLOGY

Physicochemical treatments for the removal of PAH compounds are costly and arduous (Zhang et al., 2014; Sibi et al., 2015), whereas bioremediation using microbes to improve the cleanup of hydrocarbons and having no other toxic pollutants has been proven as an effective technology for PAH removal (Aislabie et al., 1998). A bioaugmentation study by amending nutrients at contaminated Arctic soil was demonstrated, and the maximum stimulus in microorganisms action was found at the lowest level of nutrients (Braddock et al., 1997).

An indigenous bacterial consortium has been applied to petroleum-contaminated sites containing hydrocarbon, and survival rates of microorganisms were assessed (Mishra et al., 2001). In microcosm studies at the laboratory scale, PAH degradation was positively attained by indigenous microbes. Contaminated Kuwaiti soil was studied for the degradation of PAH compounds in the soil microcosm in which 20–45% PAH removal was reported (Horinouchi et al., 2000). The inadequate attainment rates for PAH degradation may be due to heavy contamination of sampling sites and microorganisms unable to survive after the first inoculation. The use of genetically engineered microorganisms (GEMs) may influence heavily contaminated sites for bioremediation of toxic pollutants (Samanta et al., 2002). Lahkar and Deka (2016) reported anthracene degradation by *Aspergillus* sp. strain HJ1 amended with anthracene at 20 ppm in 20 mL of minimum salt medium (MSM) broth in shake flask culture. Anthracene was degraded up to 25.4%, 39.2%, 40.2%, 41.8%, 48%, and 55%, observed on 3, 6, 9, 12, 15, and 18 days of incubation, respectively. Similarly, biodegradation studies of PAH compounds in broth cultures were carried out. The degradation rate of a single PAH compound (200 ppm phenanthrene, 200 ppm pyrene, and 50 ppm benzo [α] pyrene) by microbial consortia showed the maximum degradation efficiency (Kuppusamy et al., 2016).

Lors et al. (2012) used a fungal bioremediation strategy for soils contaminated with PAH compounds. Though several studies deal with observing PAH-polluted soils in laboratory, in situ or ex situ but in both the conditions, the difficulty is whether ex situ studies can be similar to in situ bioremediation was hardly revealed (Ahtiainen et al., 2002; Robinson et al., 2003; Dandie et al., 2010; Lors et al., 2010); hence there is a pressing requirement for field-scale authentication of laboratory methods (Diplock et al., 2009; Lors et al., 2012).

4.9.1 Bacterial Remediation

Microbes use many different types of metabolic strategies by which they acquire nutrition and growth multiplication. In biodegradation studies, bacteria use PAHs as a sole carbon source for metabolism. The minimum aqueous solubility and maximum octanol-water partition coefficient (log K_{ow}) of PAH compounds results in the partition into the cell wall structure of bacteria (Figure 4.3). The partition of PAHs to cell walls is carried out by reflexive transport to a concentration gradient from the surrounding to the cell membrane (Miyata et al., 2004). Such transport mainly depends on PAH concentration and availability of contaminants to microbes. Maximum bioavailability of a PAH leads to higher transportation to the microbial cells. Several bacteria could use this mechanism to facilitate fast degradation of PAH compounds, and accumulation of some PAH compounds may cause disruption to the cell wall and membrane proteins that eventually cause cell death (Sikkema et al., 1995). Yet some bacteria appear to adapt to minimum availability and PAH concentration. The adaptation mechanism might include the direct contact to solid phase PAHs (Wick et al., 2002). Bugg et al. (2000) revealed that an active efflux mechanism is present in *Pseudomonas fluorescens* LP6a for transference of complex PAH compounds. Partial research studies have

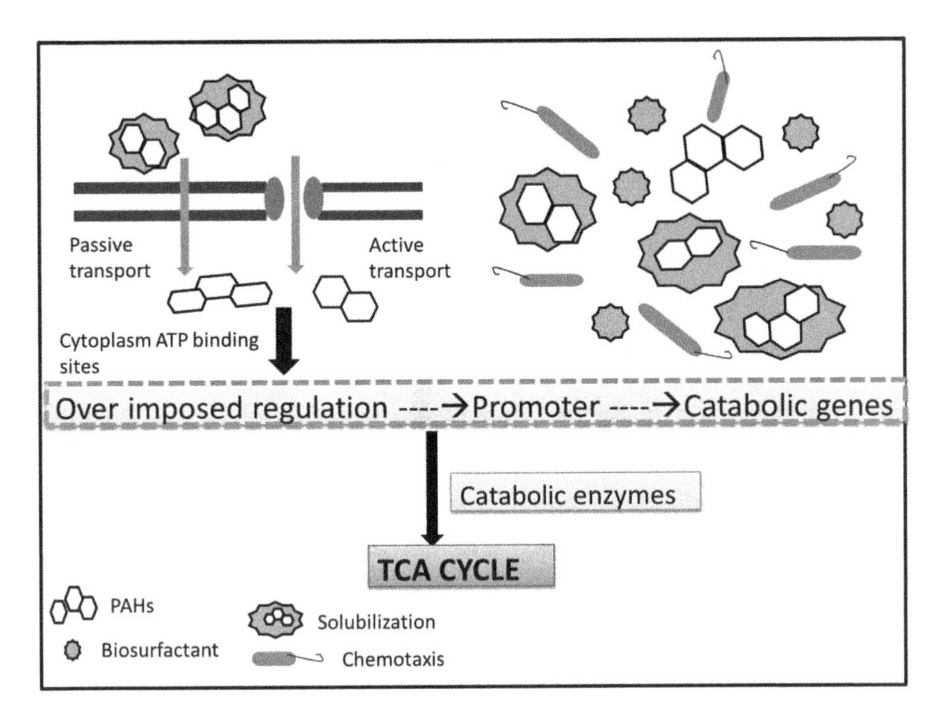

FIGURE 4.3 Major approach adapted by bacteria for PAH degradation.

been conducted on the cellular binding and transport of PAH compounds in aqueous medium by microbes.

Aerobic PAH transformation primarily includes dioxygenase or monoxygenase activity (Figure 4.3), which involves the incorporation of molecular oxygen into the aromatic nucleus that leads to the oxidation reaction (Patel and Ainsley, 1980; Samanta et al., 1998, 2002). The biodegradation and transformation of PAH compounds into different metabolites depend on the substituent breakdown reaction that may be initiated by two hydroxyl groups (–OH). The *cis*-dihydrodiols produced may further oxidize to dihydroxy compounds (Johnsen et al., 2005; Chauhan et al., 2008).

Additional reactions result in precursors of tricarboxylic acid cycle intermediate compounds. While PAHs undergo the biodegradation, the efficiency rate of the degradation pathway and kind of reaction metabolites production depend on the number of fused benzene rings in a PAH compound.

4.9.2 Fungal Remediation

Various fungi have been studied for PAH transformation. PAH-degrading fungi are classified into ligninolytic and non-ligninolytic fungi (Cerniglia, 1997). Key mechanisms included in PAH fungal degradation involve the activity of the cytochrome

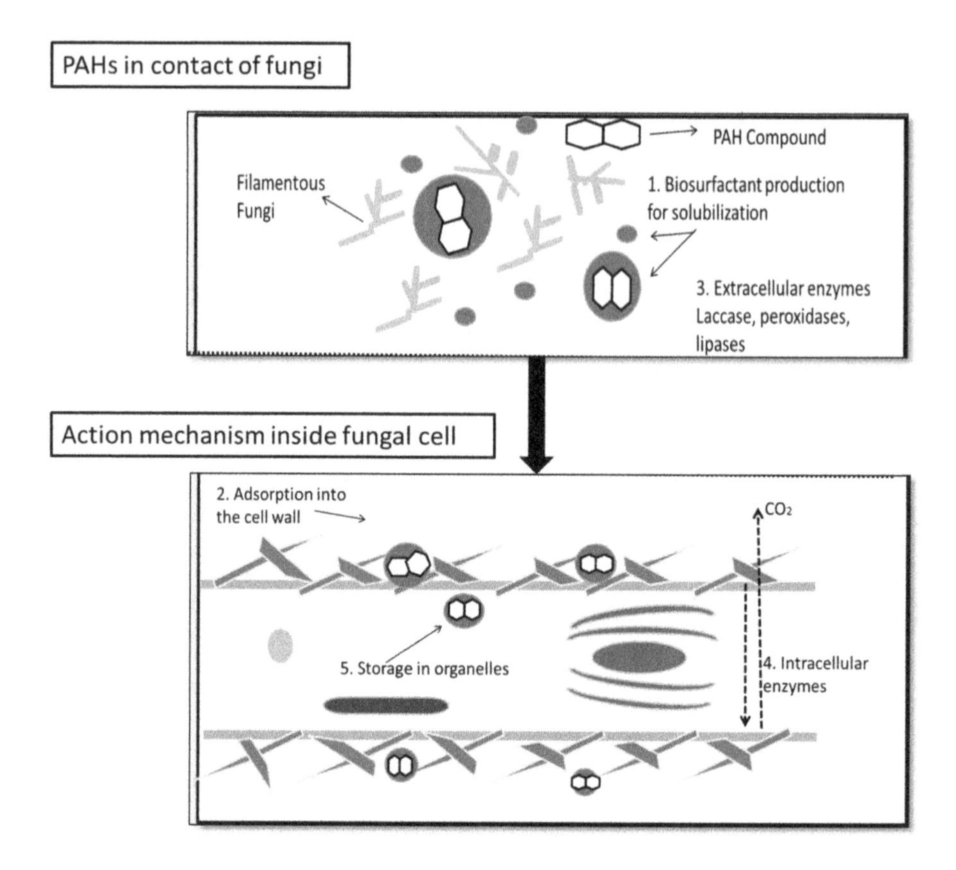

FIGURE 4.4 Major strategies adapted by fungi for PAHs degradation.

P450 (CYT P450) monooxygenase intracellular enzyme, which is mainly produced by non-ligninolytic enzymes, and second are extracellular enzymes for lignin metabolism, i.e., lignin peroxidase, manganese peroxidase, and laccase (Tortella and Diez, 2005). Fungi catabolize PAH compounds to intermediates analogous to those formed by a mammalian enzymatic system (Figure 4.4).

The principal path for oxidation of PAH is CYT P450 monooxygenase and involvement of O_2 in order to produce arene oxide, and water is encompassed in the first conversion of PAH by non-ligninolytic fungi that transform PAH into *trans*-dihydrodiols (Cerniglia, 1997; Sevcan et al., 2017). Ligninolytic fungi such as *Pleurotus ostreatus* and *Cunninghamella elegans* metabolized fluoranthene into fluoranthene *trans*-2,3-dihydrodiol (Tortella and Diez, 2005). Similarly, in the case of high-molecular-weight PAH, *P. ostreatus* (ligninolytic fungi) were able to break down pyrene into pyrene *trans* 4,5-dihydrodiol and anthracene into 9,10-anthraquinone (Bezalel et al., 1996a, 1996b). However, variability in the regiospecificity and stereoisomerism provides insight of significant enzyme which aids in the transformation of complex PAH compounds by an enzymatic mechanism.

4.9.2.1 Non-Ligninolytic Fungi

Non-ligninolytic fungi mainly involve two mechanism: one is the CYT P450 system (Bezalel et al., 1997) and other are soluble extracellular enzymes such as LiP, MnP, and laccase (Bonugli-Santos et al., 2010). Fungi also show some nonenzymatic mechanisms such as bioadsorption, bioprecipitation, and bioremediation interceded by enzymatic activity (Harms et al., 2011). Filamentous fungi also show the bioadsorption mechanism by the chitosan or chitin that exists in the cell wall of fungi (Gadd and Pan, 2016). An enzyme involved during lignin degradation comprises extracellular enzymes, such as MnP, LiP, versatile peroxidase, and laccases, which might have a key role in the conversion of PAHs (Hofrichter et al., 2015). However, only 12 CYT P450 bands have been recognized in the conversion of hydrocarbons such as cytochrome P52, cytochrome P53, cytochrome P505, and cytochrome P55 (Moktali et al., 2012). Ascomycota includes the most diverse division, with >65% of the currently described fungi belonging to this phylum (Harms et al., 2011).

Fungi are highly adaptable to diversified environmental conditions, and their ability to chelate metal ions results in detoxification of metals, which are commonly found in industrial wastewater (Gadd et al., 2014; Tigini et al., 2014). Research gaps in enzymatic conversion remain unclear, which hinders the use of fungi in bioremediation. Genetic engineering could be beneficial and may represent a vital step to developing a biodegradation study by fungi (Olicón-Hernández et al., 2017).

4.9.3 Bacterial and Fungal Enzymatic Interactions

Major enzymes that play an important role in PAH transformation are listed in Table 4.5 and in Figure 4.5. Dioxygenase enzymes are accountable for the formation of *cis*-dihydrodiols from PAH arene, which seems to most pervasive among bacteria species. Dioxygenases commonly comprise multicomponent enzymatic systems (Peng et al., 2008). Dioxygenase enzymes encompassing microorganisms have been

TABLE 4.5
Major Enzymes Action on PAHs Breakdown

Major Enzymes	Oxidized Product	References
Cytochrome P450 monooxygenases	Epoxides	Milo and Casto (1992), Marston et al. (2001)
Dioxygenases	*Cis* or *Trans* dihydrodiols	Sutherland et al. (1995); Moody et al. (2004)
Peroxidase	Quinones	Cerniglia and Sutherland (2010)
Laccase	Phenols	Cerniglia and Sutherland (2010)
Hydrolases	Fatty acids	-

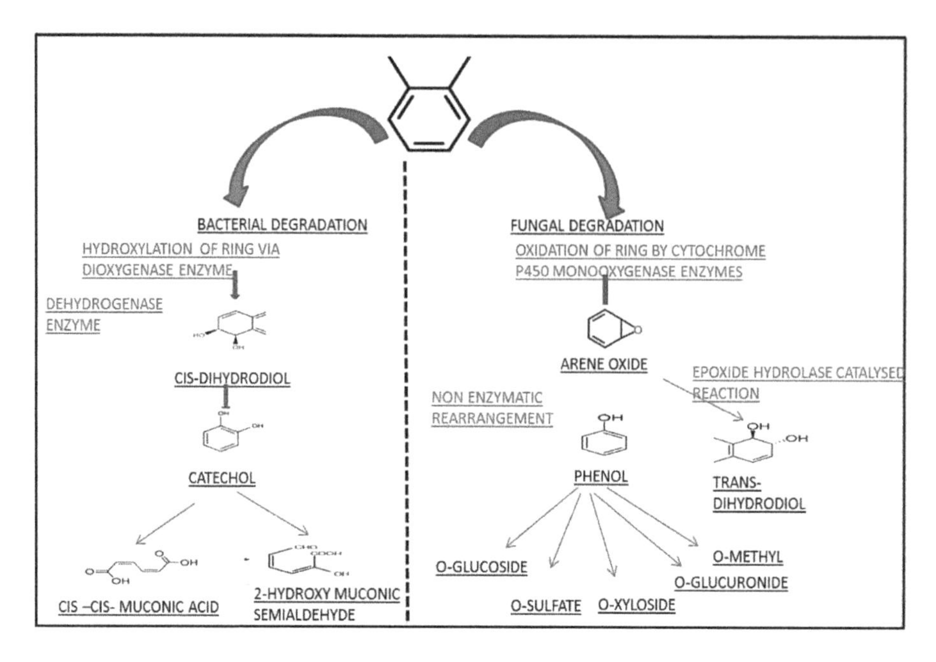

FIGURE 4.5 Bacterial and fungal metabolism of PAHs.

grown on different ranges of carbon as a sole source and classified as toluene dioxygenase, naphthalene dioxygenase, and biphenyl dioxygenase. For the breakdown of larger PAH compounds, both naphthalene dioxygenase and biphenyl dioxygenase are capable of activating dihydroxylation and are able to metabolize complex substrates (Boyd and Sheldrake, 1998; Peng et al., 2008).

However, for non-ligninolytic fungi, CYT P450 may also oxidize numerous PAH compounds into phenols (Pothuluri et al., 1996). Woody components containing a high-lignin content are mainly fragmented by extracellular enzymes produced by ligninolytic fungi/white rot fungi; similarly, PAHs may be oxidized by ligninolytic fungi to generate transient PAH diphenols and oxidized into quinones. PAH compounds are oxidized by lignin peroxidase in the presence of H_2O_2 and manganese peroxidase through Mn-dependent peroxidation (Liado et al., 2013). Lignin is a polymer with phenylpropane subunits that include linkages such as aryl ether and carbon-carbon bonds, which resemble with PAH compounds. Thus, ligninolytic fungi have been perceived as imminent PAH-degrading organisms. The ability of lignin-degrading fungi to break down PAH compounds depends on the production of multiple enzymatic systems such as lignin-peroxidase, manganese peroxidase, and laccase (Sevcan et al., 2017).

4.10 ROLE OF BIOSURFACTANT PRODUCING MICROBES: PAH DEGRADATION

Due to the poor solubility and hydrophobic nature of PAHs, there are some alternative extraction agents that have been studied to increase their solubility and degradation. The use of synthetic surfactants is resisted due to poor decomposition and environmental noxiousness (Reis et al., 2013). In addition, use of briskly exhausting petroleum products that are sources for production of chemical surfactants is a base environmental issue. These issues have compelled the exploration of ecological alternatives of low toxicity acquired from natural sources in place of synthetic surfactants.

Biosurfactant is a biologically formed surfactant revealed as a possible alternative for chemical surfactants due to its easy degradability and low level of toxicity (Irorere et al., 2017). The biosurfactant structure comprises a hydrophilic moiety possessing anionic/cationic amino acids, peptides, monosaccharaides, disaccharides, and polysaccharides. The hydrophobic moiety is frequently composed of saturated, unsaturated, or hydroxylated fatty acids or hydrophobic peptides (Banat, 1995). Low-molecular-weight biosurfactants are common glycolipids or lipopeptides, which are more active in lowering interfacial tensions in a liquid medium (Lin, 1996). Reduction in the surface tension of noble biosurfactants is less than 30 m/Nm (Wang and Mulligan, 2004; Elliot et al., 2011). PAH degradation rates depend on the mass transfer rates of the organic compounds from the solid phase to the liquid phase by increasing the surface area substantial to the maximum suspension rate (Volkering et al., 1992).

Many studies have reported on the production of biosurfactant and its role in enhancing the efficiency of the biodegradation of PAHs. Dong et al. (2016) reported that a new biosurfactant-producing bacterial, *Acinetobacter junii* BD, was isolated from a reservoir for the reduction of surface tension and emulsification of crude oil. Some examples of bacterial strains have been mentioned in Table 4.6. The biosurfactant produced by *Pseudomonas aeruginosa* CB1 enhanced the degradation of PAH compounds up to 86.5% (by adding biosurfactant) and 57% in the control (without biosurfactant and nutrient addition) during 45 days of experimental setup (Bezza and Chirwa, 2016).

Fungi could also produce a structurally and functionally versatile amphiphilic surfactant resulting in a decline in surface tension (Günther, 2017). Fungal biosurfactants based on their structure also include sophorolipid, polysaccharide, glycolipid, and glycolipoproteins (Bhardwaj et al., 2013). The major type of biosurfactants

TABLE 4.6
Biosurfactant Produced by Bacteria

S. No.	Bacteria	Type of Biosurfactant	References
1	*Pseudomonas aeruginosa*	Rhamnolipids	Elliot et al. (2011)
2	*Mycobacterium* sp	Trehalolipids	Lang and Wagner (1987)
3	*Pseudomonas fluorescens*	Viscosin (lipopeptide)	Banat (1995)
4	*Bacillus licheniformis*	Lipopeptides and lipoproteins Peptide-lipid	Joshi et al. (2015)
5	*Bacillus subtilis*	Surfactin	Gurjar and Sengupta (2015)
6	*Bacillus cereus*	Rhamnolipids	Sidkey et al. (2012)
7	*Acinetobacter sp.*	Phospholipids	Scott et al. (1976)
8	*Enterobacter cloacae*, *Pseudomonas* sp.	Rhamnolipids	Darvishi et al. (2011)

TABLE 4.7
Biosurfactant Produced by Fungi

S. No.	Fungi	Type of Biosurfactants	References
1	*Torulopsis bombicola*	Sophorose lipid	Kim et al. (1997)
2	*Candida bombicola*	*Sophorolipid*	Cavalero and Cooper (2003)
3	*Candida ingens*	Not mentioned	Amézcua-Vega et al. (2007)
4	*Candida lipolytica*	Protein-lipid polysaccharide complex	Rufno et al. (2007)
5	*Candida ishiwadae*	Glycolipid	Thanomsub et al. (2004)
6	*Candida batistae*	Sophorolipid	Konishi et al. (2008)
7	*Aspergillus ustus*	Glycolipoprotein	Kiran et al. (2009)
8	*Ustilago maydis*	Glycolipid	Alejandro et al. (2011)
9	*Trichosporon asahii*	Sophorolipid	Chandran and Das (2010)
10	*Fusarium* sp. BS-8	Lipopeptides	Qazi et al. (2014)
11	*Aspergillus* sp. MSF3	Glycolipoprotein	Kiran et al. (2009)

produced by fungi is mentioned in Table 4.7. Furthermore, fungi yield a better quantity of biosurfactant than bacteria, which may be due to the presence of the rigid cell wall in fungi (Bhardwaj et al., 2013).

4.11 FACTORS AFFECTING BIODEGRADATION

The accomplishment of the bioremediation process depends on a several factors, such as the (i) physicochemical and microbial characteristics of the contaminated sampling site, (ii) ability of isolated and screened microorganisms to mineralize the hazardous pollutants, and (iii) availability of the toxic compounds to microbes (Alexander, 1994; Pointing, 2001; Bhatt et al., 2002; Semple et al., 2003). Here, major environmental factors are discussed that affect the rate of biodegradation.

4.11.1 pH

The pH value mainly affects the enzymatic activity; degradation of PAHs in sewage sludge is affected by the influence of pH value and can impact the microbial enzymatic mechanism. pH is an essential parameter that may alter the dissolution of PAH compounds in a broth medium and the metabolic activity of microbes—the optimum pH range for bacteria to degrade organic pollutants ranges between 5.5 and 7.8 (Bossert and Bartha 1984; Wong et al., 2001). Oleszczuk (2006) found the pH range of 6.1–6.9 is suitable for the survival of bacteria and the simultaneous removal of PAH compounds.

4.11.2 Temperature

Temperature affects the capacity of the microbes to degrade PAH and the solubility of PAH rises with an increase in temperature, which stimulates the availability of

PAH to microorganisms. An increase in temperature decreases the amount of dissolved oxygen, resulting in a reduction in the metabolic activity of microbes. Most of the microbes are able to degrade hydrocarbons at a particular temperature range that also regulates the enzymatic mechanism. Hydrocarbon utilizing and mineralizing microbes are active at the range of 20–35°C. At a low temperature range, the volatility of the hydrocarbons decreases and becomes more toxic to degrading microbes (Sihag et al., 2014).

4.11.3 Bioavailability

The bioavailability of hydrocarbons at the contaminated site is a crucial phase in the bioremediation process. The availability of hydrocarbons could be enhanced by using surfactants, which has recently received significant focus in soil bioremediation (Chung et al., 2000). *Phanerochaete chrysosporium* could degrade PAHs in artificially contaminated soil in addition with Tween 80 (Zheng and Obbard, 2001). PAHs can be degraded in artificially contaminated soil by using *Pleurotus ostreatus* in the presence of Tween 40 and Triton X-100 (Marquez-Rocha et al., 2000).

4.11.4 Salinity

PAH degradation is positively correlated with salt concentrations in estuarine sediments as reported by Shiaris (1989). The optimum biodegradation of PAH was reported in 1% NaCl. A high NaCl concentration diminishes the biodegradation of hydrocarbons in a nutrient broth medium (Diaz et al., 2002).

4.11.5 Co-substrate

Co-substrate mainly helps in increasing the efficiency of the degradation process. Beef extract has been reported as a suitable co-substrate for the PAH compound degradation by using bacterial inoculum (99.1% degradation of phenanthrene in 7 days). Zhu et al., 2016 reported that inoculating endophytic rhizobacteria at PAHs contaminated sites enhanced the PAHs removal rate.

4.11.6 Heavy Metal as Co-contaminant

Heavy metals are of major concern and present at a high ratio in contaminated sites (Thavamani et al., 2011). PAHs frequently occur with heavy metals at contaminated sites, which may cause hindrance during the biodegradation of PAH compounds (Thavamani et al., 2013). High concentrations of these metals can damage cell membranes and prevent enzymatic activity (Bruins et al., 2000). Isolation of metal-tolerant PAH compounds for degrading microorganisms is achievable to avoid metal inhibition during PAH degradation. Microbial degradation of phenanthrene co-contaminated with zinc in soil has been found to be somewhat restricted (Lloyd and Lovely, 2001).

4.12 NANOREMEDIATION AS AN ADVANCED TECHNOLOGY

Nanoremediation technology using nanoparticles has been widely applied in the fields of the chemical industry and agriculture sector due to its small size and large surface-to-volume proportion (Dimkpa et al., 2011; Xu et al., 2017). Synthesis of metal nanoparticles and their application as a photocatalyst in remediation fields has received significant attention (Aiken and Finke, 1999; Daniel and Astruc, 2004). There are some research studies available with respect to the use of metal nanoparticles for degradation of PAH compounds (Nelkenbaum et al., 2007; Jacinto et al., 2009; Zhou et al., 2009). In biological synthesis of nanoparticles, microbes and plants are utilized. Microorganisms aggregate inorganic materials within or outside the cell to form nanoparticles, and some microbial species are capable of producing metal nanoparticles. Biosynthesis of nanoparticles such as gold, silver, selenium, tellurium, platinum, palladium, and silica using extracellular extracts of bacteria, actinomycetes, fungi, and different plant parts have been reported (Narayanan and Sakthivel, 2010). Bacterial and extracellular enzymes of fungi react with metal ions and reduce the metal into nanoparticles (Alghuthaymi et al., 2015).

The fungal synthesis (mycosynthesis) of metal nanoparticles includes fungal synthesis of nanoparticles in the nanotechnology field (Meyer, 2008; Rai et al., 2009). Various fungal strains such as *Fusarium* sp., *Aspergillus* sp., *Verticillium* sp., and *Penicillium* sp. have been recognized as promising biological sources for metal nanoparticle synthesis. Interaction of fungal enzymes with metal ions leads to reduction of metals into nanoparticles. Fungi could also function as an initiator in the bacterial degradation of pollutants by improving the growth of microorganisms or through immobilization of remediating biological agents and promoting microbial enzyme production. Nanoparticles may also improve the production of biosurfactant by microorganisms and aid to increase the solubility of PAHs in an aqueous medium (Kumari et al., 2016). Nevertheless, a suitable selection of nanomaterials and method of synthesis for the remediation process is an important phase (Rizwan et al., 2014). Table 4.8 presents the ability of different synthesized nanoparticles to degrade PAH compounds.

4.13 FUTURE PROSPECTS

Some microorganisms can be used for the effective cleanup of hazardous waste such as PAH compounds generated by refinery industries. Moreover, these microbes could be used for the efficient degradation of complex PAHs and for the remediation of PAH-contaminated sites. However, to strengthen the fungal remediation technology, it is important to find out more members—in particular, indigenous fungal species—for PAH biodegradation. A detailed understanding of the degradation genes, as well as of their respective roles in the degradation of toxic compounds, might deliver a platform for the development of bioremediation approaches in the near future.

TABLE 4.8
Synthesized Nanoparticles and Their PAH Degradation Efficiency

S. No	Source of Synthesis	Synthesized Nanoparticles	PAH Compounds	Degradation Percentage	References
1	Chemical method (use of solvent)	Poly(ethylene) glycol-modified urethane acrylate (PMUA)	Phenanthrene	30%, 12 days	Tungittiplakorn et al. (2005)
2	*Green sapindus-mukorossi* as natural surfactant	Iron hexacyanoferrate (FeHCF)	Fluorene, chrysene and benzo[*a*]pyrene, anthracene	70–80%	Jassal et al. (2015)
3	*Coriandrum sativum* leaf extract	ZnO	Anthracene	96%, 240 minutes, 100 μg	Hassan et al. (2015)
4	Chemical method	TiO_2	Anthracene	25 ppm	Karam et al. (2014)
5	Chemical (precipitation method)	Goethite (α-FeOOH) and magnetite (Fe_3O_4) FeO	Anthracene	3.21 and 4.39 hours half-lives	Gupta et al. (2017)
6	Chemical method	Curcumin-conjugated ZnO	Fluoranthene	with 93% removal after 10 minutes	Moussawi and Patra (2016)
7	Green synthesis	AgO	Phenanthrene	85%	Abbasi et al. (2014)
8	Green synthesis	Fe NPs	Naphthalene	96 hours, 100%	Jin et al. (2016)

REFERENCES

Abbasi, Maryam, Saeed, Fatima, Rafique, Uzaira. (2014). Preparation of silver nanoparticles from synthetic and natural sources: remediation model for PAHs. IOP Conference Series: Materials Science and Engineering. 60:012061. doi:10.1088/1757-899X/60/1/012061.

Abdel-Shafy, Hussein I., Mansour, Mona S. M. (2016). A review on polycyclic aromatic hydrocarbons: source, environmental impact, effect on human health and remediation. Egyptian Journal of Petroleum. 25:107–123.

Ahtiainen, J., Valo, R., Järvinen, M., Joutti, A. (2002). Microbial toxicity tests and chemical analysis as monitoring parameters at composting of creosote-contaminated soil. Ecotoxicol Environ Saf. 53(2):323–329, doi:10.1006/eesa.2002.2225.

Aiken, III J. D., Finke, R. G. (1999). A review of modern transition-metal nanoclusters: their synthesis, characterization, and applications in catalysis. Journal of Molecular Catalysis A: Chemical. 145:1–44. http://dx.doi.org/10.1016/S1381-1169(99)00098-9.

Aislabie, J., McLeod, M., Fraser, R. (1998). Potential for biodegradation of hydrocarbons in soil from the Ross dependency, Antarctica. Applied Microbiology and Biotechnology. 49:210–214.

Albers, Peter H. (2003). Petroleum and Individual Polycyclic Aromatic Hydrocarbons, Chap. 14, Lewis Publishers, Boca Raton, FL, pp. 341–371.

Alejandro, C. S., Humberto, H. S., María, J. F. (2011). Production of glycolipids with antimicrobial activity by Ustilago maydis FBD12 in submerged culture. African Journal of Microbiology Research. 5:2512–2523.

Alexander, M. 1994. Biodegradation and Bioremediation. Academic Press, San Diego.

Alghuthaymi, Mousa A., Almoammar, Hassan, Rai, Mahindra, Said-Galiev, Ernest, Abd-Elsalam, Kamel A. (2015). Myconanoparticles: synthesis and their role in phytopathogens management. Biotechnology & Biotechnological Equipment. 29(2):221–236, http://dx.doi.org/10.1080/13102818.2015.1008194.

Alkino, M., Tabuchi, T. M., Wang, X., Colon-Carmona, A. (2005). Stress response to polycyclic aromatic hydrocarbons in Arabidopsis include growth inhibition and hypersensitive response-like symptoms. Journal of Experimental Botany. 56:2983–2994.

Amaral, P. F. F., Silva, J. M., Lehocky, M., Barros-Timmons, A. M. V., Coelho, M. A. S., Marrucho, I. M., Coutinho, J. A. P. (2006). Production and characterization of a bioemulsifier from *Yarrowia lipolytica*. Process Biochemistry. 41:1894–1898.

Amézcua-Vega, C., Poggi-Varaldo, H. M., Esparza-García, F., Ríos-Leal, E., Rodríguez-Vázquez, R. (2007). Effect of culture conditions on fatty acids composition of a biosurfactant produced by *Candida ingens* and changes of surface tension of culture media. Bioresource Technology. 98:237–240.

Banat, I. M. (1995). Biosurfactants production and possible uses in microbial enhanced oil recovery and oil pollution remediation: a review. Bioresource Technology. 51:1–12.

Bezalel, L., Hadar, Y., Cerniglia, C. E. (1997). Enzymatic mechanisms involved in phenanthrene degradation by the white rot fungus Pleurotus ostreatus. Applied Environment Microbiology. 63:2495–2501.

Bezalel, L., Hadar, Y., Fu, P. P., Freeman, J. P., Cerniglia, C. E. (1996a). Initial oxidation products in the metabolism of pyrene, anthracene, fluorene, and dibenzothiophene by the white rot fungus *Pleurotus ostreatus*. Applied and Environmental Microbiology. 62:2554–2559.

Bezalel, L., Hadar, Y., Fu, P. P., Freeman, J. P., Cerniglia, C. E. (1996b). Metabolism of phenanthrene by the white rot fungus *Pleurotus ostreatus*. Applied and Environmental Microbiology. 62:2547–2553.

Bezza, Fisseha Andualem, Chirwa, Evans M. Nkhalambayausi. (2016). Biosurfactant-enhanced bioremediation of aged polycyclic aromatic hydrocarbons (PAHs) in creosote contaminated soil. Chemosphere. 144:635–644.

Bhardwaj, G., Cameotra, S. S., Chopra, H. K. (2013). Biosurfactants from fungi: a review. Journal of Petroleum & Environmental Biotechnology. 4:160, doi:10.4172/2157-7463.1000160.

Bhatt, M., Cajthaml, T., Sasek, V. (2002). Mycoremediation of PAH-contaminated soil. Folia Microbiologica. 47:255–258.

Bonugli-Santos, Rafaella Costa, Durranta, Lucia Regina, Manuelad, Da Silva, Durães Sette, Lara. (2010). Production of laccase, manganese peroxidase and lignin peroxidase by Brazilian marine-derived fungi. Enzyme and Microbial Technology. 46(1):32–37.

Bossert, I., Bartha, R. (1984). The fate of petroleum in soil ecosystems. In Atlas, R. M. (ed.), Petroleum Microbiology. Macmillan, New York, pp. 441–4473.

Boyd, D. R., Sheldrake, G. N. (1998). The dioxygenase-catalysed formation of vicinal *cis*-diols. Natural Product Reports. 15:309–324.

Braddock, J. F., Ruth, M. L., Catterall, P. H., Walworth, J. L., McCarthy, K. A. (1997). Enhancement and inhibition of microbial activity in hydrocarbon-contaminated arctic oils: implications for nutrient-amended bioremediation. Environmental Science &Technology. 31:2078–2084.

Bruins, M. R., Kapil, S., Oehme, F. W. (2000). Microbial resistance to heavy metals in the environment. Ecotoxicology and Environmental Safety. 45:198–207.

Bugg, T., Foght, J. M., Pickard, M. A., Gray, M. R. (2000). Uptake and active efflux of polycyclic aromatic hydrocarbons by Pseudomonas fluorescens LP6a. Applied Microbiology and Biotechnology. 66:5387–5392.

Casillas, R. P., Crow, S. A., Heinze, T. M., Deck, J., Cerniglia, C. E. (1996). Initial oxidative and subsequent conjugative metabolites produced during the metabolism of phenanthrene by fungi. Journal of Industrial Microbiology. 16:205e–215. http://dx.doi.org/10.1007/BF01570023.

Cavalero, D. A., Cooper, D. G. (2003). The effect of medium composition on the structure and physical state of sophorolipids produced by *Candida bombicola* ATCC 22214. Journal of Biotechnology. 103:31–41.

Cerniglia, C. E. (1997). Fungal metabolism of polycyclic aromatic hydrocarbons: past, present and future applications in bioremediation. Journal of Industrial Microbiology & Biotechnology. 19:324–333.

Cerniglia, C. E., Sutherland, J. B. (2010). Degradation of polycyclic aromatic hydrocarbons by fungi. In: Timmis, K. N., McGenity, T. J., van der Meer, J. R., de Lorenzo, V. (eds.), Handbook of Hydrocarbon and Lipid Microbiology. Springer, Berlin, pp. 2080–2110.

Chandran, P., Das, N. (2010). Biosurfactant production and diesel oil degradation by yeast species *Trichosporon asahii* isolated from petroleum hydrocarbon contaminated soil. International Journal of Engineering Science and Technology. 2:6942–6953.

Chauhan, Archana, Fazlurrahman, Oakeshott, John G., Jain, Rakesh K. (2008). Bacterial metabolism of polycyclic aromatic hydrocarbons: strategies for bioremediation. Indian Journal of Microbiology. 48:95–113.

Chouychai, Waraporn, Thongkukiatkul, Amporn, Upatham, Suchart, Lee, Hung, Pokethitiyook, Prayad, Kruatrachue, Maleeya. (2007). Phytotoxicity assay of crop plants to phenanthrene and pyrene contaminants in acidic soil. Wiley Periodicals, Inc, Environmental Toxicology. 597–604. doi 10.1002/tox.

Chung, N. H., Lee, I. S., Song, H. S., Bang, W. G. (2000). Mechanisms used by white-rot fungus to degrade lignin and toxic chemicals. Journal of Microbiology and Biotechnology. 10:737–752.

Crimi, M., Quickel, M., Ko, S. (2009). Enhanced permanganate in situ chemical oxidation through MnO2 particle stabilization: evaluation in 1-D transport systems. Journal of Contaminant Hydrology. 105(1–2):69–79. https://doi.org/10.1016/j.jconhyd.2008.11.007.

Dandie, C. E., Weber, J., Aleer, S., Adetutu, E. M., Ball, A. S., Juhasz, A. L. (2010). Assessment of five bioaccessibility assays for predicting the efficacy of petroleum hydrocarbon biodegradation in aged contaminated soils. Chemosphere. 81(9):1061–1068. https://doi.org/10.1016/j.chemosphere.2010.09.059.

Daniel, M. C., Astruc, D. (2004). Gold nanoparticles: assembly, supramolecular chemistry, quantum-size-related properties, and applications toward biology, catalysis, and nanotechnology. Chemical Review. 104(1):293–346.

Darvishi, Parviz, Ayatollahi, Shahab, Mowla, Dariush, Niazi, Ali (2011). Biosurfactant production under extreme environmental conditions by an efficient microbial consortium, ERCPPI-2. Colloids and Surfaces B: Biointerfaces. 84(2):292–300.

Diaz, M. P., Boyd, K. G., Grigson, S. J. W., Burgess, J. G. (2002). Biodegradation of crude oil across a wide range salinities by an extremely halotolerant bacterial consortium MPD-M, immobilized ontopolypropylene fibers. Biotechnology and Bioengineering. 79:145–153. doi: 10.1002/bit.10318.

Dimkpa, C. O., Calder, A., Gajjar, P., Merugu, S. (2011). Interaction of silver nanoparticles with an environmentally beneficial bacterium, pseudomonas chlororaphis. Journal of Hazardous Materials. 188:428–435.

Diplock, E. E., Mardlin, D. P., Killham, K. S., Paton, G. I. (2009). Predicting bioremediation of hydrocarbons: laboratory to field scale. Environmental Pollution. 157:1831–1840.

Dong, H., Xia, W., Dong, H., She, Y., Zhu, P., Liang, K., Zhang, Z., Liang, C., Song, Z., Sun, S., Zhang, G. (2016). Rhamnolipids produced by indigenous acinetobacter junii from petroleum reservoir and its potential in enhanced oil recovery. Frontiers in Microbiology. 7:1710-1–1710-13.

Eisler, R. (2000). Polycyclic aromatic hydrocarbons. In: Handbook of Chemical Risk Assessment, Chap. 25, Vol. 2, Lewis Publishers, Boca Raton, pp. 1343–1411.

Elliot, Roy, Singhal, Naresh, Swif, Simon. (2011). Surfactants and bacterial bioremediation of polycyclic aromatic hydrocarbon contaminated soil—unlocking the targets, Critical Reviews in Environmental Science and Technology, 41:78–124, ISSN: 1064-3389 print / 1547–6537. online, doi: 10.1080/00102200802641798.

EPRI (Electric Power Research Institute). (2000). Literature review of background polycyclic aromatic hydrocarbons. Final report.

Gadd, G. M., Bahri-Esfahani, J., Li, Q., Rhee, Y. J., Wei, Z., Fomina, M., et al. (2014). Oxalate production by fungi: significance in geomycology, biodeterioration and bioremediation. Fungal Biological Review. 28:36–55. 10.1016/j.fbr.2014.05.001.

Gadd, G. M., Pan, X. (2016). Biomineralization, bioremediation and biorecovery of toxic metals and radionuclides. Geomicrobiology Journal. 33:175–178. doi.10.1080/014904 51.2015.1087603.

Gan, S., Lau, E. V., Ng, H. K. (2009). Remediation of soils contaminated with polycyclic aromatic hydrocarbons (PAHs). Journal of Hazardous Materials. 172(2–3):532–549. https://doi.org/10.1016/j.jhazmat.2009.07.118.

Gan, S., Ng, H. K. (2012a). Inorganic chelated modified- Fenton treatment of polycyclic aromatic hydrocarbon (PAH)-contaminated soils. Chemical Engineering Journal. 180:1–8. https://doi.org/10.1016/j.cej.2011.10.082.

Gan, S., Ng, H. K. (2012b). Modified Fenton oxidation of polycyclic aromatic hydrocarbon (PAH)-contaminated soils and the potential of bioremediation as post-treatment. Science of the Total Environment. 419:240–249. https://doi. org/10.1016/j.scitotenv.2011.12.053.

Gitipour, Saeid, Sorial, George A., Ghasemi, Soroush, Bazyari, Mahdieh. (2018). Treatment technologies for PAH-contaminated sites: a critical review. Environmental Monitoring and Assessment. 190:546. https://doi.org/10.1007/s10661-018-6936-4.

Günther, M. (2017). Fungal glycolipids as biosurfactants. Current Biotechnology. 5:1–13. doi .10.2174/2211550105666160822170256.

Gupta, Himanshu, Kumar, Rahul, Park, Hyun-Sun, Jeon, Byong-Hun. (2017). Photocatalytic efficiency of iron oxide nanoparticles for the degradation of priority pollutant anthracene. Journal Geosystem Engineering. 20(1).

Gurjar, Jigar, Sengupta, Bina. (2015). Production of surfactin from rice mill polishing residue by submerged fermentation using *Bacillus subtilis* MTCC 2423. Bioresource Technology. 189:243–249.

Harms, H., Schlosser, D., Wick, L. Y. (2011). Untapped potential: exploiting fungi in bioremediation of hazardous chemicals. Nature Review Microbiology. 9:177–192. doi.10.1038/ nrmicro2519.

Hassan, S. S. M., El Azab, W. E., Ali, H. R., Mansour, M. S. M. (2015). Green synthesis and characterization of ZnO nanoparticles for photocatalytic degradation of anthracene. Advances in Natural Science: Nanoscience and Nanotechnology .6:045012.

Hofrichter, M., Kellner, H., Pecyna, M. J., Ullrich, R. (2015). Fungal unspecific peroxygenases: heme-thiolate proteins that combine peroxidase and cytochrome P450 properties. In: Hrycay, E. G. (eds.), Monooxygenase, Peroxidase and Peroxygenase Properties and Mechanisms of Cytochrome P450. Springer International Publishing AG, Cham, Switzerland

Horinouchi, M., Youshi, N., Eriko, S., Sugina, R., Kanchana, J., et al. (2000). Removal of polycyclic aromatic hydrocarbon from oil-contaminated Kuwaiti soil. Biotechnology Letters. 22(8):687–691.

Hussain, K., Hoque, R. R., Balachandran, S., Medhi, S., Hussain, F. L., Idris, M. G., Rahman, M. (2018). Monitoring and risk analysis of PAHs in the environment. In: Hussain, C. M. (ed.), Handbook of Environmental Materials Management. doi.10.1007/978-3-319-58538-3_29-2.

IARC. (1973). Certain polycyclic aromatic hydrocarbons and heterocyclic compounds. Monographs on the Evaluation of Carcinogenic Risk of the Chemical to Man, Vol. 3. World Health Organization, International Agency for Research on Cancer, Lyon, France.

International Agency for Research on Cancer (IARC). (1987). Monographs on the evaluation of carcinogenic risk of chemicals to humans. Supplement No. 7. Overall evaluations of carcinogenicity: an updating of IARC monographs volumes 1 to 42. Lyon: International Agency for Research on Cancer. http://monographs.iarc.fr/ENG/Monographs/suppl7/

Irorere, Victor U., Tripathi, Lakshmi, Marchant, Roger, McClean, Stephen, Banat, Ibrahim M. (2017). Microbial rhamnolipid production: a critical re-evaluation of published data and suggested future publication criteria. Applied Microbiology and Biotechnology. 101:3941–3951. doi 10.1007/s00253-017-8262-0.

Jacinto, M. J., Santos, O. H. C. F., Landers, R., Kiyohara, P. K., Rossi, L. M. (2009). On the catalytic hydrogenation of polycyclic aromatic hydrocarbons into less toxic compounds by a facile recoverable catalyst. Applied Catalysis, B. 90:688–692. doi.10.1016/j.apcatb.2009.04.031.

Jassal, V., Shanker, U., Shankar, S. (2015). Synthesis, characterization and applications of nano-structured metal hexacyanoferrates: A Review. Journal of Environmental Analytical Chemistry 2:128. doi:10.4172jreac.1000128.

Jin, Xiaoying, Yu, Bing, Lin, Jiajiang, Chen, Zuliang. (2016). Integration of biodegradation and nano-oxidation for removal of PAHs from aqueous solution. ACS Sustainable Chemistry & Engineering. 4(9). doi.10.1021/acssuschemeng.6b00933.

Johnsen, A. R., Lukas, Y., Hauke, H. (2005). Principles of microbial PAH-degradation in soil. Environmental Pollution. 1:71–84.

Joshi, S. J., Geetha, S. J., Desai, A. J. (2015). Characterization and application of biosurfactant produced by Bacillus licheniformis R2. Applied Biochemical Biotechnology. 177(2):346–361.

Karam, Faiq F., Hussein, Falah H., Baqir, Sadiq J., Halbus, Ahmed F., Dillert, Ralf, Bahnemann, Detelf. (2014). Photocatalytic degradation of anthracene in closed system reactor. International Journal of Photoenergy. Article ID 503825, 6 pages, http://dx.doi.org/10.1155/2014/503825.

Khairy, M. A., Kolb, M., Mostafa, A. R., El-Fiky, A., Bahadir, M. (2009). Risk assessment of polycyclic aromatic hydrocarbons in a Mediterranean semi-enclosed basin affected by human activities (Abu Qir Bay, Egypt). Journal of Hazardous Materials. 170(1):389–397.

Khodadoust, A. P., Bagchi, R., Suidan, M. T., Brenner, R. C., Sellers, N. G. (2000). Removal of PAHs from highly contaminated soils found at prior manufactured gas operations. Journal of Hazardous Materials. B80:159–174.

Kim, S. Y., Oh, D. K., Lee, K. H., Kim, J. H. (1997). Effect of soybean oil and glucose on sophorose lipid fermentation by *Torulopsis bombicola* in continuous culture. Applied Microbiology and Biotechnology. 48:23–26.

Kiran, G. S., Hema, T. A., Gandhimathi, R., Selvin, J., Thomas, T. A., et al. (2009). Optimization and production of a biosurfactant from the sponge-associated marine fungus *Aspergillus ustus* MSF3. Colloids and Surface B, Biointerfaces. 73:250–256.

Konishi, M., Fukuoka, T., Morita, T., Imura, T., Kitamoto, D. (2008). Production of new types of sophorolipids by Candida batistae. Journal of Oleo Science. 57:359–369.

Kumari, B., Singh, S. N., Singh, D. P. (2016). Induced degradation of crude oil mediated by microbial augmentation and bulking agents. International Journal of Environmental Science and Technology. 13:1029–1042. doi 10.1007/s13762-016-0934-2.

Kuo, C. Y., Chien, P. S., Kuo, W. C., Wei, C. T., Jui-Yeh, R. J. Y. (2013). Comparison of polycyclic aromatic hydrocarbon emissions on gasoline- and diesel-dominated routes. Environmental Monitoring and Assessment. 185(7):5749–5761. doi: 10.1007/s10661-012-2981-6.

Kuppusamy, Saranya, Palanisami, Thavamani, Mallavarapu, Megharaj, Naidu, Ravi. (2016). Biodegradation of polycyclic aromatic hydrocarbons (PAHs) by novel bacterial consortia tolerant to diverse physical settings e Assessments in liquid- and slurry-phase systems. International Biodeterioration & Biodegradation. 108:149–157.

Lahkar, Jiumoni, Deka, Hemen. (2016). Isolation of polycyclic aromatic hydrocarbons (PAHs) degrading fungal candidate from oil-contaminated soil and degradation potentiality study on anthracene. Polycyclic Aromatic Compounds. 37:141–147. doi.10.1080/10406638.2016.1220957.

Lang, S., Wagner, F. (1987). Structure and properties of biosurfactants. In: Kosaric, N., Cairns, W. L., Gray, N. C. C. (eds.), Biosurfactants and Biotechnology. Marcel Dekker, Inc., New York, pp. 21–47.

Lau, Ee Von, Gan, Suyin, Kiat Ng, Hoon, Poh, Phaik Eong. (2014). Extraction agents for the removal of polycyclic aromatic hydrocarbons (PAHs) from soil in soil washing technologies. Environmental Pollution. 184:640–649.

Lee, P., Ong, S., Golchin, J., Nelson, G. L. (2001). Use of solvents to enhance PAH biodegradation of coal tar-contaminated soils. Water Research. 16:3941–3949.

Liado, S., Covino, S., Solanas, A. M., et al. (2013). Comparative assessment of bioremediation approaches to highly recalcitrant PAH degradation in a real industrial polluted soil. Journal of Hazardous Materials. 248–249:407–414. http://dx.doi.org/10.1016/j.jhazmat.2013.01.020.

Lin, S. C. (1996). Biosurfactants: recent advances. Journal of Chemical Technology and Biotechnology. 66:109–120.

Lippman, S. M., Hawk, E. T. (2009). Cancer prevention: from 1727 to milestones of the past 100 years. Cancer Research. 69(13):5269–5284.

Lloyd, J. R., Lovely, D. R. (2001). Microbial detoxification of metals and radionuclides. Current Opinion in Biotechnology. 12:248–253.

Lors, Christine, Damidot, Denis, Ponge, Jean-François, Périé, Frédéric. (2012). Comparison of a bioremediation process of PAHs in a PAH-contaminated soil at field and laboratory scales. Environmental Pollution. 165:11–17.

Lors, C., Ryngaert, A., Périé, F., Ludo Diels, L., Damidot, D., (2010). Evolution of bacterial community during bioremediation of PAHs in a coal tar contaminated soil. Chemosphere. 81:1263–1271.

Maliszewska-Kordybach, B., Smreczak, B. (2000). Ecotoxicological activity of soils polluted with polycyclic aromatic hydrocarbons (PAHs)—effect on plants. Environmental Technology. 21:1099–1110.

Marquez-Rocha, F. J., Hernandez-Rodriguez, V. Z., Vazquez-Duhalt, R. (2000). Biodegradation of soil-adsorbed polycyclic aromatic hydrocarbons by the white rot fungus *Pleurotus ostreatus*. Biotechnology Letters. 22:469–472.

Marston, C. P., Pereira, C., Ferguson, J., Fischer, K., Hedstrom, O., Dashwood, W. M., et al. (2001). Effect of a complex environmental mixture from coal tar containing polycyclic aromatic hydrocarbons (PAH) on the tumor initiation, PAH-DNA binding and metabolic activation of carcinogenic PAH in mouse epidermis. Carcinogenesis. 22:1077–1086. 10.1093/carcin/22.7.1077.

Masih, A., Taneja, A. (2006). Polycyclic aromatic hydrocarbons (PAHs) concentrations and its related carcinogenic potencies in soil at a semi-arid region of India. Chemosphere. 65:449–456.

Meyer, V. (2008). Genetic engineering of filamentous fungi progress, obstacles and future trends. Biotechnology Advances. 26:177–185.

Milo, G. E., Casto, B. C. (1992). Events of tumor progression associated with carcinogen treatment of epithelium and fibroblast compared with mutagenic events. In: Milo, G. E., Casto, B. C., Shuler, C. F. (eds.), Transformation of Human Epithelial Cells: Molecular and Oncogenetic Mechanisms. CRC Press, Boca Raton, FL, 261–284.

Mishra, S., et al. (2001). Evaluation of inoculum addition to stimulate in situ bioremediation of oily-sludge-contaminated soil. Applied and Environmental Microbiology. 67:1675–1681.

Miyata, N., Iwahori, K., Foght, J. M., Gray, M. R. (2004). Saturable, energy-dependent uptake of phenanthrene in aqueous phase by *Mycobacterium* sp. strain RJGII-135. Applied and Environmental Microbiology. 70:363–369.

Moktali, V., Park, J., Fedorova-Abrams, N. D., Park, B., Choi, J., Lee, Y.-H., et al. (2012). Systematic and searchable classification of cytochrome P450 proteins encoded by fungal and oomycete genomes. BMC Genomics. 13:525. doi.10.1186/1471-2164-13-525.

Moody, J. D., Freeman, J. P., Fu, P. P., Cerniglia, C. E. (2004). Degradation of benzo[a]pyrene by *Mycobacterium vanbaalenii* PYR-1. Applied and Environmental and Microbiology. 70:340–345. 10.1128/AEM.70.1.340-345.2004.

Moussawi, R. N., Patra, D. (2016). Nanoparticle self-assembled grain like curcumin conjugated ZnO: curcumin conjugation enhances removal of perylene, fluoranthene, and chrysene by ZnO. Science Report. 6:24565. doi.10.1038/srep24565.

Narayanan, K. B., Sakthivel, N. (2010). Biological synthesis of metal nanoparticles by microbes. Advances in Colloid ans Interface Science. 156:1–13.

Neff, J. M. (1985). Polycyclic aromatic hydrocarbons. In: Rand, G. M., Petrocilli, S. R. (eds.), Fundamentals of Aquatic Toxicology, Chap. 14. Hemisphere, New York.

Nelkenbaum, E., Dror, I., Berkowitz, B. (2007). Reductive hydrogenation of polycyclic aromatic hydrocarbons catalyzed by metalloporphyrins. Chemosphere. 68:210–217. 10.1016/j.chemosphere.2007.01.034.

NIOSH (National Institute for Occupational Safety and Health). (1977). Criteria for a Recommended Standard: Occupational Exposure to Coal Tar Products. US Department of Health and Human Services, Department of Health and National Institute for Occupational Safety and Health, Cincinnati, OH.

Oleszczuk, Patryk. (2006). Influence of different bulking agent on the disappearance of polycyclic aromatic hydrocarbons (PAHs) during sewage sludge composting. Water, Air, and Soil Pollution. 175:15–32.

Olicón-Hernández, Darío R., González-López, Jesús, Aranda, Elisabet (2017). Overview on the biochemical potential of filamentous fungi to degrade pharmaceutical compounds. Frontiers in Microbiology. 8, Article 1792, Review. doi.10.3389/fmicb.2017.01792.

Occupational Safety & Health Administration (OSHA). Coal Tar Pitch Volatiles [Available at https://www.osha.gov/SLTC/coaltarpitchvolatiles/hazards.html; accessed 14.11.10].

OSHA. (1983). Occupational Safety and Health Administration. Code of Federal Regulations. 29 CFR 1910.1002.

Patel, T. R., Ainsley, E. A. (1980). Naphthalene metabolism by pseudomonads: purification and properties of 1,2-dihydroxynaphthalene oxygenase. Journal of Bacteriology. 143:668–673.

Peng, Ri-He, Xiong, Ai-Sheng, Xue, Yong, Fu, Xiao-Yan, Gao, Feng, Zhao, Wei, Tian, Yong-Sheng, Yao, Quan-Hong. (2008). Microbial biodegradation of polyaromatic hydrocarbons. FEMS Microbiology Review. 32:927–955.

Pointing, S. B. (2001). Feasibility of bioremediation by white-rot fungi. Applied Microbiology Biotechnology. 57:20–33.

Pothuluri, J. V., Evans, F. E., Heinze, T. M., Cerniglia, C. E. (1996). Formation of sulfate and glucoside conjugates of benzo[e]pyrene by *Cunninghamella elegans*. Applied Microbiology and Biotechnology 45:677–683.

Qazi, Muneer A., Kanwal, Tayyaba, Jadoon, Muniba, Ahmed, Safia. (2014). Isolation and characterization of a biosurfactant-producing *Fusarium* sp. BS-8 from oil contaminated soil. Biotechnology Progress. 30(5): 1065–1075. doi.10.1002/btpr.1933.

Radding, S. B., Mill, T., Gould, C. W., Liu, H. D., Jhonson, H. L., Bomberger, D. C., Fojo, C. V. (1976). The Environmental Fate of Selected Polynuclear Aromatic Hydrocarbons. Office of Toxic Substances, EPA, Washington DC.

Radzi M., Nur-Aainaa-Syafini, Tay, Kheng-Soo, Bakar, Nor-Kartini Abu, Emenike, Chijioke Uche, Krishnan, Shamini, Hamid, Fauziah Shahul, Abas, Mhd-Radzi. (2015). Degradation of polycyclic aromatic hydrocarbons (pyrene and fluoranthene) by bacterial consortium isolated from contaminated road side soil and soil termite fungal comb. Environmental Earth Science. 74:5383–5391.

Rai, M., Yadav, P., Bridge, P., Gade, A. (2009). Myconanotechnology (NT) a new and emerging science. In: Rai, B. (ed.), Applied Mycology. CAB International, London, UK, pp. 258–267.

Ravindra, K., Sokhi, R., Grieken, R. V. (2008). Atmospheric polycyclic aromatic hydrocarbons: source attribution, emission factors and regulation. Atmospheric Environment. 42:2895–2921.

Reis, R. S., Pacheco, G. J., Pereira, A. G., Freire, D. M. G. (2013). Biosurfactants: production and applications. In: Rolando, C. (ed.), Biodegradation—Life of Science. InTech. doi.10.5772/56144. [Avialble at https://www.intechopen.com/books/biodegradation-life-of-science/biosurfactants-production-and-applications; accessed 31.10.19]

Rengarajan, Thamaraiselvan, Rajendran, Peramaiyan, Nandakumar, Natarajan, Lokeshkumar, Boopathy, Rajendran, Palaniswami, Rajkapoor, Balasubramanian. (2015). Exposure to polycyclic aromatic hydrocarbons with special focus on cancer. Asian Pacific Journal of Tropical Biomedicine. 5(3):182–189.

Rivas, F. J. (2006). Polycyclic aromatic hydrocarbons sorbed on soils: a short review of chemical oxidation based treatments. Journal of Hazardous Materials. 138(2):234–251. https://doi.org/10.1016/j.jhazmat.2006.07.048.

Rizwan, M. D., Singh, M., Mitra, C. K., Morve, R. K. (2014). Review: ecofriendly application of nanomaterials: nanobioremediation. Journal of Nanoparticles. 7, Article ID 431787.

Robinson, S. L., Novak, J. T., Widdowson, M. A., Crosswell, S. B., Fetterolf, G. J. (2003). Field and laboratory evaluation of the impact of tall fescue on polyaromatic hydrocarbon degradation in an aged creosote-contaminated surface soil. Journal of Environmental Engineering. 129:232–240.

Rufno, R. D., Sarubbo, L. A., Campos-Takaki, G. M. (2007). Enhancement of stability of biosurfactant produced by *Candida lipolytica* using industrial residue as substrate. World Journal of Microbiology and Biotechnology. 23:729–734.

Samanta, S. K., Rani, M., Jain, R. K. (1998). Segregation and structural instability of a recombinant plasmid carrying genes for naphthalene degradation. Letters in Applied Microbiology. 26:265–269.

Samanta, S. K., Singh, O. V., Jain, R. K. (2002). Polycyclic aromatic hydrocarbons: environmental pollution and bioremediation. Trends Biotechnology. 20:243–248.

Santodonato, S., Howard, P., Basu, D. (1981). Health and ecological assessment of polynuclear aromatic hydrocarbons. Journal of Environmental Pathology Toxicology. Special Issue, 5:1–364.

Scott, C. C., Makula, S. R., Finnerty, W. R. (1976). Isolation and characterization of membranes from a hydrocarbon-oxidizing Acinetobacter sp. Journal of Bacteriology, 127(1):469–480.

Semple, K. T., Morriss, A. W. J., Paton, G. I. (2003). Bioavailability of hydrophobic organic contaminants in soils fundamental concepts and techniques for analysis. European Journal of Soil Science. 54, 809–818.

Sevcan, Aydin, Karacy, Hatice Aygun, Shahi, Aiyoub, Gokc, Selen, Ince, Bahar, Ince, Orhan. (2017). Aerobic and anaerobic fungal metabolism and Omics insights for increasing polycyclic aromatic hydrocarbons biodegradation. Fungal Biology Reviews. 31:61–72.

Shiaris, M. P. (1989). Seasonal biotransformation of naphthalene, phenanthrene and benzo[a] pyrene in surficial estuarine sediments. Applied and Environmental Microbiology. 55:1391–1399.

Sibi, G, Simaria, C, Pant, G. (2015). Characterization and evaluation of polycyclic aromatic hydrocarbon (PAH) degrading bacteria isolated from oil contaminated soil. Applied Microbiology: Open Access. 1:104. doi.10.4172/2471-9315.1000104.

Sidkey, Nagwa M., Hadry, Al, Eman, A. (2012). Biosurfactant production by *Bacillus cereus*, B7 from lubricant oil waste. International Journal of Science and Research (IJSR). ISSN (Online): 2319–7064.

Sihag, Shallu, Pathak, Hardik, Jaroli, D. P. (2014). Factors affecting the rate of biodegradation of polyaromatic hydrocarbons. International Journal of Pure & Applied Bioscience. 2(3):185–202. ISSN: 2320–7051.

Sikkema, J., de Bont, J. A. M., Poolman, B. (1995). Mechanisms of membrane toxicity of hydrocarbons. Microbiology Review. 59:201–222.

Song, W., Li J., Zhang, W., Hu, X., Wang, L. (2012). An experimental study on the remediation of phenanthrene using ultrasound and soil washing. Environmental Earth Science. 66, 1487–1496.

Sutherland, J. B., Rafii, F., Khan, A. A., Cerniglia, C.E. (1995). Mechanisms of polycyclic aromatic hydrocarbon degradation. Microbial Transformation and Degradation of Toxic Organic Chemicals, 15:269.

Sverdrup, L. E., Krogh, P. H., Nielsen, T., Kjaer, C., Stenessen, J. (2003). Toxicity of eight polycyclic aromatic hydrocarbons to red clover (Trifolium pretense) ryegrass (Lolium perenne) and mustard (Sinapsis alba). Chemosphere. 53:993–1003.

Thanomsub, B., Watcharachaipong, T., Chotelersak, K., Arunrattiyakorn, P., Nitoda, T. et al. (2004). Monoacylglycerols: glycolipid biosurfactants produced by a thermotolerant yeast, *Candida ishiwadae*. Journal of Applied Microbiology. 96:588–592.

Thavamani, P., Megharaj, M., Krishnamurti, G. S. R., McFarland, R., Naidu, R. (2011). Finger printing of mixed contaminants from former manufactured gas plant (MGP) site soils: implications to bioremediation. Environment International. 37:184–189.

Thavamani, P., Megharaj, M., Venkateswarlu, K., Naidu, R. (2013). Mixed contamination of polyaromatic hydrocarbons and metals at manufactured gas plant sites: toxicity and implications to bioremediation. In: Wong, M. H. (ed.), Environmental Contamination: Health Risks, Bioavailability and Bioremediation. Taylor and Francis, New York, pp. 347–368.

Tigini, V., Prigione, V., Varese, G. C. (2014). Mycological and ecotoxicological characterisation of landfill leachate before and after traditional treatments. The Science of the Total Environment. 487:335–341. doi.10.1016/j.scitotenv.2014.04.026.

Tiwari, Jitendra Nath, Chaturvedi, Prashant, Ansari, Nasreen Gazi, Patel, Devendra Kumar, Jain, Sudhir Kumar, Murthy, Ramesh Chandra. (2011). Assessment of polycyclic aromatic hydrocarbons (PAH) and heavy metals in the vicinity of an oil refinery in India. Soil and Sediment Contamination: An International Journal. 20(3):315–328.

Tortella, G. R., Diez, M. C. (2005). Fungal diversity and use in decomposition of environmental pollutants. Critical Review in Microbiology. 31:197–212.

Tudoran, M. A., Putz, M. V. (2012). Polycyclic aromatic hydrocarbons: from in cerebro to in silico eco-toxicity fate. Chemical Bulletin "Politehnica" University of Timisoara. 57(71):1.

Van Brummelen, Timco C., Van Gestel, Cornelis A. M., Verweij, Rudo A. (1996). Long-term toxicity of five polycyclic aromatic hydrocarbons for the terrestrial isopods *Oniscus asellus* and *Porcellio scaber*. Environmental Toxicology and Chemistry. 15(7):1199–1210.

Varanasi, U. (ed.) (1989). Metabolic activation of PAH in subcellular fractions and cell cultures from aquatic and terrestrial species. In: Metabolism of Polycyclic Aromatic Hydrocarbons in the Aquatic Environment, Chap. 6, CRC Press, Boca Raton, FL.

Veignie, E., Rafin, C., Landy, D., Fourmentin, S., Surpateanu, G. (2009). Fenton degradation assisted by cyclodextrins of a high molecular weight polycyclic aromatic hydrocarbon benzo[a]pyrene. J Hazard Mater. 15168(2–3):1296–1301. doi: 10.1016/j.jhazmat.2009.03.012.

Volkering, F., Breure, A. M., Sterkenburg, A., van Andel, J. G. (1992). Microbial degradation of polycyclic aromatic hydrocarbons: effect of substrate availability on bacterial growth kinetics. Applied Microbiology and Biotechnology. 36:548–552.

Wang, S., Mulligan, C. N. (2004). An evaluation of surfactant foam technology in remediation of contaminated soil. Chemosphere. 57:1079–1089.

Wang, Z., Ren, P., Sun, Y., Ma, X., Liu, X., Na, G. (2013). Gas/particle partitioning of polycyclic aromatic hydrocarbons in coastal atmosphere of the north Yellow Sea, China. Environmental Science and Pollution Research International. 20:5753–5763.

Tungittiplakorn, Warapong, Cohen, Claude, Lion, Leonard W. (2005). Engineered polymeric nanoparticles for bioremediation of hydrophobic contaminants. Environmental Science & Technology. 39:1354–1358.

WHO. (1971). International Standards for Drinking Water, 3rd ed. World Health Organization, Geneva, Switzerland, p. 37.

Wick, L. Y., Ruiz de Munain, A., Springael, D. (2002). Responses of *Mycobacterium* sp. LB501T to the low bioavailability of solid anthracene. Applied Microbiology and Biotechnology. 58:378–385.

Wong, J. W. C., Lai, K. M., Wan, C. K., Ma, K. K., Fang, M. (2001). Isolation and optimization of PAHs-degradative bacteria from contaminated soil for PAHs bioremediation. Water, Air, and Soil Pollution. 13:1–13.

Xu, Jiangbing, Luo, Xiaosan, Wang, Yanling, Feng, Youzhi. (2017). Evaluation of zinc oxide nanoparticles on lettuce (Lactuca sativa L.) growth and soil bacterial community. Environmental Science and Pollution Research. https://doi.org/10.1007/s11356-017-0953-7.

Zhang, Q., Wang, D., Li, M., Xiang, W. N., Achal, V. (2014). Isolation and characterization of diesel degrading bacteria, *Sphingomonas* sp. and *Acinetobacter junii* from petroleum contaminated soil. Frontiers of Earth Science. 8:58–63.

Zheng, Z. M., Obbard, J. P. (2001). Effect of non-ionic surfactants on elimination of polycyclic aromatic hydrocarbons (PAHs) in soil–slurry by *Phanerochaete chrysosporium*. Journal of Chemical Technology and Biotechnology. 76:423–429.

Zhou, X., Wu, T., Hu, B., Jiang, T., Han, B. (2009). Ru nanoparticles stabilized by poly (N-vinyl-2-pyrrolidone) grafted onto silica: Very active and stable catalysts for hydrogenation of aromatics. Journal of Molecular Catalysis A: Chemical. 306:143–148.

Zhu, Xuezhu, Ni, Xue, Gatheru Waigi, Michael, Liu, Juan, Sun, Kai, Gao, Yanzheng. (2016). Biodegradation of mixed PAHs by PAH-degrading endophytic bacteria. International Journal of Environmental Research and Public Health. 13:805. doi.10.3390/ijerph13080805.

5 Relative Influence of "Physicochemical Variables" and "Indigenous Bacterial Diversity" on the Efficiency of Bioaugmentation-Mediated In Situ Bioremediation

*Janmejay Pandey**

CONTENTS

* Corresponding author: E-mail: Janmejay@curaj.ac.in

5.1 INTRODUCTION

Soil xenobiotic pollution caused by the injudicious production, usage, and disposal of anthropogenic xenobiotic pollutants (e.g., dyes, explosives, pesticides) is one of the major environmental concerns in agriculture and industry. Bioremediation has been defined as one of the most efficient, economic, and environmentally friendly approaches for the decontamination and restoration of ecological niches contaminated with such pollutants [1–6]. Among different bioremediation procedures, bioaugmentation, i.e., exogenous introduction of selected microorganism(s) showing specific degradation capabilities on the contaminated habitats, is regarded as one of the most potent approaches for in situ bioremediation [7–12]. Consequently, the majority of the studies in the field of microbial bioremediation have focused on isolation and characterization of microbial isolates capable of mineralizing or co-metabolically transforming the toxic pollutants. Furthermore, a large number of studies have reported cloning, sequencing, and characterization of genes, gene clusters, and proteins involved in the degradation of diverse xenobiotic compounds [13–23]. On the contrary, only a very few reports have shown successful application of degradative bacterial isolates in bioaugmentation-based in situ bioremediation [10, 24–28]. It is often suggested that a number of factors, e.g., strain selection, environmental heterogeneity, microenvironment of contaminated sites, physical-chemical nature of the contaminating pollutant, humidity, pollutant bioavailability, and environmental engineering procedures, may influence the outcome of the bioaugmentation process [7–8, 29–30]. Apparently, there is no information available on how indigenous microbial communities influence the efficiency of bioaugmentation-mediated in situ bioremediation. Therefore, a systematic study focusing on the influence of native bacterial/microbial community structure on the efficiency of bioaugmentation-mediated bioremediation process could be extremely useful for both basic understanding and technology development process.

With this rationale in mind, a feasibility study was carried out using a model based on the biodegradation of *p*-nitrophenol (PNP) by bioaugmentation of a degradative actinobacterial isolate, viz., *Arthrobacter protophormiae* strain RKJ100. PNP, an auto-hydrolysis product of organophosphate pesticides (parathion and methyl parathion) [31], is one of the most important and widely distributed xenobiotic compounds. It has been used for chemical synthesis of pesticides, dyes, plastics, explosives, etc. [32]. Furthermore, it has been categorized as a priority pollutant by environmental monitoring agencies due to its acute toxicity and mutagenicity to most forms of life [33–34]. In previous studies, strain RKJ100 was successfully used for in situ bioremediation of PNP-contaminated soils at different experimental scales, ranging from microcosms (20 g of soil) to small-scale field studies (300 kg of soil in standardized plots) [35–37]. However, the potential influence of biotic and abiotic characteristics of soil on efficiency of PNP degradation was not evaluated.

During the present study, comparative effects of soil characteristics and indigenous bacterial community structures on the efficiency of RKJ100 bioaugmentation-mediated PNP bioremediation process was assessed experimentally in a microcosm-scale study. For this, microcosms were prepared with soil samples collected from seven geographically separated provinces of India. Soil samples were compared for their respective physicochemical characteristics and native bacterial community structures. Soil samples were spiked with known concentrations of the target pollutant (PNP) prior to preparation of the microcosm. Subsequently, microcosms were bioaugmented with pre-grown cells of strain RKJ100. Pollutant degradation was determined by estimating the residual concentration of the pollutant at different time points after bioaugmentation. The results obtained from a preliminary analysis of PNP degradation were subjected to numerical ecology and statistical analysis to determine any possible correlation between PNP degradation kinetics, bacterial community structures, and physicochemical characteristics of soil samples. The results obtained from the statistical analyses clearly demonstrated that successful in situ bioremediation of PNP could be achieved in different soil types through bioaugmentation of pre-grown and enriched cultures of strain RKJ100; however, the efficiency of biodegradation may be differentially influenced by native bacterial community structure. Therefore, it is critical to systematically characterize the optimal performance parameters.

5.2 MATERIALS AND METHODS

5.2.1 Bacterial Strain, Culture Conditions, and Inoculums Preparation for Bioaugmentation

Arthrobacter protophormiae strain RKJ100 was previously isolated and characterized for its ability to utilize a number of nitroaromatic compounds (NACs), including PNP, 4-nitrocatechol (4-NC), and *o*-nitrobenzoate (ONB), as sole sources of carbon and energy [38–41]. This strain was tested successfully for its capability to degrade PNP in laboratory-scale, microcosm, and small-scale field studies [36–37]. During the preset study, this strain was used to decontaminate synthetically contaminated soil samples collected from seven geographically isolated provinces of India in microcosm studies (see later). Strain RKJ100 was grown either in shake flask culture or high-density fermentations using nutrient broth (NB) supplemented with 0.5–1 mM of PNP. High-density fermentations were carried out in a laboratory-scale bioreactor (10 L, New-Brunswick Scientific, Edison,

NJ, USA). For preparation of the inoculums used in the bioaugmentation, a bacterial cell mass was harvested by centrifugation at 5,000 × *g*, washed twice with 50 mL sterile saline, and immobilized on UV-sterilized "corn-cob powder" according to the procedure described earlier [36].

5.2.2 Microcosm Bioaugmentation Studies

Soil samples were collected from the agriculture fields of Assam, Andhra Pradesh, Gujarat, Karnataka, Maharashtra, Rajasthan, and Tamil-Nadu as a function of preexisting information on their different physicochemical characteristics. Collected soil samples were transported to the laboratory in cooled boxes and examined for their major physicochemical characteristics using standard geochemistry methods. Additionally, each soil sample was examined for indigenous concentration of PNP to rule out interference of preexisting PNP contamination. Soil samples were then spiked with PNP according to the protocol described earlier [37]. The indigenous bacterial community structures of soil samples were determined with terminal restriction fragment length polymorphism (T-RFLP) analysis carried out on a pool of eubacterial 16S rRNA genes amplified from total soil metagenomics DNA according to the standard procedure (described later). For establishing soil microcosm, 30 g of spiked soils were distributed per sterile glass beakers in triplicate and bioaugmented with pre-grown and PNP-induced RKJ100 cells (10^6 cells/g of soil). Control microcosms were established using (i) non-bioaugmented soil microcosms, (ii) bioaugmented sterilized soil microcosms, and (iii) non-bioaugmented sterilized soil microcosms. The non-bioaugmented microcosms added comparable amounts of "corn-cob" powder. For sterilization, 500 g of each soil sample was packed in autoclaving compatible plastic bags and subjected to five cycles of autoclaving. Sterilization of the soil samples was ascertained before using them for subsequent study with serial dilution and spread plate culturing. Each microcosm was covered with perforated aluminum foil and incubated at 30°C for 30 days. During the incubation period, the microcosms were sprinkled with sterilized distilled water at regular time intervals to compensate for the loss of water via evaporation. Soil samples (3 g) were collected from microcosms on a regular timescale interval to analyze it for (i) degradation of spiked pollutant; and (ii) survival of bioaugmented strain. The procedures used for these analyses are described later.

5.2.3 Total Soil DNA Isolation and T-RFLP Procedure

Soil DNA from collected soil samples was isolated using Mega-Prep Total Soil DNA Isolation Kit (Mo-Bio Laboratories, Inc., Salana-Beach, CA, USA) as per the manufacturer's instructions. Isolated DNA was quantified spectrophotometrically at 260 nm using NanoDrop-1000 (NanoDrop Technologies, Wilmington, DE, USA). Partial 16S rRNA gene was amplified for T-RFLP analysis using universal eubacterial 16S primers: 6-FAM-labeled 27F (5′-GAG TTT GAT CCT GGC TCA G-3′) and unlabeled 926R (5′-CCG TCA ATT CCT TTG AGT T-3′). Primers used during this study were procured from Bio-Basic, Inc., Markham, ON, Canada. The

T-RFLP polymerase chain reaction (PCR) cocktail consisted of 150 ng of total soil DNA, 1X ThermoPol Buffer, 200 μM dNTPs, and 100 pmol of each primer and 0.5 U of Deep Vent DNA polymerase (New England BioLabs, Inc., Ipswich, MA, USA) to form the final volume of 50 μL. The PCR program used for amplification was as follows: initial denaturation and enzyme activation at 95°C for 3 min, followed by 30 cycles of denaturation at 95°C for 30 s, annealing at 49°C for 30 s, extension at 75°C for 2 min, and a final extension for 5 min at 75°C on a personal thermocycler (Eppendorf, Germany). The PCR reaction was carried out in triplicate for each sample and the resulting products were pooled, ethanol precipitated, and resuspended in 20 μL of water prior to restriction digestion using 5 units of *Hae*III (New England BioLabs) and 10 μL of DNA in a final reaction volume of 20 μL for 3 h. The digested PCR products were ethanol precipitated, washed, and dissolved in a Tris-EDTA (where EDTA is ethylenediaminetetraacetic acid) buffer. The dissolved digestion product was mixed with 10 μL of deionized formamide and 0.5 μL GeneScan 1000-bp TAMRA size standard (Applied Biosystems, Warrington, UK), denatured by thermal denaturation, and analyzed with capillary electrophoresis using ABI prism 310 DNA Sequencer (Applied Biosystems). The output data were subjected to T-RF fragment size correction and binning according to the standard procedure.

5.2.4 T-RFLP Data Analyses

The output data were processed for binning with TREEFLAP (the T-RFLP peak sorting function for Microsoft Excel; available from http://www.sci.monash.edu.au/wsc/staff/walsh/treeflap.xls). The mismatched peaks were assigned their correct positions by manual correction. Second, all the output data were normalized for the determination of relative percentage abundance of identified fragments. Afterwards, T-RFLP profiles for all the samples were matched with the computer-simulated *in silico* digestion pattern of the 16S rRNA gene database (RDB, version 8.0) with the help of "T-RFLP Fragment Sorter 4.0" (available from http://www.oardc.ohio-state.edu/T-RFLPfragsort). This program compares the base pair length of experimentally generated terminal restriction fragments (T-RFs) with those generated by an *in silico* digestion of all the 16S rRNA gene sequences as submitted to the Ribosome Database Project. The database comparison was carried out with a restriction digestion profile generated with the same restriction endonucleases that were used for the experimental digestion and the T-RF matches were carried out with one base as fixed sizing error. The database match leads to the generation of a primary list of microorganisms (cultured, uncultured, environmental clones). Since the program makes use of multiple T-RFLP profiles (at least three profiles generated with different restriction endonuclease) for each sample, the confidence of the database match is significantly high.

For application in the numerical ecology and statistics, T-RFLP profiles for each soil sample were tabulated with Microsoft Excel to generate input files compatible with "R package." For comparative representation, T-RFLP profiles were plotted in form of an XY bar graph showing the T-RF base pair size on X axis and the percentage of relative abundance of respective T-RFs on the Y axis.

5.2.5 Quantification of PNP Degradation in Microcosms

Soil samples collected at different time points during bioaugmentation were subjected to residual PNP extraction and quantification according to the procedure reported earlier [36, 37]. Briefly, 1 g of soil sample was suspended in 10 mL of 5% NaOH, vortexed thoroughly for 5 min at ambient temperature, and followed by centrifugation at 1,500 rpm for 10 min. The supernatant was extracted with double volumes of ethyl acetate at neutral and acidic pH (after acidification to pH 2.0 with HCl). The aqueous phases from both neutral and acidic extractions were pooled, evaporated to dryness under vacuum in a Rotavapor (BUCHI, Switzerland), and finally dissolved in 1 mL methanol. Quantification of PNP was performed by high-performance liquid chromatography (HPLC) using a Waters 600 model equipped with a Waters 996 photodiode array detector operating at 315 nm. Separation was carried out using Waters Spherisorb 5 μm C8 column as the stationary phase and acetonitrile: water (80:20 v/v) containing 0.1% trifluoroacetic acid with a flow rate of 1.0 mL/min as the mobile phase. A standard plot generated with known concentrations of PNP stocks (prepared in HPLC grade methanol) was used to determine the concentration of residual PNP. Control-normalized residual concentrations of PNP determined in different samples collected during the bioaugmentation process were subsequently analyzed with regression analysis using Sigma Plot (SPSS, Inc., CA, USA) according to the exponential decay equation to assess the kinetics and efficiency of degradation process.

5.2.6 Survival of Bioaugmented Strain in Microcosms

The soil samples collected were also analyzed for survival of the bioaugmented strain. For this culture-dependent (colony-forming unit [CFU] counting and colony hybridization analyses) and culture-independent (slot blot analysis) approaches were implemented.

5.2.6.1 CFU Counting and Colony Hybridization

For CFU counting and colony hybridization analyses, 0.50 g of collected soil samples were suspended in 5 mL of sterile saline. The suspension was serially diluted and plated on selective medium, i.e., minimal salt medium agar (containing: 4.0 g/L $NaH_2PO_4 \times 2\,H_2O$, 4.0 g/L K_2HPO_4, 0.8 g/L $(NH_4)_2SO_4$, and 0.8 g/L $MgSO_4$, 1/1000 × trace element solution, and 2 % w/v ultrapure agar) supplemented with 0.5 mM PNP as a sole carbon, energy, and nitrogen source. Utilization of PNP was monitored by observing the decolorization of the yellow color of the PNP on these selective medium plates. All the colonies included within the decolorized regions of the PNP plate were counted. Colony hybridization was performed to further substantiate the results of the CFU counts. Bacterial colonies obtained on the earlier CFU counting on selective plates were screened using a 300 bp DNA fragment that is amplified as a partial amplicon to form a benzenetriol dioxygenase (*btd*) gene of strain RKJ100. The enzyme encoded by this gene has been shown to be involved in the PNP metabolic pathway of *A. protophormiae* strain RKJ100 and used to monitor cell survival of strain RKJ100 during earlier in situ biodegradation studies [36]. The colony hybridization process used was as follows: bacterial colonies were transferred onto a HybondN+ nylon membrane

(GE HealthCare, Amersham, UK). Membranes were then treated sequentially for 10 min each with cell lysis buffer (10% SDS, 0.5 N NaOH), denaturation buffer (1.5 M NaOH, 0.5 M Tris-Cl; pH 7.4), transfer buffer (1.5 M NaCl), and 2× SSC buffer. Afterwards, membranes were air-dried and UV cross-linked before probing them with the P^{32}-labeled 300-bp fragment of the *btd* gene. After hybridization, the membrane was exposed to film for 2 hours and hybridization signals were scanned with Phosphor-Imager (Bio-Rad Laboratories, Hercules, CA, USA).

5.2.6.2 Dot Blot Analysis

For quantitation of strain RKJ100 with culture-independent approach; dot blot analysis was carried out according to the standard procedure described earlier [36]. Briefly, total soil DNA was isolated from 1 g of each soil sample using Ultra-Clean Mega-Prep Soil DNA isolation kit (Mo-Bio Laboratories) according to the manufacturer instructions. Resulting total soil DNA was quantified spectrophotometrically using NanoDrop-1000 (NanoDrop Technologies). Normalized concentrations of total soil DNA were vacuum blotted onto HybondN+ nylon membrane (GE HealthCare) using slot-blot vacuum blotter (GE HealthCare) according to procedure described by the manufacturer. Afterwards, the membrane was air-dried and subjected to hybridization using P^{32}-labeled 300-bp fragment of the *btd* gene as described above. Hybridization signals were scanned Phosphor-Imager (Bio-Rad Laboratories). Quantitation of hybridization signals was carried out with Gel-Quant Software version 2.0 (Bio-Rad Laboratories). Standard plot generated with known concentrations of RKJ100 genomic DNA (each isolated from 10^2, 10^3, 10^4, 10^5, and 10^6 cells) was used to determine the concentration of RKJ100-specific DNA and corresponding cell number of strain RKJ100 within soil sample collected at different time points during bioaugmentation.

5.2.7 Numerical Ecology Statistical Analysis

The PNP degradation patterns obtained during bioaugmentation were subjected to numerical ecology statistical analysis using matrices generated with (i) T-RFLP data for different soil samples, (ii) physicochemical characteristics of different soil samples, and (iii) degree of survival of the bioaugmented strain in different microcosms during bioaugmentation. This analysis was carried out with Mantel tests (using "R" software; http://www.r-project.org/). The goal of this analysis was to evaluate potential correlations among the previously noted factors and PNP degradation kinetics. The influence of each of these factors on PNP degradation kinetics was measured in terms of correlation index, wherein values of 0.00 represented no correlation (no influence) and 1.00 represented perfect correlation (complete interdependence).

5.3 RESULTS AND DISCUSSIONS

5.3.1 Physicochemical Characteristics of Soil Samples

PNP degradation by bioaugmentation by of *A. protophormiae* strain RKJ100 was studied in microcosm studies using soil samples collected from seven geographically distinct provinces of India. The selection of soil samples was carried out on purpose to assess performance and applicability of the *A. protophormiae* RKJ100 bioaugmentation for in situ PNP degradation in different soil types. In the hindsight, it was expected that the soil samples would have divergent physicochemical characteristics and indigenous bacterial community structures; therefore, the results from the study could also provide important insight into the influence of these factors on kinetics and efficiency of pollutant removal during bioaugmentation process. As expected, all the soil samples were found to have significant differences in their physicochemical characteristics. Notably, the soil sample taken from Assam showed an acidic pH (6.65), whereas all others were slightly alkaline. Another strong contrast was observed with the total sand content of different soil samples; as soils collected from Rajasthan and Gujarat showed large amounts of sand content (78% and 35%, respectively), compared to the others (7–13%). Other physicochemical characteristics showed comparable values. Table 5.1 presents a

TABLE 5.1
Physicochemical Variables of the Soil Samples Used During the Study

Physicochemical Variable	Geographical Origin of Soil Sample						
	Andhra	Assam	Gujarat	Karnataka	Maharashtra	Rajasthan	Tamil Nadu
pH	8.36	6.65	7.73	7.5	8.98	9.04	8.91
Electrical conductivity (Siemens. m^{-1})	0.186	0.027	0.041	0.046	0.084	0.094	0.088
Organic matter content (ppm)	1.63	1.42	1.01	0.92	1.76	0.78	1.13
Nitrogen content (ppm)	0.056	0.037	0.023	0.033	0.029	0.022	0.041
Phosphorous content (ppm)	11.1	8.5	12.6	8.4	13.5	7.8	9.0
Potassium content (ppm)	157	124	161	101	182	139	108
Chloride content (ppm)	0.106	0.109	0.0561	0.067	0.87	0.07	0.011
Sodium absorbance ratio	4.03	2.88	2.04	3.1	1.28	1.59	1.64
Water holding capacity (mL* g $soil^{-1}$)	0.34	0.25	0.58	0.41	0.36	0.716	0.27
Clay content (%)	52	55	29	37	40	8	48
Silt content (%)	38	38	36	51	48	14	39
Sand content (%)	10	7	35	12	12	78	13
Native PNP (ppm)	2.5	8.05	5.6	4.2	3.7	0.42	5

summary of some of the important physicochemical characteristics analyzed during the study. These results clearly establish the distinct physicochemical nature of soil samples used during the present study.

The soil samples were also analyzed for their indigenous PNP concentration before using them into microcosm studies. The results from this analysis showed presence of PNP concentration in a range of 0.42–8.05 ppm (Table 5.1). The highest concentration of PNP was found in soil samples collected from Assam, whereas the soil sample from Rajasthan had minimum indigenous concentrations. The values were found to be comparable for other soil samples (Table 5.1). These results further established that soil samples selected for the present study were significantly diverse from each other in terms of both physicochemical characteristics as well as their relative residual PNP contamination.

5.3.2 Bacterial Community Structures of Soil Samples

The eubacterial community structures of different soil samples were analyzed with T-RFLP analysis of pool of 16S rRNA gene amplified from respective total soil DNA. Results from this analysis clearly demonstrated large differences in community structures of different soil samples, showing variations in diversity (measured in terms of total T-RF numbers). For instance, the bacterial community structure of Assam soil was found to be most diverse and showed the presence of 78 T-RFs (each T-RF corresponding to a ribotype), whereas the Rajasthan soil had only 28 distinct T-RFs, indicating that the microbial community in Rajasthan soil sample was significantly less diverse. Apart from differences observed with the diversity of T-RFs, their relative contributions to the total bacterial diversity were also found to be appreciably diverse among different soil samples. Figure 5.1 shows a graphical representation of percentage normalized T-RFLP profiles of all seven soil samples. Database comparison of T-RFLP profiles also corroborated the divergent nature of native bacterial structures. These results clearly established that the native bacterial diversity of different soil types were significantly different from each other.

5.3.3 PNP Degradation During Microcosm Study

Controls carried out on non-bioaugmented unsterilized microcosms showed that a limited amount of PNP was lost from different microcosms during the course of experiment; this observation indicated the presence of an indigenous PNP degradation potential. However, the extent of PNP degradation in non-bioaugmented microcosms was only nominal. The quantification of residual PNP concentrations in bioaugmented microcosm demonstrated that the exogenous addition of strain RKJ100 cells in the soil microcosms led to significant improvement of PNP degradation rates; five bioaugmented microcosms (Andhra Pradesh, Assam, Karnataka, Maharashtra, and Tamil Nadu soil samples) showed more than 60% of PNP degradation during 30 days of incubation. The residual PNP concentration

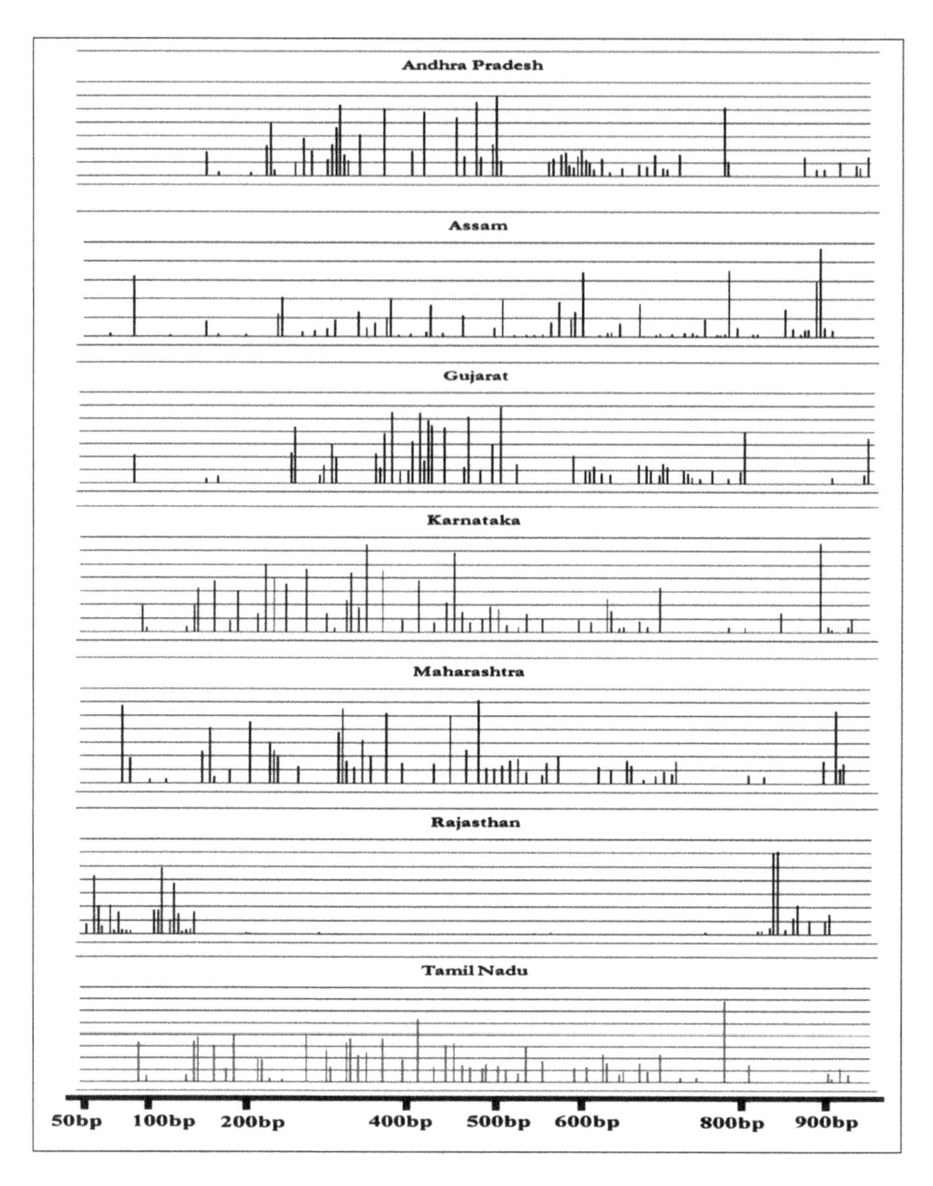

FIGURE 5.1 Representative percentage of normalized terminal restriction fragment length polymorphism (T-RFLP) analysis profile of eubacterial 16S rRNA gene pools amplified from total soil DNA of soil samples and with digested with *Hae*III restriction endonuclease.

decreased to less than 40% of initial pollutant concentration. Noticeably, the maximum degradation was attained in microcosm prepared with soil sample from Andhra Pradesh.

The efficiency of degradation was rather poor in microcosms prepared with Gujarat and Rajasthan soil samples, as only less than 30% PNP degradation was observed in them during the same time course. Inefficient PNP degradation in microcosms prepared with these soil samples suggested either the presence of unfavorable environmental conditions (soil texture, mineralogy, and other physicochemical characteristics) or a possible antagonistic influence of native microbial communities on the bioaugmented strain or a combination thereof.

At this stage, it was important to make the distinction between the respective influence of physicochemical conditions of soil samples and the possible antagonistic effects induced by activity of native bacterial communities on the survival capacity of the strain RJK100 and kinetics of PNP degradation. Therefore, a comparative evaluation was performed with results obtained for PNP degradation in unsterilized and sterilized soil microcosms. Autoclaving of soil samples had no to minor impact on efficiency of PNP degradation in five of the seven tested soil samples (data not shown).

5.3.4 Cell Survival of Bioaugmented Strain

Cells survival of bioaugmented strain was monitored as an alternative gauge for the efficiency of bioaugmentation process. Monitoring the survival of bioaugmented strain is also important for determining the environmental fate of the bioaugmented population. As observed with results from culture-dependent and culture-independent methods, the bioaugmented strain showed different survival pattern in different microcosms. The colony hybridization signal count per gram soil on 30th day of incubation in microcosms showed that RKJ100 survived much better in microcosms prepared with soil samples of Andhra Pradesh, Assam, Karnataka, Maharashtra, and Tamil Nadu as compared to that made with Rajasthan or Gujarat soil; survival of bioaugmented strain was also three to five time better in the samples obtained from Andhra Pradesh, Assam, Karnataka, Maharashtra, and Tamil Nadu. In slot blot analysis, the signal strength for RKJ100-specific DNA within microcosms on 30th day of incubation also showed similar results. This result was in close agreement with approximately two to five times improved cell survival observed in colony hybridization analysis. Conspicuously, this observation corresponded well with improvement in efficiency of pollutant degradation in microcosm prepared with sterilized Rajasthan soil as mentioned earlier. These observations clearly demonstrate that survival of bioaugmented strain has a significant effect on the efficiency of pollutant degradation.

5.3.5 Numerical Ecology and Statistics Analyses

The results discussed earlier provided a strong indication for the possible correlation between the survival of the bioaugmented strain and PNP degradation during bioaugmentation. However, with this analysis, it was not possible to determine the quantitative parameters for such an influence and for the role of abiotic and biotic factors on the efficiency of pollutant degradation. Therefore, a four-way "Mantel tests" analysis was performed for quantitative statistical assessment of correlations among (i) PNP degradation kinetics, (ii) physicochemical variables of soil samples, (iii) native microbial community structure of soil samples, and

(iv) strain RKJ100 survival. The correlation was determined in terms of Mantel statistics (r). The empirical upper limit for each combination is estimated at 90%, 95%, 97.5%, and 99% correlation. Subsequently, the obtained "r" value is compared. Also, the significance of analyses was estimated with p-value of the statistical test ($p \leq 0.05$ indicates a strong probability of a significant correlation between the two data matrices).

The first noticeable result from this analyses showed that native bacterial community structures (T-RFLP profiles) were adequately correlated with the physicochemical characteristics (soil variables) of their respective habitats as indicated by values of 0.459, which falls between 95% and 97.5 % correlation (Table 5.2; row 1). This observation confirmed the potential role of the soil variable in determining the structure and diversity of indigenous bacterial population.

The other noticeable result was that survival of bioaugmented strain (RKJ100) was found to have high correlation ($r = 0.440$) with soil variables and indigenous microbial structure ($r = 0.483$) (Table 5.2; row 2). These values correspond to more than empirical values possible at 97.5% and 95%, respectively. This observation clearly suggests that both biotic and abiotic factors contribute toward the differential survival of the bioaugmented strain.

The last set of results from Mantel test analysis showed PNP degradation to have the highest correlation with survival of the bioaugmented strain ($r = 0.764$; importantly, this value is very close to the maximum possible empirical value at 99.5% correlation) (Table 5.2, row 3). PNP degradation was also found to be greatly correlated with native microbial community structure ($r = 0.521$) and soil variables ($r = 0.558$), both showing possible correlation between 97.5% and 99%.

Together these results clearly established that "survival of bioaugmented strain" contributes as the most important factor in determining the efficiency of PNP degradation within different soil types. Further, the results obtained during the present study strongly indicated that both "physicochemical variables of soil samples" and "indigenous bacterial community structures" may selectively influence "survival of bioaugmented strain" and in turn they may influence the final outcome and efficiency of bioaugmentation-mediated biodegradation process.

To conclude, results obtained during present study clearly established that RKJ100 bioaugmentation could be successfully used for PNP bioremediation in diverse contaminated niches and different soil types. Only a few other studies have reported successful bioaugmentation-mediated bioremediation of NACs. J. R. Simplot *Ex Situ* Bioremediation Technology reported bioremediation of tri-nitrotoluene (TNT) contaminated soil with a consortium of anaerobic soil microorganisms. Notably, this technology was used for bioremediation of TNT-contaminated soil and it was performed in a bioreactor. The bioremediation process resulted in ~99% pollutant removal during 9 months of incubation. In other study, Strong et al., reported field scale remediation of atrazine-contaminated soil using atrazine hydrolase expressing Escherichia *coli* recombinant [42]. A few other studies have reported enhanced biodegradation of xenobiotic pollutants during bioaugmentation studies [24–26, 28, 43–44]. These studies have usually shown improved biodegradation

TABLE 5.2
Statistical Analysis of Results Obtained During Bioaugmentation-Mediated In Situ Degradation of PNP-Contaminated Soil Samples

	Soil Characteristics	T-RFLP Profile	RKJ100 Survival	PNP Degradation
T-RFLP profile	Mantel statistic, $r = \mathbf{0.459}$ Significance : 0.034 Empirical upper confidence limit for r 90% = 0.365 95% = 0.427 97.5% = 0.472 99% = 0.531			
RKJ100 survival	Mantel statistic, $r = \mathbf{0.440}$ Significance : 0.016 Empirical upper confidence limit for r 90% = 0.331 95% = 0.376 97.5% = 0.424 99% = 0.460	Mantel statistic, $r = \mathbf{0.483}$ Significance : 0.031 Empirical upper confidence limit for r 90% = 0.388 95% = 0.449 97.5% = 0.494 99% = 0.519		
PNP degradation	Mantel statistic, $r = \mathbf{0.558}$ Significance : 0.013 Empirical upper confidence limit for r 90% = 0.362 95% = 0.448 97.5% = 0.491 99% = 0.570	Mantel statistic, $r = \mathbf{0.521}$ Significance : 0.015 Empirical upper confidence limit for r 90% = 0.388 95% = 0.445 97.5% = 0.478 99% = 0.532	Mantel statistic, $r = \mathbf{0.764}$ Significance: 0.012 Empirical upper confidence limit for r 90% = 0.234 95% = 0.322 97.5% = 0.728 99% = 0.775	

Note: r = correlation coefficient; p = p-value of the statistical test ($p \leq 0.05$ indicates a strong probability of a significant correlation between the two data matrices).
Mantel tests were carried out on T-RFLP data used as a Hellinger-transformed dissimilarity matrix and PNP degradation, strain survival, and soil variables data sets used as Euclidean distances matrices using 1000 permutations.

only under highly controlled and optimized environments. Therefore, it is difficult to justify whether their results could be extrapolated to diverse environmental niches or not. On the contrary, results obtained during the present study clearly demonstrate that bioaugmentation of *A. protophormiae* strain RKJ100 can stimulate PNP bioremediation in a number of soil samples with diverse geographical origins, physicochemical variables, and indigenous microbial communities. To the best of our knowledge, this is one of the first experimental demonstrations for applicability of a bioaugmentation process for successful pollutant decontamination in different soil types characterized by diverse physicochemical characteristics and indigenous bacterial community structures. Furthermore, this study also presents first experimental evidence for relative influence of physicochemical variable of soil and its native bacterial community structure on efficiency of decontamination process. This study clearly demonstrated that "survival of bioaugmented strain" acts as the most important factor determining the efficiency of the bioaugmentation process. Based on these findings, we propose that bioaugmentation patterns, e.g., inoculums size, inoculums frequency, etc. could be optimized for different soil type to ensure optimal performance of bioaugmentation-mediated bioremediation procedure.

ACKNOWLEDGMENTS

This study was partly funded by Council for Scientific and Industrial Research (CSIR), India, Department of Biotechnology (DBT), and Indo-Swiss Collaboration in Biotechnology (ISCB). The study was performed in the Laboratory of Late Dr. R. K. Jain (IMTECH-CSIR, Chandigarh) in Scientific Collaboration with Prof. Christof Holliger (EPFL-Lausanne, Switzerland). Experiments related to numerical ecological analyses and statistical assessments were carried out under guidance of Prof. C. Holliger and Dr. P. Rossi (EPFL-Lausanne, Switzerland).

REFERENCES

1. Crawford, R. L. Bioremediation of groundwater pollution. Curr. Opin. Biotechnol. 2 (1991):436–439.
2. Bouwer, E. J., Zehnder, A. J. Bioremediation of organic compounds--putting microbial metabolism to work. Trends Biotechnol. 11 (1993):360–367.
3. Parales, R. E., Haddock, J. D. Biocatalytic degradation of pollutants. Curr. Opin. Biotechnol. 15 (2004):374–379.
4. Lloyd, J. R., Renshaw, J. C. Bioremediation of radioactive waste: radionuclide-microbe interactions in laboratory and field-scale studies. Curr. Opin. Biotechnol. 16 (2005):254–260.
5. Lovley, D. R. Cleaning up with genomics: applying molecular biology to bioremediation. Nat. Rev. Microbiol. 1 (2003):35–44.
6. Paul, D., Pandey, G., Pandey, J., Jain, R. K. Accessing microbial diversity for bioremediation and environmental restoration. Trends Biotechnol. 23 (2005):135–142.
7. Tyagi, M., da Fonseca, M. M., de Carvalho, C. C. Bioaugmentation and biostimulation strategies to improve the effectiveness of bioremediation processes. Biodegradation. 22 (2011):231–241.
8. Vogel, T. M. Bioaugmentation as a soil bioremediation approach. Curr. Opin. Biotechnol. 7 (1996):311–316.
9. Jiao, Y., Zhao, Q., Jin, W., Hao, X., You, S. Bioaugmentation of a biological contact oxidation ditch with indigenous nitrifying bacteria for in situ remediation of nitrogen-rich stream water. Bioresour. Technol. 102 (2011):990–995.

10. Jitnuyanont, P., Sayavedra-Soto, L. A., Semprini, L. Bioaugmentation of butane-utilizing microorganisms to promote cometabolism of 1,1,1-trichloroethane in groundwater microcosms. Biodegradation. 12 (2001):11–22.
11. Weyens, N., van der Lelie, D., Artois, T., Smeets, K., Taghavi, S., Newman, L., Carleer, R., Vangronsveld, J. Bioaugmentation with engineered endophytic bacteria improves contaminant fate in phytoremediation. Environ. Sci. Technol. 43 (2009):9413–9418.
12. Barathi, S., Vasudevan, N. Bioremediation of crude oil contaminated soil by bioaugmentation of *Pseudomonas fluorescens* NS1. J. Environ. Sci. Health A Tox. Hazard. Subst. Environ. Eng. 38 (2003):1857–1866.
13. Kube, M., Beck, A., Meyerdierks, A., Amann, R., Reinhardt, R., Rabus, R. A catabolic gene cluster for anaerobic benzoate degradation in methanotrophic microbial black sea mats. Syst. Appl. Microbiol. 28 (2005):287–294.
14. Tsoi, T. V., Plotnikova, E. G., Cole, J. R., Guerin, W. F., Bagdasarian, M., Tiedje, J. M. Cloning, expression, and nucleotide sequence of the pseudomonas aeruginosa 142 ohb genes coding for oxygenolytic ortho dehalogenation of halobenzoates. Appl. Environ. Microbiol. 65 (1999):2151–2162.
15. Nordin, K., Unell, M., Jansson, J. K. Novel 4-chlorophenol degradation gene cluster and degradation route via hydroxyquinol in *Arthrobacter chlorophenolicus* A6. Appl. Environ. Microbiol. 71 (2005):6538–6544.
16. Jesenska, A., Bartos, M., Czernekova, V., Rychlik, I., Pavlik, I., Damborsky, J. Cloning and expression of the haloalkane dehalogenase gene dhmA from Mycobacterium avium N85 and preliminary characterization of dhmA. Appl. Environ. Microbiol. 68 (2002):3724–3730.
17. Chauvaux, S., Chevalier, F., Le Dantec, C., Fayolle, F., Miras, I., Kunst, F., Beguin, P. Cloning of a genetically unstable cytochrome P-450 gene cluster involved in degradation of the pollutant ethyl tert-butyl ether by Rhodococcus ruber. J. Bacteriol. 183 (2001):6551–6557.
18. Bohuslavek, J., Payne, J. W., Liu, Y., Bolton, H., Xun, Jr., L. Cloning, sequencing, and characterization of a gene cluster involved in EDTA degradation from the bacterium BNC1. Appl. Environ. Microbiol. 67 (2001):688–695.
19. Jeon, C. O., Park, W., Padmanabhan, P., DeRito, C., Snape, J. R., Madsen, E. L. Discovery of a bacterium, with distinctive dioxygenase, that is responsible for in situ biodegradation in contaminated sediment. Proc. Natl. Acad. Sci. USA. 100 (2003):13591–13596.
20. Belchik, S. M., Xun, L. Functions of flavin reductase and quinone reductase in 2,4,6-trichlorophenol degradation by *Cupriavidus necator* JMP134. J. Bacteriol. 190 (2008):1615–1619.
21. Sanchez, M. A., Gonzalez, B. Genetic characterization of 2,4,6-trichlorophenol degradation in Cupriavidus necator JMP134. Appl. Environ. Microbiol. 73 (2007):2769–2776.
22. Zhang, S., Sun, W., Xu, L., Zheng, X., Chu, X., Tian, J., Wu, N., Fan, Y. Identification of the *para*-nitrophenol catabolic pathway, and characterization of three enzymes involved in the hydroquinone pathway, in pseudomonas sp. 1-7. BMC Microbiol. 12 (2012):27.
23. Zhong, W. H., Sun, M., He, G. Q., Feng, X. S., Yu, Z. N. Isolation of 2,4-dichlorophenol degrading bacterium strain and cloning and expression of its 2,4-dichlorophenol hydroxylase gene. Sheng Wu Gong Cheng Xue Bao. 20 (2004):209–214.
24. Tribedi, P., Sil, A. K. Bioaugmentation of polyethylene succinate-contaminated soil with pseudomonas sp. AKS2 results in increased microbial activity and better polymer degradation. Environ. Sci. Pollut. Res. Int. 20(3) (2012):1318–1326.

25. Li, X., Lin, X., Li, P., Liu, W., Wang, L., Ma, F., Chukwuka, K. S. Biodegradation of the low concentration of polycyclic aromatic hydrocarbons in soil by microbial consortium during incubation. J. Hazard Mater. 172 (2009):601–605.
26. Van Dillewijn, P., Caballero, A., Paz, J. A., Gonzalez-Perez, M. M., Oliva, J. M., Ramos, J. L. Bioremediation of 2,4,6-trinitrotoluene under field conditions. Environ. Sci. Technol. 41 (2007):1378–1383.
27. Smith, A. E., Hristova, K., Wood, I., Mackay, D. M., Lory, E., Lorenzana, D., Scow, K. M. Comparison of biostimulation versus bioaugmentation with bacterial strain PM1 for treatment of groundwater contaminated with methyl tertiary butyl ether (MTBE). Environ. Health Perspect. 113 (2005):317–322.
28. Payne, R. B., May, H. D., Sowers, K. R. Enhanced reductive dechlorination of polychlorinated biphenyl impacted sediment by bioaugmentation with a dehalorespiring bacterium. Environ. Sci. Technol. 45 (2011):8772–8779.
29. Thompson, I. P., van der Gast, C. J., Ciric, L., Singer, A. C. Bioaugmentation for bioremediation: the challenge of strain selection. Environ. Microbiol. 7 (2005):909–915.
30. El Fantroussi, S., Agathos, S. N. Is bioaugmentation a feasible strategy for pollutant removal and site remediation? Curr. Opin. Microbiol. 8 (2005):268–275.
31. Rubin, C., Esteban, E., Hill, R. H., Pearce, Jr., K. Introduction---the methyl parathion story: a chronicle of misuse and preventable human exposure. Environ Health Perspect. 110(Suppl 6) (2002) 1037–1040.
32. Imtiaz, R., Haugh, G. Analysis of environmental and biologic methyl parathion data to improve future data collection. Environ. Health Perspect. 110(Suppl 6) (2002):1071–1074.
33. Wissiack, R., Rosenberg, E. Universal screening method for the determination of US environmental protection agency phenols at the lower ng l(-1) level in water samples by on-line solid-phase extraction-high-performance liquid chromatography-atmospheric pressure chemical ionization mass spectrometry within a single run. J. Chromatogr. A. 963 (2002):149–157.
34. Spain, J. C., Pritchard, P. H., Bourquin, A. W. Effects of adaptation on biodegradation rates in sediment/water cores from estuarine and freshwater environments. Appl. Environ. Microbiol. 40 (1980):726–734.
35. Paul, D., Pandey, G., Meier, C., van der Meer, J. R., Jain, R. K. Bacterial community structure of a pesticide-contaminated site and assessment of changes induced in community structure during bioremediation. FEMS Microbiol. Ecol. 57 (2006):116–127.
36. Labana, S., Pandey, G., Paul, D., Sharma, N. K., Basu, A., Jain, R. K. Pot and field studies on bioremediation of *p*-nitrophenol contaminated soil using *Arthrobacter protophormiae* RKJ100. Environ. Sci. Technol. 39 (2005):3330–3337.
37. Labana, S., Singh, O. V., Basu, A., Pandey, G., Jain, R. K. A microcosm study on bioremediation of *p*-nitrophenol-contaminated soil using *Arthrobacter protophormiae* RKJ100. Appl. Microbiol. Biotechnol. 68 (2005):417–424.
38. Pandey, G., Paul, D., Jain, R. K. Branching of *o*-nitrobenzoate degradation pathway in *Arthrobacter protophormiae* RKJ100: identification of new intermediates. FEMS Microbiol. Lett. 229 (2003):231–236.
39. Bhushan, B., Chauhan, A., Samanta, S. K., Jain, R. K. Kinetics of biodegradation of *p*-nitrophenol by different bacteria. Biochem. Biophys. Res. Commun. 274 (2000):626–630.
40. Chauhan, A., Chakraborti, A. K., Jain, R. K. Plasmid-encoded degradation of *p*-nitrophenol and 4-nitrocatechol by *Arthrobacter protophormiae*. Biochem. Biophys. Res. Commun. 270 (2000):733–740.
41. Chauhan, A., Jain, R. K. Degradation of *o*-nitrobenzoate via anthranilic acid (o-aminobenzoate) by *Arthrobacter protophormiae*: a plasmid-encoded new pathway. Biochem. Biophys. Res. Commun. 267 (2000):236–244.

42. Strong, L. C., McTavish, H., Sadowsky, M. J., Wackett, L. P. Field-scale remediation of atrazine-contaminated soil using recombinant *Escherichia coli* expressing atrazine chlorohydrolase. Environ. Microbiol. 2 (2000):91–98.
43. Larsen, S. B., Karakashev, D., Angelidaki, I., Schmidt, J. E. Ex-situ bioremediation of polycyclic aromatic hydrocarbons in sewage sludge. J. Hazard. Mater. 164 (2009):1568–1572.
44. Sun, G. D., Xu, Y., Jin, J. H., Zhong, Z. P., Liu, Y., Luo, M., Liu, Z. P. Pilot scale ex-situ bioremediation of heavily PAHs-contaminated soil by indigenous microorganisms and bioaugmentation by a PAHs-degrading and bioemulsifier-producing strain. J. Hazard. Mater. 233–234 (2012):72–78.

6 Role of Enzymes in Bioremediation of Organic Pollutants

Smita Chaudhry and Rashmi Paliwal*

CONTENTS

6.1 INTRODUCTION

Contamination by organic pollutants is a critical issue that places serious pressure on the global environment. Unfortunately, this pressure is continuously increasing with technological and industrial revolutions fulfilling the demand of growing populations. Organic pollutants such as pesticides, polycyclic aromatic hydrocarbons (PAHs), dyes, polychlorinated biphenyls (PCBs), BTEX (benzene, toluene, ethylbenzene, and xylene), plastics, biopolymers, phenols, chlorophenols,

* Corresponding author: Email: smitachaudhry11@gmail.com

nitrocompounds, polyethene, etc., from various industrial activities are continuously being released into the adjoining land and/or water bodies. These toxic pollutants are characterized as ubiquitous compounds with both natural and anthropogenic sources. Human activities that are related to the release of such persistent organic pollutants (POPs) include not only the production and application but also the unsafe disposal of these compounds in nature. These substances are also present in the workplace environment, and the general population may get exposed to these chemicals even through food. Exposure to such toxic compounds can cause harmful effects on human health, which include neurological disorders; liver dysfunction; reproductive system problems; behavioral, immune, and endocrine disorders; and carcinogenic effects (Stojić et al., 2018).

Over the centuries, the work on the development of an effective methodology to clean the contaminated sites has resulted in conventional techniques that in turn disrupt and cause an abrupt change in the soil and water quality. The search for effective techniques has led to the evolution of biological agents (bacteria, fungi, algae, and plants) for the remediation of the polluted site. Application of biological agents such as microbes for the breakdown of organic contaminants is known as biodegradation. These contaminants are usually considered a food source that the microbes utilize as a substrate and degrade through enzymatic activity. Therefore, biodegradation of any organic contaminant involves a series of degradation steps that ultimately result in degradation of the toxic parent compound. Biodegradation of organic pollutants occurs through two mechanisms (i.e., aerobic and anaerobic). Aerobic degradation usually involves incorporation of an oxygen atom into the compound structure with the help of the enzyme oxygenase, whereas the anaerobic transformation of organic contaminants involves the modification of certain functional groups of the compound with the help of specific enzymes, such as transferase, carboxylase, dehydrogenase, etc. Thus, each step in aerobic and anaerobic degradation pathways is catalyzed by a specific enzyme that is produced and released within the microbial cell or released by the cells into the environment for initiation of the degradation process. Enzymes that are released into the environment by the cell are called extracellular enzymes. Application of different enzymes from various biological agents can provide a better tool for the removal of contaminants of diverse nature from the affected environment. Application of enzymatic remediation on a large scale is limited by the fact that production of enzymes in bulk is a tedious task. However, advanced technologies such as enzyme immobilization, enhanced enzymatic system, and nanotechnology have received considerable attention in recent years and can overcome the problem associated with the application of enzymatic remediation on a large scale. This chapter describes the role of various organic pollutants and the interaction of enzymes from various sources (microbes and plants) with the contaminants. It also highlights the recent development in their application for bioremediation of organic contaminants.

6.2 ORGANIC POLLUTANTS' STRUCTURE AND CHARACTERISTICS

Toxic organic chemicals are ubiquitous in nature and have resulted from various industrial activities. Organic compounds that are potentially hazardous in nature with a tendency to resist degradation and persist in ecosphere for a long period of

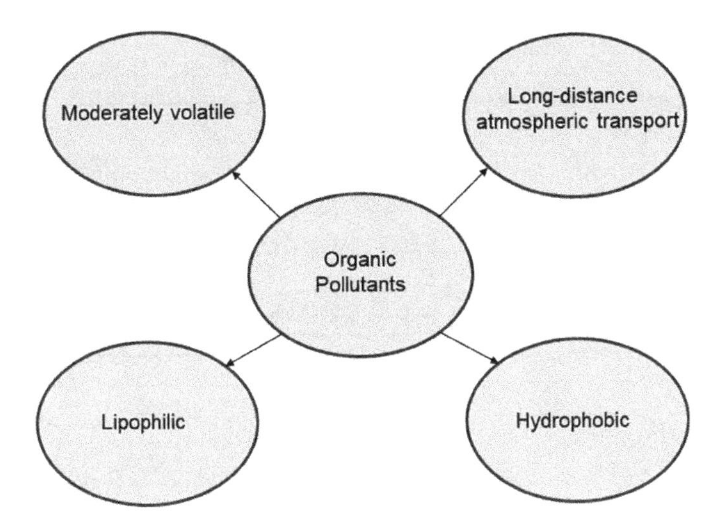

FIGURE 6.1 Characteristics of organic pollutants.

time are known as POPs. POPs are moderately volatile in nature that causes their long-distance atmospheric transportation. Certain characteristics of POPs, such as poor solubility in water (hydrophobicity) and high solubility in fats (lipophilicity), make these contaminants biologically harmful. Some important characteristics of POPs are shown in Figure 6.1.

As these compounds are lipophilic in nature, they can easily pass through biological membranes and get deposited in lipid-containing tissues. All these properties ensure the high mobility of POPs and their widespread atmospheric distribution, even in the remote regions of the earth. The structure of organic contaminant molecules is mostly based on one of the three basic structures: aliphatic, alicyclic, or aromatic (Figure 6.2).

Aliphatic hydrocarbons: Aliphatic hydrocarbons share several common sources and contaminate the environment. These include straight-chain and branched-chain structures, e.g., linear alkylbenzene sulfonate (LAS) detergents, trichloroethylene (TCE), perchloroethene (PCE), etc. TCE and PCE are halogenated compounds and are extensively used as industrial solvents. The improper use and disposal of these compounds are responsible for groundwater contamination.

Alicyclic hydrocarbons: Alicyclic hydrocarbons constitute a major component of crude oil and are also found as common components in microbial lipids, plant oils, paraffins, and pesticides. These include cyclopentane and cyclohexane simple compounds, as well as trimethylcyclopentane and various cycloparaffins among the complex structures (Figure 6.2).

Aromatic hydrocarbons: Compounds with at least one unsaturated ring system form the basic structure of aromatic hydrocarbons. These compounds contain benzene as the parent hydrocarbon. Certain aromatic compounds containing the two or more fused unsaturated rings are called polyaromatic hydrocarbons (PAHs). Some aromatic hydrocarbons are chlorinated and extensively used as preservatives and pesticides,

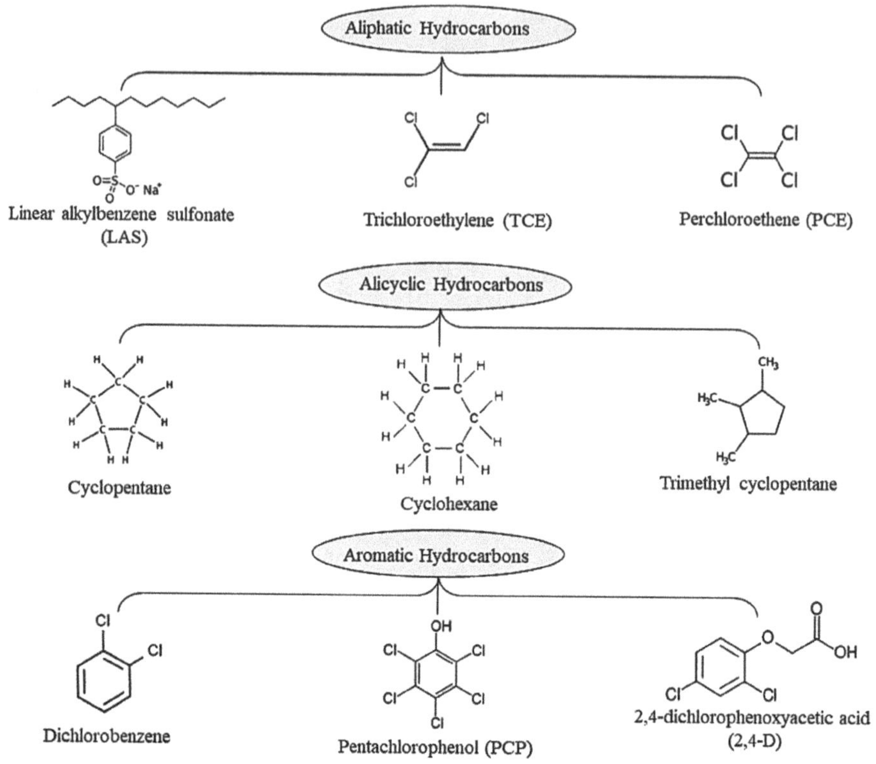

FIGURE 6.2 Basic structures of organic compounds.

such as dichlorobenzene, pentachlorophenol (PCP), 2,4-dichlorophenoxyacetic acid (2,4-D), dichlorodiphenyltrichloroethane (DDT), etc. (Figure 6.2). Aromatic hydrocarbons form a major class of organic contaminants that are thought to cause severe health issues in humans. These issues include carcinogenicity and teratogenicity, and noncancer risks include diabetes, reduced IQ, and behavioral impacts.

6.3 INTENTIONALLY AND UNINTENTIONALLY PRODUCED POPs

These days, special attention is being given to controlling and managing the POPs due to their potent toxic nature. The Stockholm Convention on Persistent Organic Pollutants (2001) included 12 initial and 16 new POPs that can be placed into two categories: intentionally and unintentionally produced POPs (Figure 6.3). Intentionally produced POPs include the organochlorine pesticides such as DDT, aldrin, chlordane, etc., and industrial chemicals such as PCBs. Almost all the POP pesticides originate from anthropogenic activities and are associated with the manufacturing, application, and disposition of such chemicals. Despite the awareness of their potent toxicities and promotion of integrated pest management, these recalcitrant organochlorine pesticides are still being applied in various developing countries.

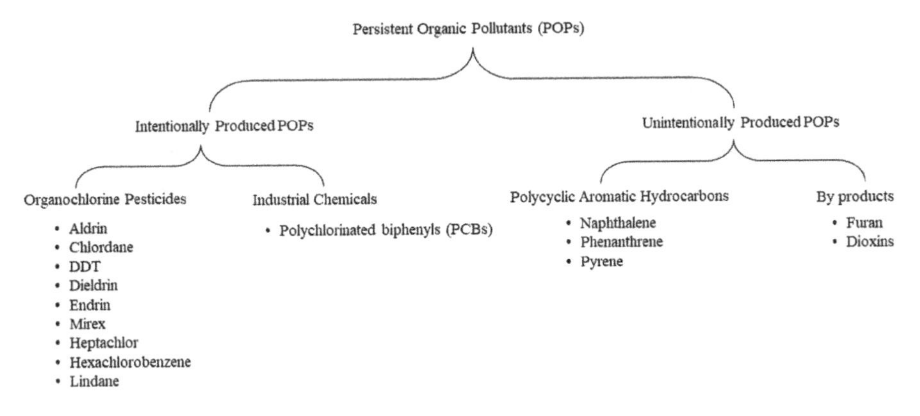

FIGURE 6.3 Persistent organic pollutants (POPs) classification.

The long-distance transportation capability of POP pesticides is causing their associated problems in the remote regions where these pesticides have never been used, such as in the Antarctic region (Kang et al., 2012). The presence of pesticide residues in vertebrates such as fish species, Adeliepenguins, Weddell seals, and skuas in Antarctica has already been confirmed by scientists in various studies (George and Frear, 1966; Stojić et al., 2018).

Among the other intentionally produced POPs, PCBs are extremely dangerous, as their concentration is much higher in workplaces that have electric devices and in the air of hazardous and toxic waste landfills (Stojić et al., 2014). Biphenyl is an unchlorinated parent compound of the PCBs, which were first described in 1881 (Waid, 1986). Their toxicity is dependent upon the number and position of each chlorine atom on the biphenyl ring. PCB congeners with a higher number of chlorine atoms are highly toxic in nature. Similarly, PCBs with chlorines in the ortho positions of each ring are less toxic compared to the non-ortho or mono-ortho PCBs. PCBs can also travel long distances, and their atmospheric migration depends on the number of chlorine atoms. PCBs with zeroto one chlorine atom remain in the atmosphere, those with one to four chlorine atoms tend to transport toward polar latitudes, those with four to eight chlorines remain in mid latitudes, and those with eight to nine chlorines remain close to the contamination source (Wania and Mackay,1996; Stojić et al., 2018).

PAHs are unintentionally produced toxic compounds derived from the incomplete combustion of organic materials and are released into the environment by anthropogenic as well as natural geological processes. PAHs mainly consist of carbon and hydrogen atoms and are poorly soluble, hydrophobic compounds. PAH compounds generally contain two or more fused benzene rings in a linear, angular, or clustered arrangement and can be classified into "small" and "large" PAHs on the basis of the number of fused aromatic rings present (Di-Toro et al., 2000; Arey and Atkinson, 2003). Small PAHs contain six or fewer fused aromatic rings, whereas those containing more than six aromatic rings are known as "large" PAHs (Rengarajan et al., 2015). The general characteristics of PAHs include high melting and boiling points, low vapor pressure, very low aqueous solubility, and highly lipophilic

TABLE 6.1
List of Potentially Toxic PAHs

PAHs	Molecular Weight (g/mol)
Acenaphthylene ($C_{12}H_8$)	152.19
Acenaphthene ($C_{10}H_6(CH_2)_2$)	154.2
Fluorene ($C_{13}H_{10}$)	166.22
Phenenthrene ($C_{14}H_{10}$)	178.23
Anthracene ($C_{14}H_{10}$)	178.23
Fluoranthene ($C_{16}H_{10}$)	202.25
Pyrene ($C_{16}H_{10}$)	202.25
Benz[*a*]anthracene ($C_{18}H_{12}$)	228.3
Chrysene ($C_{18}H_{12}$)	228.3
Benzo[*b*]fluoranthene ($C_{20}H_{12}$)	252.3
Benzo[k]fluoranthene ($C_{20}H_{12}$)	252.3
Benzo[*a*]pyrene ($C_{20}H_{12}$)	252.31
dibenz[*a,h*]anthracene ($C_{22}H_{14}$)	278.3
Benzo[*g,h,i*] perylene ($C_{22}H_{12}$)	276.3
Indeno[1,2,3-*c,d*]pyrene ($C_{22}H_{12}$)	276.3
Benzo[*j*]fluoranthene ($C_{20}H_{12}$)	252.3
Benzo[*e*]pyrene ($C_{20}H_{12}$)	252.3

nature, making them highly soluble in organic solvents. There are more than 100 different PAH compounds. Although the health effects of individual PAHs are not the same, 17 PAHs have been identified by United States Environmental Protection Agency (U.S. EPA) as being of greatest concern with regard to potential exposure and adverse health effects on humans. The profiles of 17 PAHs issued by the Agency for Toxic Substances and Disease Registry (ATSDR) are given in Table 6.1.

Dioxins and furans are the toxic by-products generated from the waste incineration and technological activities and also make up part of the released smoke stack effluent. Dioxins are also water insoluble and highly fat soluble, and hence tend to accumulate in higher animal species. The most prominent and most toxic dioxin compound is 2,3,7,8-tetrachlorodibenzo-*p*-dioxin (TCDD), which is associated with the manufacture of 2,4-D and 2,4,5 trichlorophenoxyacetic acid (2,4,5-T), hexachlorophene, and other pesticides that have 2,4,5-T as a precursor. The organic contaminants like dioxins and furans are carcinogenic and are reported to cause the following health issues:

- problems in reproduction and development
- destruction of the immune system
- interference with the hormonal system

Their decomposition is a very slow process that causes their accumulation in the body. Therefore, the chronic exposure to dioxins is particularly dangerous for human health (Yonemoto, 2000; Radenkova-Saeva, 2009).

6.4 ROLE OF ENZYMES IN THE DEGRADATION OF ORGANIC CONTAMINANTS

Enzymes act as biological catalysts, or biocatalysts, facilitating the biological reactions that involve the transformation of substrates into products. Enzymes can be proteins or glycoproteins consisting of at least one polypeptide moiety. The protein or glycoprotein moiety in an enzyme is called the apoenzyme, while the nonprotein moiety is called the prosthetic group, and the combination of an apoenzyme and a prosthetic group is known as the holoenzyme. The active sites in enzymes are the regions that are directly involved in the process of conversion of a substrate (Karigar and Rao, 2011). Enzymes possess several beneficial characteristics and present advantages over traditional remediation technologies (Gianfreda and Bollag, 2002), some of which are given next:

- Enzymes possess both narrow and broad specificity towards the substrate.
- They can participate in extensive transformations of structurally complex and toxic compounds and result in innocuous end products.
- They remain functional under extreme environmental conditions, which are limiting for the growth and activities of most microbes.
- Enzyme activities are not inhibited by inhibitors of microbial metabolism and remain active even in the presence of microbial predators.
- The enzymatic transformation is independent of pollutant concentrations.

These characteristics make the enzymes and the application of enzymatic techniques environmentally sound processes. All the enzymes can be classified into six different categories. These are oxidoreductases, transferases, hydrolases, lyases, isomerases, and ligases (Figure 6.4). Enzyme-mediated biodegradation is an emerging technology that offers a more effective approach for bioremediation than the chemical treatments. The enzymes involved in the transformation of organic contaminant may be "constitutive" or "induced." The constitutive enzymes are generally expressed even in low concentration of organic compounds, whereas the induced enzymes are synthesized and their production depends upon the growth conditions of cells. Therefore,

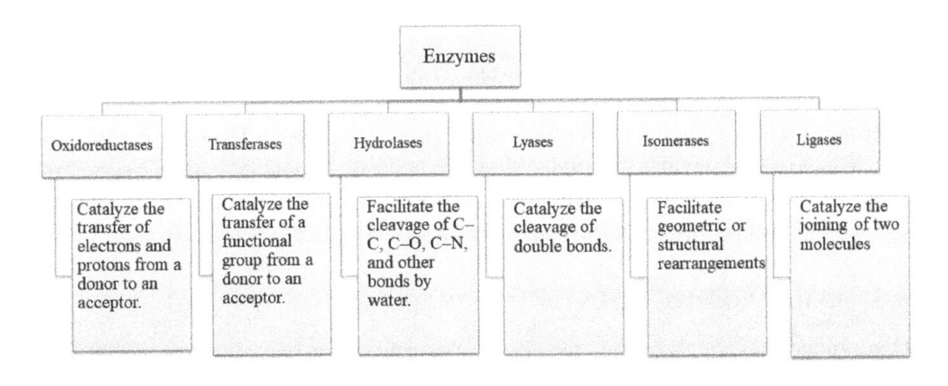

FIGURE 6.4 Classification of enzymes.

TABLE 6.2
Enzymes and Their Functions in the Process of Bioremediation

Enzyme classification	Functions
Oxidoreductase	
1. Oxygenases	Catalyze oxidation of aromatic compounds such as chlorinated biphenyls, aliphatic olefins by incorporating one or two atoms from molecular oxygen.
2. Laccases	Cleave ring present in aromatic compounds and reduce one molecule of oxygen in water and produces free radicals.
3. Peroxidases	Catalyze reduction reaction in the presence of peroxides, such as hydrogen peroxide (H_2O_2) and generate reactive free radicals after oxidation of organic compounds.
Hydrolases	
1. Lipases	Break triglycerol into glycerol and fatty acid and are widely used for wastewater treatment, polyaromatic hydrocarbon degradation, etc.
2. Cellulases	Break down complex cellulosic materials into simple sugars and are commonly used in the treatment of agricultural residues such as cotton waste, sawdust of *Khaya ivorensis*, and rice straw.
3. Carboxylesterases	Catalyze the hydrolysis of carboxyl ester bond present in synthetic pesticides such as organophosphates with addition of water.
4. Phosphotriesterases	Catalyze hydrolysis of phosphotriesters, the main components of organophosphorus compounds such as pesticides.
5. Haloalkane dehalogenases	Catalyze the degradation of halogenated aliphatic compounds such as 1,2,3-trichloropropane.

Source: Sharma et al. (2018).

the expression of induced enzymes can be modified with the presence or absence of organic compounds. Bryant and DeLuca (1990) reported an enhancement of five- to tenfold in the production of intracellular nitroreductase in *Enterobacter cloacae* by the addition of 2,4,6-trinitrotoluene (TNT) to the culture media. The plant-mediated remediation is generally catalyzed by endogenous enzymes. These enzymes generally attack the common functional groups, i.e., nitro (NO_2), amino (NH_2), hydroxyl (OH), carboxyl (COO), and halogens (Cl, Br, and I) of aromatic and aliphatic recalcitrant compounds. Microorganisms present several enzyme systems working individually or in groups for the degradation of recalcitrant organic compounds. The most representative classes of the enzymes that play a vital role in the degradation of organic contaminants are oxidoreductases and hydrolases, which can further be classified into different classes. Different enzymes of the two major classes are discussed in Table 6.2.

6.5 OXIDOREDUCTASES

Oxidoreductases catalyze the transfer of electrons from one molecule, the donor (reductant), to another compound known as the electron acceptor (oxidant). This group of enzymes usually utilizes NADP or NAD+ as cofactors. The contaminants are thus oxidized to harmless compounds during the oxidation-reduction reactions (Karigar and Rao, 2011). Oxidoreductases are further classified into 22 subclasses of enzymes. The oxidoreductase enzymatic system includes peroxidases (EC 1.11; act on the peroxide as an acceptor), dioxygenases (EC 1.14; act on the paired donor with the incorporation of molecular oxygen), and laccases (EC 1.15; act on the superoxide radical as an acceptor). These subclasses of oxidoreductases play important role in transformation of the organic compounds (Ali et al., 2017). Oxygenases are the EC 1 group of enzymes classified under oxidoreductase, catalyze the oxidation of organic compounds by transferring oxygen from molecular oxygen (O_2), and utilize FAD/NADH/NADPH as the cosubstrate (Karigar and Rao, 2011). Oxygenases thus cleave the aromatic ring and increase the reactivity and water solubility of organic compounds (Arora et al., 2009). These enzymes may transfer one or both the oxygen atoms of the O_2 molecule to the substrate, on the basis of which these are classified as monooxygenases and dioxygenases.

6.5.1 Monooxygenases

Monooxygenases contribute one atom of molecular oxygen (O_2) to the organic substrate and can be further categorized into two subclasses (i.e., flavin-dependent monooxygenases and P450 monooxygenases). Flavin-dependent monooxygenases contain flavin as the prosthetic group and NADP or NADPH as the coenzyme, whereas P450 monooxygenases are heme-containing oxygenases and are present in both eukaryotes and prokaryotes (Karigar and Rao, 2011). Monooxygenases catalyze biotransformation and biodegradation of organic contaminants through various reactions such as dehalogenation, desulfurization, denitrification, ammonification, and hydroxylation (Arora et al., 2010). Pentachlorophenol 4-monooxygenase (PcpB) is a flavin monooxygenase that catalyzes the hydroxylation of PCP at the para position with the removal of the chloride ion. PcpB is reported in many pentachlorophenol degrading bacteria such as *Sphingobium*, *Sphingomonads*, and *Pseudomonas* and has been purified and characterized PcpB from *Sphingomonas chlorophenolicus* ATCC 39723(Wang et al., 2002). Urlacher et al. (2004) reported P450 monooxygenase in *Bacillus megaterium* BM3 that is able to hydroxylate a variety of alkanes, fatty acids, and aromatic compounds.

6.5.2 Dioxygenases

Dioxygenases are multicomponent enzyme systems and transfer molecular oxygen to the organic substrate. The dioxygenases enzyme system can be subdivided into two classes of enzymes (i.e., the extradiol and the intradiol dioxygenases). The extradiol dioxygenases contain non-heme iron (II) cofactor in their active site and catalyze meta-cleavage in the aromatic ring, whereas the intradiol dioxygenases contain non-heme iron (III) as cofactor in their active sites and catalyze ortho-cleavage (Arora et al., 2009). The meta-cleavage involves the breaking of the bond between the hydroxylated carbon and the adjacent non-hydroxylated carbon, whereas the ortho-cleavage involves the ring cleavage at the C—C bond between the two hydroxyl groups (Figure 6.5).

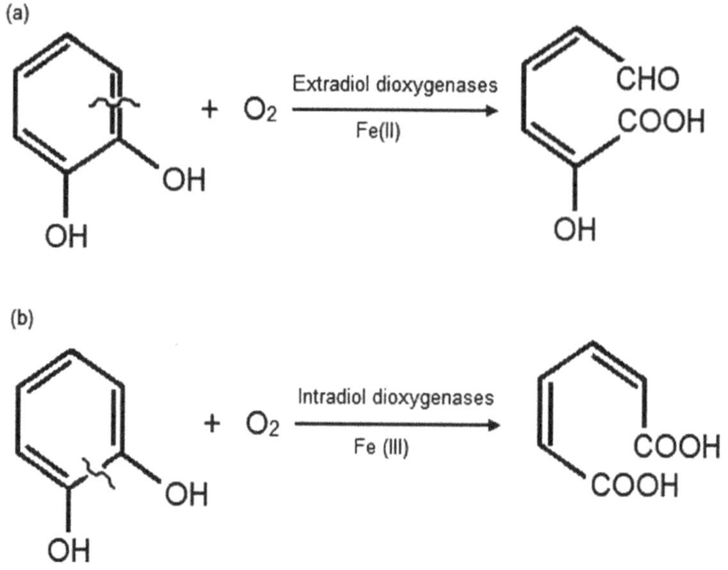

FIGURE 6.5 Extradiolring cleaving dioxygenase and intradiolring cleaving dioxygenase. (From Wang et al., 2017.)

6.5.3 Laccases

Laccases are the multicopper oxidases produced by certain plants, fungi, bacteria, and insects. Laccases catalyze the oxidation of a wide range of reduced phenolicand aromatic substrates (Gianfreda et al., 1999; Mai et al., 2000). Laccases need atmospheric oxygen as the electron acceptor to oxidize phenols, polyphenols, aromatic amines, and a range of non-phenolic substrates (Mai et al., 2004). Almost all the fungi examined secrete more than one laccase isozyme. Two LAC isozymes, LacI and LacI, I have been identified in *Physisporinusrivulosus* T241i, *Trametestrogii*, *Cerrena unicolor* 137, and *Panustigrinus* (Cadimaliev et al., 2005; Lorenzo et al., 2006; Zouari-Mechichi et al., 2006; Mäkelä et al., 2006). These microorganisms produce both intra- and extracellular laccases that are capable of catalyzing the degradation of a wide range of organic compounds such as ortho- and paradiphenols, aminophenols, polyphenols, polyamines, lignins, aryl diamines, etc. Laccases are also identified in some bacterial strains such as the *Bacillus sphaericus* strain (Claus and Filip, 1997); *Marinomonasmediterranea* (Sanchez-Amat et al.,2001), *Streptomyces cyaneus* (Arias et al., 2003), *Streptomyces griseus* (Endo et al., 2003), *Streptomyces lavendulae* (Suzuki et al., 2003), and *Streptomyces coelicolor* (Machczynski et al., 2004; Skálová et al., 2007; Koschorreck et al., 2008).

6.5.4 Peroxidases

Peroxidases are heme-containing enzymes and are ubiquitously distributed among the microorganisms, as well as plants. Plant peroxidases are considered multifunctional enzymes responsible for lignification and crosslinking of biopolymers in the cell wall, thus building the defense system in plants. Peroxidases are also induced and expressed in certain plant tissues catalyzing the oxidation of aromatic substrates using hydrogen peroxidase as the cosubstrate. Peroxidases on the basis of preferred substrates include different classes such as lignin peroxidase (LiP), manganese-dependent peroxidase (MnP), and versatile peroxidase (VP). These enzymes have been reported to possess high potential to degrade toxic substances in nature (Karigar and Rao, 2011).

6.6 HYDROLASES

Hydrolytic enzymes catalyze the degradation of pesticides and insecticides into less toxic compounds. Hydrolases disrupt the major chemical bonds such as esters, peptide bonds, and carbon-halide bonds present in the complex structure of toxicants (Sharma et al., 2018). Extracellular hydrolases include cellulases, proteases, lipases, xylanases, DNAses, and amylases.

6.6.1 PHOSPHATASE

Phosphatases catalyze the transformation of organophosphate compounds by hydrolyzing the ester and anhydride bonds resulting in the production of phosphoric acid. On the basis of pH optima, the enzymes can be classified as alkaline phosphatase (EC 3.1.3.1) and acid phosphatase (EC 3.1.3.2), whereas on the basis of the structure of the substrate, the phosphatase enzyme can further be classified into three groups that include phosphoric monoester hydrolases (EC 3.1.3), phosphoric diester hydrolases (EC 3.1.4), and triphosphoric monoester hydrolases (EC 3.1.5). Alkaline phosphatase catalyzes the hydrolysis of a variety of organophosphorus substrates and need Zn^{2+} or Cr^{2+} for their activity. The enzyme has been purified and characterized in halotolerant bacterium *Halomonaselongata* (Bylund et al., 1990). Acid phosphatases are classified as monophosphoric-monoester-phosphohydrolases and catalyze the hydrolysis of C—O—P bonds. Phosphatases of plant and microbial origin are particularly gaining attention due to their capacity to hydrolyze the di- and triphosphate ester bond that is present in organophosphorus pesticides. The hydrolysis of an organophosphorus pesticide (methyl paraoxon) by acid phosphatase leading to the production of *p*-nitrophenol is presented in Figure 6.6.

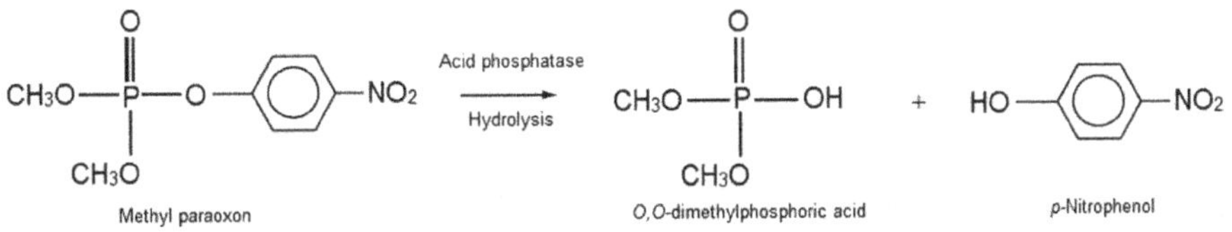

FIGURE 6.6 Hydrolysis of methyl paraoxon by acid phosphatase (EC 3.1.3.2).

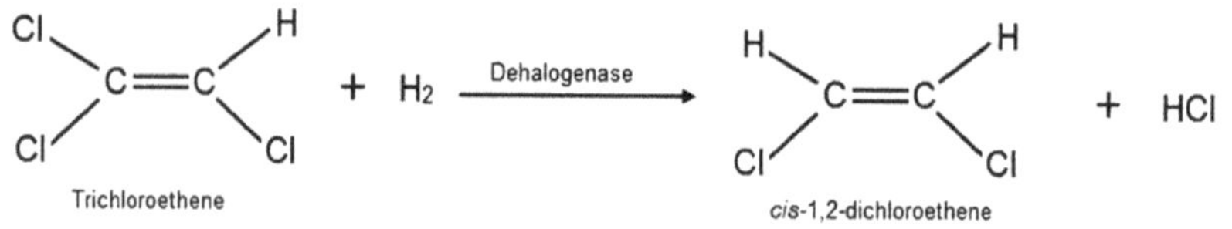

FIGURE 6.7 Dehalogenation of trichloroethene to *cis*-1,2-dichloroethene by the activity of dehalogenase. (From Wolfe and Hoehamer, 2003.)

6.6.2 Dehalogenases

Dehalogenases catalyze the hydrolysis of the carbon-halogen bond in the halogenated substrates, thereby resulting in the production of alcohol and halides (Kotik and Famerova, 2012). During the enzymatic cleavage of a carbon-halogen bond, the halogen groups such as bromine, chlorine, and iodine are removed from the substrate (aliphatic or aromatic, see Figure 6.7). Dehalogenase was first discovered in the bacterium *Xanthobacter autotrophicus* GJ10, and it is able to degrade 1,2-dichloroethane (Nagata et al., 2015).

6.7 Technologies for the Improvement of Enzymatic Degradation of Organic Pollutants

6.7.1 Metabolic Engineering

Metabolic engineering is the process of improving the production of specific cellular compounds, such as enzymes, by optimizing the genetic and regulatory systems. In the field of bioremediation or biodegradation of toxicants, the metabolic engineering involves the enhancement of metabolic capacity of microorganisms. The improvement of metabolic capability is usually done by combining the metabolic activities of different organisms in a single microbial cell. The degradation of methylphenols and methylbenzoates by combining the five different catabolic pathways from three different bacteria in a single bacterium has been reported by Rojo et al. (1987). Oxygenases are known for their important role in the degradation of a wide range of organic contaminants and are being considered as potentially targeted enzymes for metabolic engineering. Haro and deLorenzo (2001) have prepared a gene cassette by assembling dioxygenase of *Pseudomonas putida* F1, benzyl alcohol dehydrogenase (encoded by *xylB*), and benzaldehyde dehydrogenase (*xylC*) of *P.putida* mt-2 in separate mini-Tn5 transposon vectors. This gene cassette has been used to construct *Pseudomonas* strains (*Pseudomonas aeruginosa* PA142 and *P. aeruginosa* JB) genetically designed for the degradation of the recalcitrant compound 2-chlorotoluene. Lei et al. (2017) have isolated the carbendazim hydrolyzing enzyme gene from *Microbacterium*sp. djl-6F and cloned into *Escherichia coli* BL21 (DE3) to increase the levels of the enzyme. The enzyme was found to hydrolyze carbendazim into 2-aminobenzimidazole.

6.7.2 Immobilization

Application of immobilized enzymes can overcome the problems associated with the long-term utilization of suspended enzymes for the degradation of toxic compounds. Immobilization is the process of confining any biological agent or enzyme to a defined support or matrix. It resists the movement of biological agents in the space while retaining their activities in the medium. Therefore, immobilized systems can offer several advantages and improve the efficiency of the decontamination process (Engade and Gupta, 2010; Chaudhry and Paliwal, 2018). Enzymes catalyze a reaction either intracellularly (i.e., within their originating cells) or extracellularly (i.e., in the environment) when released outside by the originating cells. In the free state of activity (solublein the reaction solution), the process of catalysis is homogenous, whereas in the immobilized state, when the enzymes are retained on a solid matrix, the catalysis is heterogeneous (Gianfreda and Rao, 2004). Some advantages and disadvantages of application of enzyme immobilization are summarized in Table 6.3.

A variety of support/matrices can be used for immobilization of enzymes, which are grouped into three groups: natural polymers, synthetic polymers, and inorganic materials (Figure 6.8). Enzymes can be immobilized by various methods/interactions on different matrices that include covalent attachment to solid supports, solid support adsorption methods, entrapment in polymeric gels, crosslinking with biofunctional reagents, and encapsulation within a solid support (Sharma et al., 2018). The physical confinement of the enzyme can be developed by different interactions occurring between the enzyme and the selected matrix, or by self-aggregation. The main advantages and limitations of different immobilization techniques are

TABLE 6.3
Advantages and Disadvantages Associated with the Application of Immobilized Enzymes

Advantages	Disadvantages
• Enhanced retention period of biocatalyst • Increased process stability • Reusability of biological system • Convenience in continuous operation of system and tolerance to shock loadings • Higher hydraulic loading rates possible • Generation of relatively less biological sludge • Maintenance of activities of slow growing microbes • Development of multienzyme activity possible	• Loss or reduction in the catalytic activity of some cells or enzymes possible • Diffusion limitation for enzymes • Additional cost for isolation, purification, and recovery of active biocatalysts (enzymes)

Source: Chaudhry and Paliwal (2018).

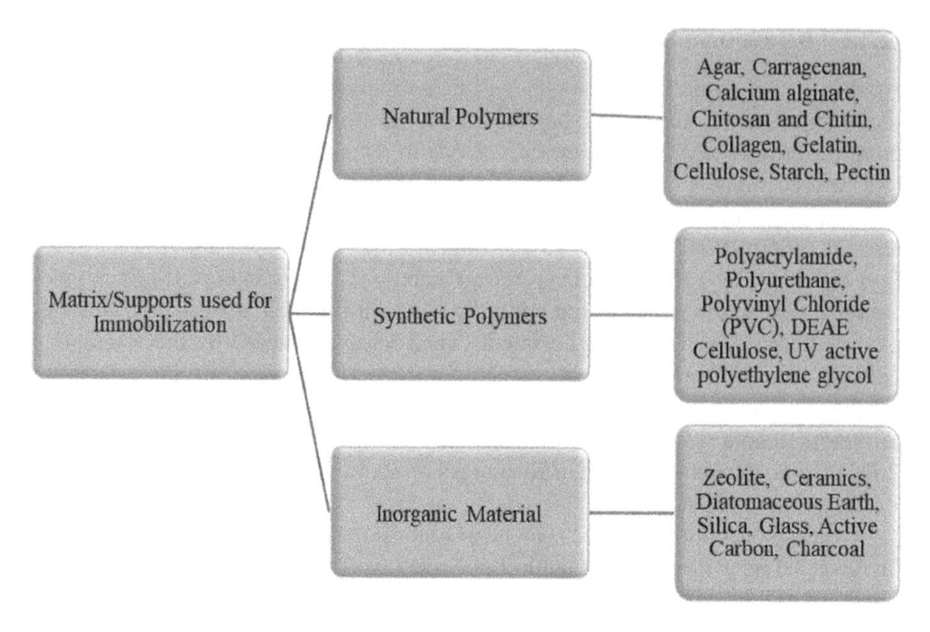

FIGURE 6.8 Different matrices/supports used for immobilization. (From Chaudhry and Paliwal, 2018.)

described by Cacicedo et al. 2019, (Table 6.4). Many reports have confirmed that the application of immobilized enzymes for degradation of organic contaminants is economical due to their stability and repeated use. Dodor et al. (2004) have reported that the immobilization of laccase from *Trametes versicolor* on kaolinite increases stability and resistant power of laccase for the degradation of anthracene and benzo[a]pyrene as compared to free enzyme.

6.7.3 Nanotechnology

Nanotechnology is an emerging branch of science with a wide field of application. Research in the field of nanobiotechnology is gaining attention in recent years. Nanoparticles or nanomaterials with nanoscale size have been proved to be an excellent tool for their application in decontamination of toxic pollutants. Interest in nanobiotechnology has also been increasing in developing biosensors by controlled immobilization of enzymes on different nanosized surfaces. Couto and Herrera (2006) studied the application of laccases (lignin-degrading enzyme) as a biosensor for the detection of industrial pollutants. Roy et al. (2005) reported the application of crosslinked enzyme crystals (CLECs) of laccase isolated from *Trametes versicolor* as a biosensor that has great advantage over the soluble enzyme. The application of nanoparticles for immobilization present an excellent support for the stability of enzymes in the reaction medium. Nanoparticles provide compatible environments for enzyme immobilization due to their small size and a large surface area. Thus, to overcome the problems associated with the application of enzyme in pollution abatement, such as more time requirement, highcost, enzyme inactivation, large quantity requirement, etc., application of nanoparticles for immobilization can play a significant role.

TABLE 6.4
Main Properties of Different Enzyme Immobilization Techniques

Techniques	Properties
Immobilization on matrix (carrier-bound)	
Binding	Advantageous for allosteric enzymes. The enzyme and the derivatized coenzyme can be retained in carrier surface. Drawback: shear sensitive.
Covalent	Increased operational stability and enzyme recycle. Sometimes problems with steric hindrance. Adaptable to specific processes.
Noncovalent	Simple technique and easy recovery of carrier. High immobilization yield. Drawback: enzyme activity is environmentally dependent, poor enzyme recycling.
Encapsulation	The enzyme is isolated and protected from the environment inside the matrix. Drawback: diffusional barriers for substrate/products.
Microencapsulation	The enzyme could be in the same media of the reaction (no change of physical state of substrate/products). Drawback: carrier stability.
Entrapment	Each enzyme molecule is dispersed in the matrix. Drawback: catalysis carried out at interphase enzyme/substrate.
Immobilization without support (carrier-free)	
Crosslinked enzymes (CLEs)	Poor mechanical properties and severe mass transfer limitations. No longer explored for industrial purposes.
Crosslinked enzyme crystals (CLECs)	High stability and activity under harsh environmental conditions. Drawbacks: high labor and technical skills.
Crosslinked enzyme aggregates (CLEAs)	High mechanical properties and yields than CLEs. Simple and easy technique. Do not require purification steps.

Source: Cacicedo et al. (2019).

6.8 CONCLUSION

Organic contamination in the environment is a matter of great concern. Toxic compounds from a variety of industrial activities, as well natural processes, are released into the lakes, rivers, soil, and air. Conventional approaches for the remediation of organic compounds such as benzene, chlorinated solvents, pesticides, etc., are generally laborious and may generate secondary toxic substances. Application of enzymes present environmental advantages against the chemical treatment of contaminants. Various enzymes such as monooxygenases, dioxygenases, laccases, and peroxidases significantly degrade the recalcitrant organic compound. The production of enzymes at a higher level for the industrial application can be improved by metabolic engineering that deals with the enhancement in the production of selective enzymes in a single cell. Modern techniques of immobilization and nanotechnology may play a significant role in bioremediation of organic contaminants.

REFERENCES

Ali, M.I., Khatoon, N., Jamal, A. (2017). Polymeric pollutant biodegradation through microbial oxidoreductase; a better strategy to safe environment. Int J Biol Macromol. http://dx.doi.org/10.1016/j.ijbiomac.2017.06.047.

Arey, J., Atkinson, R. (2003). Photochemical reactions of PAH in the atmosphere, In: Douben P.E.T. (ed.), PAHs: an ecotoxicological perspective. John Wiley and Sons Ltd., New York, pp. 47–63.

Arias, M. E., Arenes, M., Rodr´ıguez, J., Soliveri, J., Ball, A. S., Hernandez, M. (2003). Kraft pulp biobleaching and mediated oxidation of a nonphenolic substrate by laccase from *Streptomyces cyaneus* CECT 3335. Appl Environ Microbiol. 69:1953–1958.

Arora, P. K., Kumar, M., Chauhan, A., Raghava, G. P., Jain, R. K. (2009). OxDBase: a database of oxygenases involved in biodegradation. BMC Res Notes. 2:67.

Arora, P. K., Srivastava, A., Singh, V. P. (2010). Application of monooxygenases in dehalogenation, desulphurization, denitrification and hydroxylation of aromatic compounds. J Bioremediat Biodegrad. 1:1–8.

Bryant, C., DeLuca, M. (1990). Purification and characterization of an oxygen insensitive NAD(P)H nitroreductase from *Enterobacter cloacae*. J Biol Chem. 266:4119–4125.

Bylund, J. E., Dyer, J. K., Feely, D. E., Martin, E. L. (1990). Alkaline and acid phosphatases from the extensively halotolerant bacterium *Halomonaselongata*. Curr Microbiol. 20(2):125–131.

Cacicedo, M. L., Manzo, R. M., Municoy, S., Bonazza, H. L., Islan, G. A., Desimone, M., Bellino, M., Mammarella, E.J., Castro, G. R. (2019). Immobilized enzymes and their applications. Adv Enzyme Technol.169–200. doi:10.1016/b978-0-444-64114-4.00007-8.

Cadimaliev, D. A., Revin, V. V., Atykyan, N. A., Samuilov, V. D. (2005). Extracellular Oxidases of the Lignin-Degrading Fungus Panustigrinus. Biochemistry (Moscow). 70:703–707.

Chaudhry, S., Paliwal, R. (2018). Techniques for remediation of paper and pulp mill effluents: processes and constraints. In: Hussain C. (ed.), Handbook of Environmental Materials Management. Springer, Cham, pp. 1–9, ISBN - 978-3-319-58538-3. https://doi.org/10.1007/978-3-319-58538-3_134-1

Claus, H., Filip, Z. (1997). The evidence of a laccase-like activity in a *Bacillus sphaericus* strain. Microbiol Res. 152:209–215.

Couto, S. R., Herrera, J. L. T. (2006).Industrial and biotechnological applications of laccases: a review. Biotechnol Adv. 24:500–513.

Di-Toro, D.M., McGrath, J.A.,Hansen, D.J. (2000). Technical Basis for Narcotic Chemicals and Polycyclic Aromatic Hydrocarbon Criteria. I. Water and Tissue. Environ Toxicol Chem. 19:1951–1970.

Dodor, D.E., Hwang, H.M., Ekunwe, S.I. (2004). Oxidation of anthracene and benzo[a] pyrene by immobilized laccase from *Trametes versicolors*. Enzyme Microbial Technol. 35:210–217. https://doi.org/10.1016/j.enzmictec.2004.04.007.

Endo, K., Hayashi, Y., Hibi, T., Hosono, K., Beppu, T., Ueda, K. (2003). Enzymological characterization of EpoA, a laccase-like phenol oxidase produced by *Streptomyces grizeus*. J Biochem. 133:671–677.

Engade, K. B., Gupta, S. G. (2010). Decolorization of textile effluent by immobilized *Aspergillus terreus*. J Pet Environ Biotechnol. 1:101. doi:10.4172/2157-7463.1000101.

George, J.L., Frear, D. E. H. (1966). Pesticides in the Antarctic. J Appl Ecol. 3:155–167.

Gianfreda, L., Bollag, J.-M. (2002). Isolated enzymes for the transformation and detoxification of organic pollutants. In: Burns, R. G., Dick R., (eds.), Enzymes in the Environment: Activity, Ecology and Applications. Marcel Dekker, New York, pp. 491–538.

Gianfreda, L., Rao, M.A. (2004). Potential of extra cellular enzymes in remediation of polluted soils: a review. Enzyme Microb Technol. 35:339–354.

Gianfreda, L., Xu, F., Bollag, J.M. (1999). Laccases: a useful group of oxidoreductive enzymes. Bioremediat J. 3:1–25.
Haro, M. A., deLorenzo, V. (2001). Metabolic engineering of bacteria for environmental applications: construction of *Pseudomonas* strains for biodegradation of 2-chlorotoluene. J Biotechnol. 13:103–113.
Kang, J. H., Son, M. H., Hur, D. S., Hong, S., Motoyamad, H., Fukuie, K., Chang, Y. S. (2012).Deposition of organochlorine pesticides into the surface snow of East Antarctica. Sci Total Environ. 433:290–295.
Karigar, C. S., Rao, S. S. (2011). Role of microbial enzymes in the bioremediation of pollutants: a review. Enzyme Res. doi:10.4061/2011/805187.
Koschorreck, K., Richter, S. M., Ene, A. B., Roduner, E., Schmid, R. D., Urlacher, V. B. (2008). Cloning and characterization of a new laccase from *Bacillus licheniformis* catalyzing dimerization of phenolic acids. Appl Microbiol Biotechnol. 79:217–224.
Kotik, M., Famerova, V., (2012). Sequence diversity in haloalkane dehalogenases, as revealed by PCR using family-specific primers. J Microbiol Methods. 88:212–217. https://doi.org/10.1016/j.mimet.2011.11.013.
Lei, J., Wei, S., Ren, L., Hu, S., Chen, P. (2017). Hydrolysis mechanism of carbendazim hydrolase from the strain *Microbacterium* sp. djl-6F. J Environ Sci. 54:171–177. https://doi.org/10.1016/j.jes.2016.05.027.
Lorenzo, M., Moldes, D., Sanromán, M. A. (2006). Effect of heavy metals on the production of several laccase isoenzymes by *Trametes versicolor* and on their ability to decolourise dyes. Chemosphere. 63:912–917.
Machczynski, M., Vijgenboom, E., Samyn, B., Canters, G. W. (2004). Characterization of SLAC: a small laccase from *Streptomyces coelicolor* with unprecedented activity. Protein Sci. 13:2388–2397.
Mai, C., Kues, U., Militz, H. (2004). Biotechnology in the wood industry. Appl Microbiol Biotechnol. 63:477–494.
Mai, C., Schormann, W., Milstein, O., Huttermann, A. (2000). Enhanced stability of laccase in the presence of phenolic compounds. Appl Microbiol Biotechnol. 54:510–514.
Mäkelä, M. R., Hildén, K. S., Hakala, T. K., Hatakka, A., Lundell, T. K. (2006). Curr Genet. 50:323–333.
Nagata, Y., Ohtsubo, Y., Tsuda, M.2015. Properties and biotechnological applications of natural and engineered haloalkane dehalogenases. Appl Microbiol Biotechnol. 99:9865–9881.
Radenkova-Saeva, J. (2009). Immunotoxicity of dioxins. J Clin Med. 2(2):8–13.
Rengarajan, T, Rajendran, P., Nandakumar, N., Lokeshkumar, B., Rajendran, P., Nishigaki, I. (2015).Exposure to polycyclic aromatic hydrocarbons with special focus on cancer. Asian Pac J Trop Biomed. 5(3):182–189.
Rojo, F., Pieper, D. H., Engesser, K. H., Knackmuss, H. J., Timmis, K. N. (1987).Assemblage of *ortho* cleavage route for simultaneous degradation of chloro- and methylaromatics. Science. 238:1395–1398.
Roy, J. J., Abraham, T. E., Abhijith, K. S., Kumar, P. V. S., Thakur, M. S. (2005).Biosensor for the determination of phenols based on cross-linked enzyme crystals (CLEC) of laccase. Biosens Bioelectron. 21:206–211.
Sanchez-Amat, A., Lucas-Eilo, P., Fernandez, E., Garcia-Borron, J. C., Solano, F. (2001). Molecular cloning and functional characterization of a unique multipotent polyphenol oxidase from *Marinomonasmediterranea*. Biochim Biophys Acta. 1547:104–116.
Sharma, B., Dangi, A. K. Shukla, P. (2018). Contemporary enzyme based technologies for bioremediation: a review. J Environ Manage. 210:10–22.
Skálová, T., Dohnálek, J., Ostergaard, L. H., Ostergaard, P. R., Kolenko, P., Dusková, J., Hasek, J. (2007). Crystallization and preliminary X-ray diffraction analysis of the small laccase from *Streptomyces coelicolor.* Acta Crystallogr Sect F Struct Biol Cryst Commun. 63:1077–1079.

Stockholm Convention on Persistent Organic Pollutants, Stockholm, 22 May. (2001). http://chm.pops.int/. Accessed 08 Aug 2017.

Stojić, N., Pucarević, M., Mrkajić, D., Kecojević, I. (2014). Transformers as a potential for soil contamination. Meta. 53(4):689–692.

Stojić, N., Štrbac, S., Prokić, D. (2018). Soil pollution and remediation. Handbook of Environmental Materials Management. 1–34. doi:10.1007/978-3-319-58538-3_81-1.

Suzuki, T., Endo, K., Ito, M., Tsujibo, H., Miyamoto, K., Inamori, Y. (2003). A thermostable laccase from *Streptomyces lavendulae* REN-7: purification, characterization, nucleotide sequence, and expression. Biosci Biotechnol Biochem. 67:2167–2175.

U.S. Environmental Protection Agency (2009). Integrated Risk Information System (IRIS).http://cfpub.epa.gov/ncea/iris/index.cfm. Accessed 14 Sep 2010.

Urlacher, V. B., Lutz-Wahl, S, Schmid, R. D. (2004). Microbial P450 enzymes in biotechnology. Appl Microbiol Biotechnol. 64:317–325.

Waid, J. S. ed. (1986). PCBs and the Environment, Vol. I & II. CRC Press, Boca Raton, FL.

Wang, H., Marjomaki, V., Ovod, V., Kulomaa, M. S. (2002). Subcellular localization of pentachlorophenol 4-monooxygenase in *Sphingobiumchlorophenolicum* ATCC 39723. Biochem Biophys Res Commun. 299:703–709.

Wang, Y., Li, J., Liu, A. (2017). Oxygen activation by mononuclear nonheme iron dioxygenases involved in the degradation of aromatics. J Biol Inorg Chem. 22:395–405.

Wania, F., Mackay, D. (1996). Tracking the distribution of persistent organic pollutants. Environ Sci Technol. 30:390A–396A.

Wolfe, N., Hoehamer, C. (2003). Enzymes used by plants and microorganisms to detoxify organic compounds. In: McCutcheon, S. C., Schnoor, J. L. (eds.), Phytoremediation: transformation and control of contaminants. Wiley, Hoboken, pp. 159–187.

Yonemoto, J. (2000).The effects of dioxin on reproduction and development. Ind Health. 38(3):259–268.

Zouari-Mechichi, H., Mechichi, T., Dhouib, A., Sayadi, S., Martínez, A. T., Martínez, M. J. (2006). Enzyme Microb Technol. 39:141–148.

7 Bacterial Biodegradation of Phenolics and Derivatives of Phenolics

*Pooja Hirpara and Nikhil Bhatt**

CONTENTS

* Corresponding author: E-mail:bhattnikhil2114@gmail.com & bhatt@gujaratvidyapith.org

7.1 INTRODUCTION

Many industries are releasing highly aromatic compounds as waste. Thus, industrial effluent contains phenolics and their derivatives (El-Ashtoukhy et al., 2013). All these aromatic compounds accumulate in the environment—some of these compounds are naturally degraded from environmental flora, and the remaining materials are deposited in the natural environment (Chakraborty et al., 2010). These deposited compounds are hazardous to human, animal, and flora. Some aromatic compounds are carcinogenic, mutagenic, and teratogenic (Olaniran and Igbinosa, 2011). Nonbiodegradable aromatic compounds must be treated by biodegradable compounds or less toxic compounds (El-Nass et al., 2009. Aromatic pollutants such as phenolics and their derivatives are polluting the surface water, underground water, and soil. It causes a big concern worldwide (Krastanov et al., 2013). Phenolics and their derivatives such as phenol, *o*-cresol, *m*-cresol, *p*-cresol, nitrophenol, and resorcinol are highly toxic to human, animal, aquatic life, and others (Gonzalez et al., 2016). Some of these compounds are highly toxic and difficult to remove from the environment (El-Naas et al., 2010).

Phenolics and their derivatives are released from various industries (Table 7.1), such as resin manufacturing, steel industries, pharmaceutical, petroleum refineries, textile, dye staff, pulp and paper, oil and paint, coal processing, pesticide, fungicide, chemical fertilizers, varnish, tannery, herbicides, etc. (Gonzalez et al., 2016).

The concentration of phenolics in wastewater may vary from 10 to 300 mg/L to 4500 mg/L, which is the upper limit in highly polluted wastewater (Al-Khalid and El-Nass, 2012). Phenol rapidly penetrates the skin due to its properties that may cause irritation to the skin, eyes, respiratory tract, mucous membranes, etc. The liver and kidney are severely damaged by oral ingestion. Al-Khalid and El-Naas (2012) reported that 1 g of phenol may be lethal to human life. Phenolics and their derivatives with 5–25 mg/L concentration are proven to be toxic to fish. According to the WHO (World Health Organization), the permissible limit of phenolics is <1 μg/L and the U.S. Environmental Protection Agency also set this limit for surface water (Al-Khalid and El-Naas, 2102). The European Council Directive declared that the

TABLE 7.1
Phenolic Compounds Released from Industrial Wastewater

Industrial Wastewater	Range (mg/L)
Coal gasification	850–950
Coal processing	9–6800
Cooking plants	28–3900
Municipal sewage	0.02–0.77
Olive oil mill	0.5–80,000
Oil refineries	6–500
Petrochemical	2–1220
Pharmaceutical plants, wood products, paints, papermaking	0.1–3900
Plastics	600–2000

permissible limit of phenolic compound is 0.5 μg/L. In the UAE, 0.1 mg/L total phenol containing industrial water can be discharged to the marine environment (Basha et al., 2010).

Industrial wastewater is increasing day by day in environment. Saving soil, water, and natural habitats all over world is a primary concern. Several physicochemical and biological treatment technologies are available for the degradation of phenolics, such as chemical oxidation, ion exchange, activated carbon adsorption, and liquid-liquid extraction (Kilic, 2009). All of these methods are costly and have drawbacks, in that they are not able to degrade phenolic compounds completely but transfer or form other molecules (secondary/pollutant/amines) that may be more hazardous.

On the other hand, biological treatments are more effective, eco-friendly, and cheap methods as compared to physicochemical treatments. The biological method can degrade/mineralize phenolics, and thus it produces nontoxic or lesser toxic by-products. In the past decades, phenolics containing effluents were treated by traditional conventional methods, but those methods have some limitations. A novel treatment technology is needed that plays an important role in the form of pure, mixed, and/or consortium (Zaki, 2015). The aim of this chapter is to examine the involved microorganisms, degradation mechanisms of microbes, and degradation treatment technology.

7.2 PHENOLIC COMPOUND APPLICATION CLASSES AND THEIR CHARACTERISTICS ACCORDING TO STRUCTURE

Phenol is an organic compound with the formula C_6H_5OH. It contains a hydroxyl group (OH) that binds with a single benzene ring and is a monoaromatic hydrocarbon. It is a white crystalline solid and volatile and mildly acidic in nature. Due to its propensity to cause burn effects, it requires careful handling. The structure and properties of phenolic compounds are shown in Table 7.2.

7.3 VARIOUS EXISTING TREATMENT TECHNOLOGIES

7.3.1 Physicochemical Methods

Various processes have been used for the degradation of phenolics and their derivatives due to their high efficiency and rapidity. The physicochemical method contains a vast range of treatment options, from traditional methods such as coagulation and flocculation to recent advanced methods such as reverse osmosis, advanced oxidation, electrochemical process, UV light, and UV/H_2O_2.

7.3.1.1 Sorption

The sorption process comprises adsorption, absorption, and ion exchange method. Adsorption is a process in which pollutants accumulate at the interface between phases such as gas-liquid, liquid-liquid, and liquid-solid (Lamichhane et al., 2016). As a sorption substrate, various media are used in the removal of phenolics and their derivatives such as activated carbon (petroleum coke, sewage sludge, shell, bituminous, coal, etc.). Biochar is another sorption media recently used for removing phenolic compounds. It is produced by pyrolysis of agricultural and industrial residues, such as corn, fodder, rice

TABLE 7.2
Structure and Characteristics of Phenolic Compounds

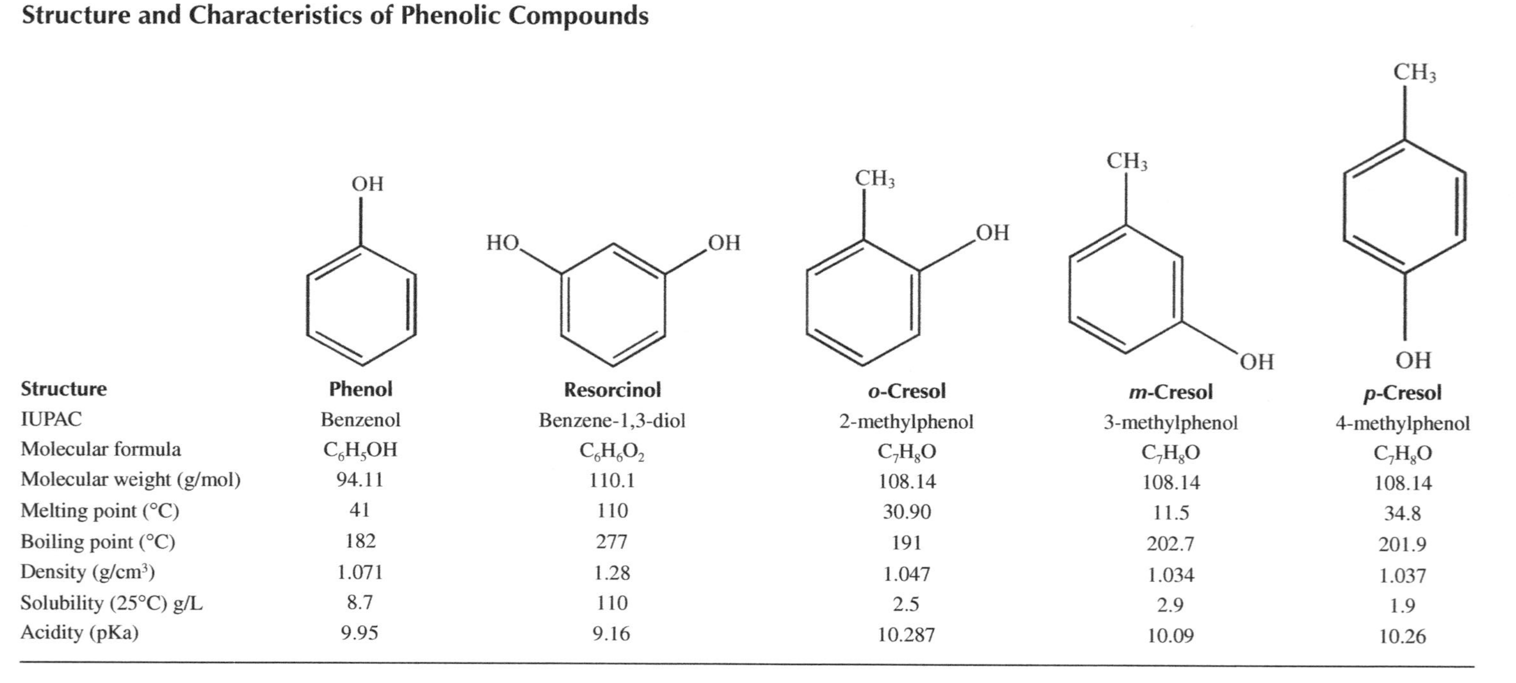

Structure	**Phenol**	**Resorcinol**	***o*-Cresol**	***m*-Cresol**	***p*-Cresol**
IUPAC	Benzenol	Benzene-1,3-diol	2-methylphenol	3-methylphenol	4-methylphenol
Molecular formula	C_6H_5OH	$C_6H_6O_2$	C_7H_8O	C_7H_8O	C_7H_8O
Molecular weight (g/mol)	94.11	110.1	108.14	108.14	108.14
Melting point (°C)	41	110	30.90	11.5	34.8
Boiling point (°C)	182	277	191	202.7	201.9
Density (g/cm^3)	1.071	1.28	1.047	1.034	1.037
Solubility (25°C) g/L	8.7	110	2.5	2.9	1.9
Acidity (pKa)	9.95	9.16	10.287	10.09	10.26

husks, peanut hulls, apricot stones, coconut shell, wheat residue, and sorghum stoves, in the absence of oxygen at 300–700°C. Modified clay minerals such as zeolite, sepiolite, and montmorillonite are used in the removal of phenolics and their derivatives.

7.3.1.2 Ion Exchange

The ion exchange process is used for the removal of ions from the water and another ion replaced from an ionic species. Very specifically designed natural and synthetic materials have been used for the removal of pollutants. Ionic resins have various specific properties, such as porosity, adsorption capacity, and density. Dowex XZ, a strong anion exchange resin, and AURIX100, a weak anion exchange resin, are used for the removal of phenol in alkaline conditions (Gonzalez et al., 2016).

7.3.1.3 Chemical Oxidation

Chemical compounds are used in the chemical oxidation process. Pollutant compounds break down or convert into less toxic compounds by chemical agents. A less toxic compound can further be degraded by the biological method. When the concentration of phenolic compounds is high, the chemical oxidation method can be used. Generally, as a chemical agent the strong oxidant hydrogen peroxidase is reported to initiate the oxidation reaction in the removal of pollutants by the oxidation process (Seetharam and Saville, 2003).

7.3.1.4 UV-Based Process

The UV-based process uses radiation for the excitation of substrates. Various types of UV/oxidizers are used, such as ozone, hydrogen peroxide, and metallic salt. In the UV/O_3 process, aqueous systems saturated with ozone are irradiated by UV light of particular nanometers. UV radiations are used for the initiation of ozone by photolysis. Many other types of UV-based processes are available, such as the UV/O_3-H_2O_2 process, Fe^{+3}/UV-VIS process, and UV/TiO_2 (Yingxu et al., 2005).

7.3.2 Biological Methods

7.3.2.1 Bioreactor for Biodegradation Study

In the last decades, the biological treatment approach is gaining attention for the degradation of phenolics and their derivatives worldwide. Presently, various types of reactor treatments are available, ranging from an activated sludge process to advanced biological treatment, hollow-fiber membrane, and microbial fuel cell. Reactors are operated in a three-way manner: continuous, semicontinuous, and batch. The general continuous reactor is used in municipal and industrial waste treatment under aerobic condition. Semibatch and batch reactors are used more in laboratory studies as compared to the continuous operated reactor. Bacterial cell applications like suspended cell culture or free microbial cell and immobilized microbial cell are playing a major role in the reactor. Bioreactor performance is enhanced by evaluating or optimizing the thermal and operation stability, retention time, pH, etc. The modified biofilm-based reactors include a fluidized bed, rotating biological contactor, trickling filter, packed bed, microporous membranes, rotating fiber discs, sequencing batch reactor (SBR), and microbial fuel cell. Many other types of bioreactors are listed in Table 7.3.

TABLE 7.3
Bioreactors Used in the Biodegradation of Phenolics and Derivatives of Phenolics

Compounds	Microorganisms	Reactors	Concentration (mg/L)	References
Phenol	*Pseudomonas putida*	Batch (free cells)	1–100	Monteiro et al. (2000)
Phenol	*Pseudomonas putida*	Bubble column bioreactor (immobilized)	100	Mordocco et al. (1999)
Phenol	*Pseudomonas putida*	Bubble column bioreactor (immobilized)	5–150	El-Naas et al. (2009)
Phenol	*Pseudomonas putida*	Spouted bed bioreactor) (immobilized)	40–190	El-Naas et al. (2008)
Phenol	*Pseudomonas putida*	Spouted bioreactor (immobilized)	10–150	El-Naas et al. (2008)
Phenol	*Pseudomonas putida*	Batch (free cell)	1000	Kumar et al. (2005)
Catechol Phenol	*Acinetobacter* sp.	Batch (free cell)	500, 100–1100	Wang et al. (2007)
Phenol	*Acinetobacter* sp. and *Sphingomonas* sp.	Bubble column bioreactor (free cell and immobilized)	200–1000	Liu et al. (2009)
Phenol	*Pseudomonas* sp.	Batch (free cell)	300–1000	Shourian et al. (2009)
Phenol	*Ewingella americana*	Batch (free cell)	0–1000	Khleifat (2006)
Phenol	*Alcaligenes faecalis*	Batch (free cell)	0–1800	Jiang et al. (2007)
Phenol	Ochrobactrum	Batch (free cell)	50–400	Kilic (2009)
Phenol	*Ralstonia eutropha*	Batch (immobilized)	25–500	Tepe and Dursun (2008)
Phenol	Indigenous mixed culture	Batch (free cell)	100–800	Saravanan et al. (2008)
Phenol	*Pseudomonas fluorescens*	Batch bioreactor (suspended cell)	100–500	Agarry et al. (2008)
Phenol	Consortium	Batch (free cell)	0–1750	Bajaj et al. (2009b)
Phenol	Mixed culture	Mixed batch reactor	25–1450	Nuhoglu and Yalcin (2005)
Phenol	Mixed culture	Continuous	400–1000	Luo et al. (2009)
Phenol	*Pseudomonas putida*	Batch (free cell)	3.18 mM	Tsai and Juang (2006)
Mixture of phenol	*Pseudomonas putida*	Batch (immobilized)	200–1000	Gonzalez et al. (2001a)
Phenol raw wastewater	*Pseudomonas putida*	Continuous stirred tank bioreactor and fluidized bed bioreactor (free and immobilized)	1000	Gonzalez et al. (2001b)

(*Continued*)

TABLE 7.3 (*Continued*)
Bioreactors Used in the Biodegradation of Phenolics and Derivatives of Phenolics

Compounds	Microorganisms	Reactors	Concentration (mg/L)	References
Phenol	*Pseudomonas putida*	Continuous fluidized bed bioreactor (immobilized)	50–250	Vinod and Reddy (2006)
Phenol	Candida *tropicalis*	Batch	2000–10,000	Varma and Gaikwad (2009)
Phenol	*Resting Candida tropicalis*	Batch	2000	Varma and Gaikwad (2009)
Phenolics	*Aspergillus awamori*	Batch	1000	Stoilova et al. (2007)
Phenol	*Cupriavidus metallidurans*	Batch (free cells) biofilm reactor	400–1400	Stehlickova et al. (2009)
Phenolics	*Halomonas campisalis*	Batch (free cell)	130	Alva and Peyton (2003)
Phenol	Aerobic granular sludge	Batch	500–5000	Ho et al. (2010)
Phenolics	Mutant strain of wild *Candida tropicalis*	Batch	4CP-400 0-800 phenol and Phenol-2500 4CP-0-30	Jiang et al. (2008)
Phenol	*Pseudomonas aeruginosa*	Packed bed bioreactor (immobilized cell)	1500	Kotresha and Vidyasagar (2017)
Phenol	*Pseudomonas stutzeri*	Continuously Internal loop air lift bioreactor	190	Viggiani et al. (2006)
Phenol	Sludge	Sequencing batch reactor (SBR)	500 mg/L rate 1.5 g phenol/L day	Jiang et al. (2006)
Phenol	*Pantoea agglomerans* + *Raoultella terrigena*	Batch culture Olive washing wastewater	150 mg/L	Maza-Marquez et al. (2013)
Phenol	*Sphingomonas* + *Rhizobiaceae*	Fixed film bioreactor	104.6 mg/L	Pozo et al. (2007)
Phenol	Sludge	SBR	500 mg/L rate 1.5 g phenol/L day	Jiang et al. (2006)
Phenol	*Pantoea agglomerans* + *Raoultella terrigena*	Batch culture Olive washing wastewater	150 mg/L	Maza-Marquez et al. (2013)
Phenol	*Azoarcus* spp. + *Microbulbifer* spp. *Pseudomonas* spp. *Thauera* spp.	Laboratory scale-activated sludge anoxic aerobic bioreactor	-	Sueoka et al. (2009)

7.3.2.2 Aerobic Biodegradation

The biodegradation of phenol is initiated by the oxygenation of an aromatic ring and then it is hydroxylated by a monooxygenase. This biodegradation is initiated by the oxygenation under aerobic conditions. First, the phenol containing an aerobic ring is hydroxylated by the monooxygenase. The ortho-cleavage forms a catechol by hydroxylation. In its main interaction process, phenol is degraded by aerobic microorganisms. Ring cleavage of catechol at the ortho-position or meta-position depends on the types of microorganisms. Microorganisms initiating ring cleavage at the ortho-position are called Co-A. Microorganisms undergo ring cleavage at the meta-position, thus initiating the meta-pathway that leads to the formation of pyruvate and acetaldehyde (Figure 7.1) (Hussain et al., 2010).

7.3.2.3 Anaerobic Biodegradation

Anaerobic biodegradation occurs by the carboxylation of phenol. Carboxylation is initiated after the phosphorylation of the phenol. Addition of the phosphate group and phenylphosphate synthase forms a phenyl phosphate molecule, which is then carboxylated to carboxybenzoate. Both enzymes are oxygen sensitive. CO_2 is used as a substrate and a metal ion as a co-catalyst.

7.4 MECHANISM OF BIODEGRADATION

Microorganisms can degrade phenolics due to a metabolic process. Microbes use pollutants as a sole source of carbon or nitrogen. It provides energy to survival. In the degradation process, a specific reaction takes place that converts the molecule into energy via responsible enzymes. The biodegradation process occurs in the presence of molecular oxygen, and the activated enzyme breaks down the aromatic ring. In the biodegradation of phenol, phenol hydroxylases initially and then ring cleavage by the ortho- or meta-pathway form catechol. When the ortho-pathway occurs by the enzyme catechol, 1,2-dioxygenase is formed, and when meta-pathway occurs by the catechol 2,3-dioxygenase is formed, which are responsible for the phenol degradation (Agarry et al., 2008; Nair et al., 2008; Hussain et al., 2010). According to Wang et al. (2007), *Acinetobacter* spp. showed the activity of catechol 1,2 dioxygenase via the ortho-pathway. *Ewingella americana* is degraded via meta-pathways, and the *Bacillus cereus* is degraded via meta-pathways (Al-Khlid and El-Naas, 2012).

7.5 ENZYME RESPONSIBLE FOR BIODEGRADATION

Microorganisms degrade toxic compounds by metabolic activity and use them as a sole source of carbon and energy. Different types of enzymes are responsible for metabolic activities in living cells. Hydroxylases, peroxidases, tyrosinase, and oxidases enzymes (Figure 7.1) are responsible for phenolic biodegradation (Haritash and Kaushik, 2009).

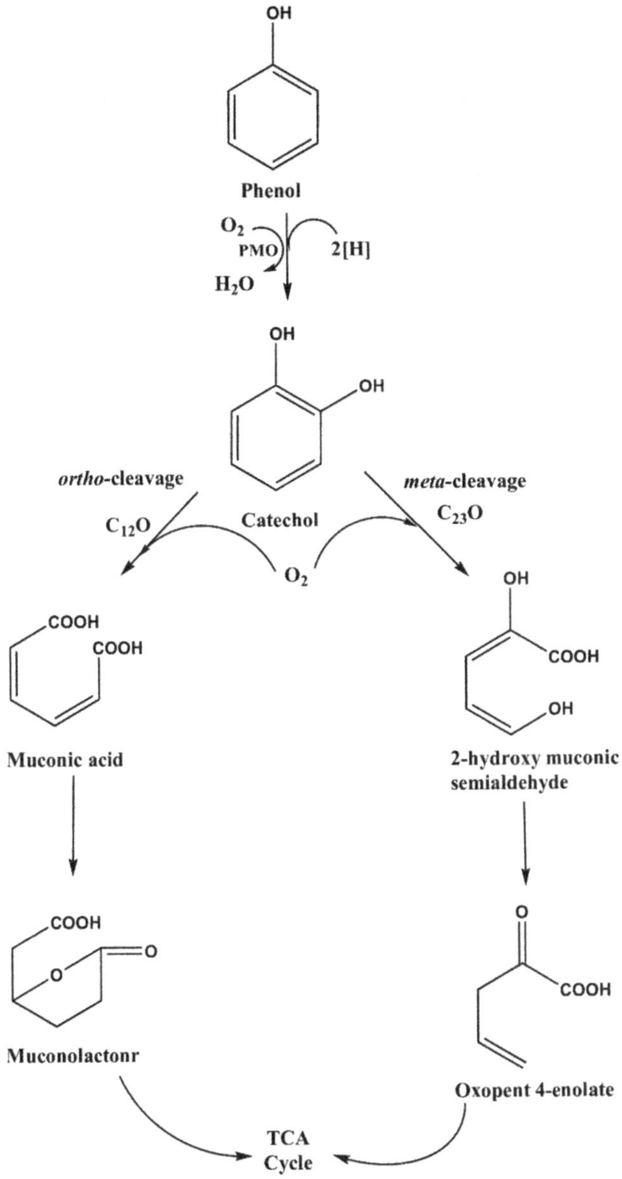

OH

Phenol

O_2 PMO H_2O 2[H]

OH OH

ortho-cleavage $C_{12}O$

Catechol

meta-cleavage $C_{23}O$

O_2

COOH COOH

Muconic acid

OH COOH OH

2-hydroxy muconic semialdehyde

COOH O O

Muconolactonr

O COOH

Oxopent 4-enolate

TCA Cycle

Enzymes involved in degration are PMO (in first two step)= Phenol monooxygenase; $C_{12}O$= Catechole 1, 2 dioxygenase $C_{23}O$= Catechole 2, 3 dioxygenase

FIGURE 7.1 Phenol biodegradation pathway.

7.5.1 Oxygenases

Oxygenase enzymes play an important role in the formation a more water-soluble compound from hydrophobic organic compounds. Due to this feature, microorganisms can break down it easily. Oxygenase enzymes have two different classes: monooxygenase and dioxygenases. Oxygenases participate in the oxidation process of various chemical (aromatic and aliphatic) compounds' metabolism.

7.5.2 Monooxygenases

The general mechanism of this class of enzyme is that the one oxygen molecule is inserted into the substrate and one molecule of oxygen is reduced to water, and, therefore, the two substrates are required. It is a more complex process in action and can also catalyze the oxygen insertion reaction. It is also called a mixed function that oxidizes the two substrates. It is also called hydroxylases because the one main substrate becomes hydroxylaselated.

7.5.3 Dioxygenases

Degradation of many chlorinated and nitro-aerobic compounds is initiated by the dioxygenases enzyme. In this enzymatic process, both oxygen molecules assimilate in substrates; dioxygenase and monooxygenase initiate the degradation and form catechol or protocatechuate. Intermediates are metabolized by ring-cleavage type to either beta-ketoadipate or 2-keto-4-hydroxyvalerate. These intermediates then enter the tricarboxylic acid cycle (TCA).

7.5.4 Hydroxylase

Phenol degradation takes place by two pathways: either meta-cleavage or ortho-cleavage. The monooxygenase of the aromatic ring can degrade various phenolic compounds. This process is initiated by flavoprotein monooxygenases that use electrons from NAD(P)H to cleave an oxygen molecule through the formation of a flavin hydroperoxide intermediate. It cannot assimilate the oxygen molecule in the substrate. These reactions can be catalyzed by a polypeptide chain or by multicomponent enzymes. It has been reported as a class of monooxygenases, consisting of a reductase component that uses NAD(P)H to reduce a flavin. It diffuses to an oxygenase large component that initiates the hydroxylation of aromatic compounds (Van Berkel et al., 2006).

7.6 MICROORGANISMS INVOLVED IN THE BIODEGRADATION OF PHENOLICS

Microorganisms can grow in specific conditions. They also vary in degradation of particular toxic compounds—some can grow aerobically and others anaerobically (Igbinosa et al., 2013). Various researchers reported that phenolics and their derivatives can degrade aerobically. Aerobic microorganisms can grow rapidly as compared to anaerobic (Agarry et al., 2008). Various types of aerobic bacteria are

TABLE 7.4
Phenolics and Derivatives of Phenolics Degrading Microorganism

Microorganisms	Experimental Conditions	Media	Concentrations	References
Acinetobacter baumannii	Aerobic	MSM	125–1000	Prasad et al. (2010)
Bacillus brevis	Aerobic	MSM	500–1750	Arutchelvan et al. (2006)
Bacillus sp.	Aerobic	MSM	-	Ali et al. (1998)
Delftia tsuruhatensis	Aerobic	MSM	-	Juarez-Jimenez et al. (2012)
Pseudomonas aeruginosa	Aerobic	MSM	200	Wang et al. (2011)
Pseudomonas stutzeri	Aerobic	MSM	450	Viggiani et al. (2006)
Acinetobacter sp. *Alcaligenes* sp.	Aerobic	Synthetic medium	2-CP: 100 Phenol: 50 *m*-Cresol: 50	Gallegos et al. (2003)
Alcaligenes odorans *Bacillus subtilis* *Corynebacterium* *Propinquum* *P. aeruginosa*	Aerobic	Synthetic medium	3.71–1.20 mg/L	Singh et al. (2013)
Comamonas sp. *Propioniferax*	Aerobic	Synthetic wastewater	250 mg/L	Jiang et al. (2006)
Cryptanaerobacter phenolicus	Anoxic/anaerobic	Synthetic medium	Phenol (3.5 mM) Hydroxybenzoate (6.5 mM)	Juteau et al. (2005)
Desulfobacterium aniline	Anoxic/anaerobic	MSM + sulfate	800–1500 µM	Ahn et al. (2009)
Natrialba sp.		Synthetic media + 25% NaCl	3%	Khemili-Talbi et al. (2015)
Sulfolobus solfataricus	Aerobic	Mineral medium	51–745	Christen et al. (2012)

reported as a pure culture or mixed culture or native consortia. *Pseudomonas* spp. is reported more than the other species. Several researchers did biodegradation study by fungi, yeast, and algae. The reported microbes shown in Table 7.4 are capable of the degradation of phenolics.

7.7 FACTORS AFFECTING THE BIODEGRADATION OF PHENOLICS

Microorganisms are affected by various physical, chemical, and biological factors like soil and water composition, nutrient availability, atmospheric conditions, such as pH, temperature, oxygen availability, and concentration of pollutants (Krastanov et al., 2013), which cause changes in the growth pattern, rate of removal, or biodegradation.

7.7.1 Nutrient Availability

Phenolics degrading microorganisms require nutrients such as carbon, nitrogen, phosphate, and potassium for biodegradation and their growth (Zaki, 2015). Sites containing phenolic compounds or wastewater rich in organic compounds have less amounts of C, N, P, and K that create starvation conditions for microbes. On the other hand, the excess amount of nutrients can create a negative effect and inhibit the growth of organisms.

7.7.2 pH

pH is a more important factor for degradation of phenolics because it helps in bacterial cell activity. Various organisms can grow under acidic and alkaline conditions (acidophiles and alkaliphiles). Some fungi reported for the degradation of phenolics at acidic pH can tolerate the acidic conditions. Generally, most organisms are pH sensitive, living in neutral pH (6.5–7.5) (El-Naas et al., 2008). The pH of a phenolic-contaminated site is acidic due of some spillage of phenolics containing compounds or leaching of coal spoil and oxidation of some compounds. However, natural microbial remediation balanced the soil pH near neutral by various chemicals, such as lime with Ca^{+2} and ammonium sulfate.

7.7.3 O_2 Availability

Biodegradation occurs in aerobic and anaerobic conditions—it depends on the availability of O_2 in the atmosphere. Researchers have reported that the phenolics and their derivatives are degraded under aerobic conditions. In the aerobic condition, O_2 plays an important role in the activation of enzymes (monoxygenase and dioxygenase), catechol, 1,2-dioxygenase, and catechol 2,3-dioxygenase. Oxygen availability or unavailability directly affects the degradation rate. Under the anaerobic condition, oxidation is carried out by nitrite, sulfate, or ferrous ions due to the absence of the oxygen molecule. Zhang et al. (2008) reported that degradation can be possible in the sulfate reducing and denitrifying conditions. Various toxic conditions also created due to the excess amount of phosphorous and ferrous ion in environments (Ho et al., 2010).

7.7.4 Pollutant Concentration

The degradation rate of the phenolics depends on the concentration of substrate interaction in two conditions: either synergistic or antagonistic (Heidari et al., 2015). Some effects are improving the rate of biodegradation by facilitating the rate of enzyme activity. It is evidenced in the synergistic condition. On the other hand, some activities are playing a negative role in biodegradation and inhibit the rate of degradation and the growth of organisms, like antagonistic interactions.

7.7.5 Temperature

Temperature is an important factor for aromatic degradation. It affects the various metabolic processes of microbes. Microbes degrade phenolics at different temperatures in various environments: 30–40°C for soil, 20–37°C for aqueous, and 15–20°C for marine environment bioremediation processes (Bajaj et al., 2009a). Temperature plays a crucial role for the biodegradation of phenolic compounds. All microorganisms can grow, and the best activities are obtained at an ambient temperature. At higher temperatures, aerobic microbial activity is inhibited due to a decreased concentration of dissolved oxygen. At lower temperatures, psychrophiles can carry out biodegradation. At higher temperatures, mesophilic organism activity is reduced due to the high concentration of pollutants because at high temperatures the solubility of phenolic compounds is increased, whereas the concentration of dissolved oxygen is decreased (Luo et al., 2009), which affects the growth of microorganisms. In some cases, at higher temperatures the compounds are transferred or produce new compounds. Newly produced compounds may be more toxic than the parental compounds. Various thermophilics can degrade phenolics and their derivatives at 60–70°C (Hasan and Jabeen, 2015).

7.8 MOLECULAR APPROACHES FOR THE BIODEGRADATION OF PHENOLICS

In recent research studies, the molecular biological technology is studied more due to its accurate result of strategies. The molecular technology gives an understanding of the responsible/catabolic gene in microorganisms from microbial community mechanisms of degradation, and using information regarding physiology, we can improve today's biodegradation technology and get more novel microorganisms and responsible enzymes for the biodegradation of pollutant compounds. Labelled DNA is used as a specific probe in the DNA hybridization process. Polymerase chain reaction (PCR) technology gave more sensitive detection. Nowadays more technologies are available as shown in Figure 7.2.

7.9 CONCLUSION AND FUTURE PROSPECTS

In the last decade, various types of microorganisms have been isolated and characterized for their capability to degrade phenolic compounds. Various enzymes are isolated from microorganisms, and different types of innovative pathways have been elucidated on the basis of ring oxidation and ring cleavage products. Recent research studies on genetic, genomics, proteomics, and metabolomics approaches are helpful in understanding the physiology, ecology, and biochemistry of phenolics degrading microorganisms. Conversely, in-depth research is required to determine exactly what is going on in phenolics and derivatives of phenolic-contaminated environments. In addition, various aspects of the bioremediation of phenolics are still unknown or have insufficient information, which requires further study. Bioremediation through enzymatic action converts phenolics to less toxic/nontoxic forms with fewer chemicals, less energy, and less time. Thus, it is a key method to degrade/remove phenolics contaminants in an eco-friendly way. These treatment strategies are used to enhance the degradation of organic pollutants.

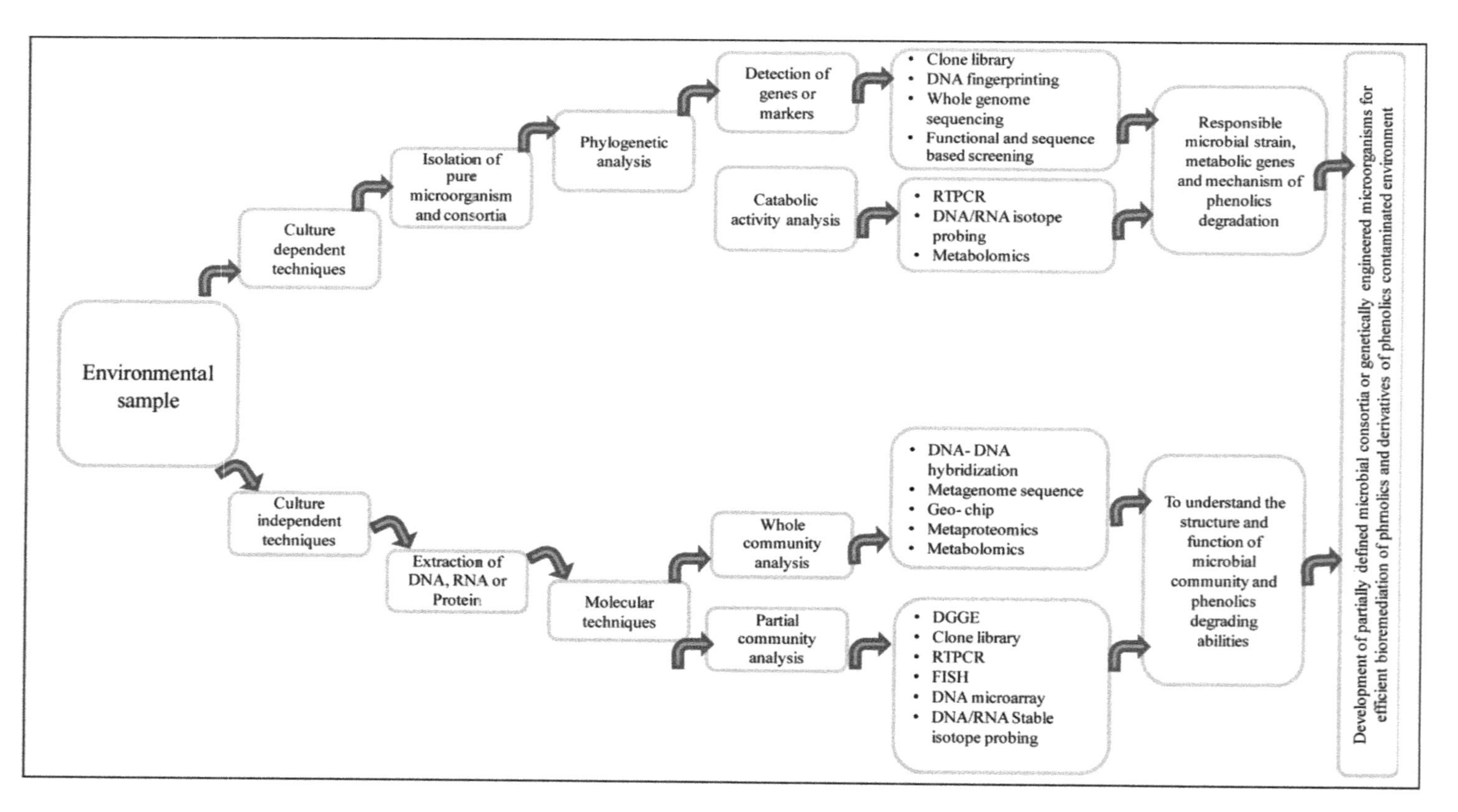

FIGURE 7.2 Recent molecular biology–based technology used in the bioremediation process.

However, there are still some persistent organic pollutants that are difficult to remove by bacteria from the environment. Genetic engineering can help solve this problem and offers a promising tool to improve the current bioremediation treatment technology and increase bacterial tolerance to these pollutants. In this chapter, we can conclude that bacteria has immense potential in the biodegradation of phenolics and their derivatives.

REFERENCES

Agarry, S. E., Solomon, B. O. 2008. Kinetics of batch microbial degradation of phenols by indigenous Pseudomonas Fluorescence. *International Journal of Environmental Science and Technology*. 5(2):223–232.

Ahn, Y. B., Chae, J. C., Zylstra, G. J., Häggblom, M. M. 2009. Degradation of phenol via phenylphosphate and carboxylation to 4-hydroxybenzoate by a newly isolated strain of the sulfate-reducing bacterium *Desulfobacterium anilini*. *Applied & Environmental Microbiology*. 75(13):4248–4253.

Ali, S., Lafuente R. L., Cowan, D. A. 1998. Meta-pathway degradation of phenolics by *thermophilic Bacilli*. *Enzymes & Microbial Technology*. 23(7–8):462–468.

Al-Khalid, T., El-Naas, M. H. 2012. Aerobic biodegradation of phenols: a comprehensive review. *Critical Reviews in Environmental Science and Technology*. 42(16):1631–1690.

Alva, V., Peyton, B. 2003. Phenol and catechol biodegradation by the haloalkaliphile *Halomonas campisalis*: influence of pH and salinity. *Environmental Science Technology*. 37:4397–4402.

Arutchelvan, V., Kanakasabai, V., Elangovan, R., Nagarajan, S., Muralikrishnan, V. 2006. Kinetics of high strength phenol degradation using *Bacillus brevis*. *Journal of Hazardous Materials B*. 129(1–3):216–222.

Bajaj, M., Gallert, C., Winter, J. 2009a. Treatment of phenolic wastewater in an anaerobic fixed bed reactor (AFBR)-recovery after shock loading. *Journal of Hazardous Materials*. 162:1330–1339.

Bajaj, M., Gallert, C., Winter, J. 2009b. Phenol degradation kinetics of an aerobic mixed culture. *Biochemical Engineering Journal*. 46:205–209.

Basha, K. M., Rajendran, A. Thangavelu, V. 2010. Recent advances in the biodegradation of resorcinol: a review. *Asian Journal of Experimental Biological Science*. 1(2):219–234.

Chakraborty, S., Bhattacharya, J., Patel, T. N., Tiwari, K. K. 2010. Biodegradation of phenol by native microorganisms isolated from cock processing wastewater. *Journal & Environmental Biology*. 31:293–296.

Christen, P., Vega, A., Casalot, L., Simon, G., Auria, R. 2012. Kinetics of aerobic phenol biodegradation by the acidophilic and hyper thermophilic archaeon *Sulfolobus solfataricus* 98/2. *Biochemical Engineering Journal*. 62:56–61.

El-Ashtoukhy, E-S. Z., El-Taweel, Y. A., Abdelwahab, O., Nassef. E. M. 2013. Treatment of petrochemical wastewater containing phenolic compounds by electrocoagulation using a fixed bed electrochemical reactor *International Journal of Electrochemical Science*. 8:1534–1550.

El-Naas, M. H., Al-Muhtaseb, S., Makhlouf, S. 2009. Biodegradation of phenol by *Pseudomonas putida* immobilized in polyvinyl alcohol (PVA) gel. *Journal of Hazardous Material*. 164:720–725.

El-Naas, M., Al-Zuhair, S., Makhlouf, S. 2008. Continuous biodegradation of phenol in a spouted bed bioreactor (SBBR). *Journal of Chemical Engineering*. 160:565–570.

El-Naas, M., Al-Zuhair, S., Makhlouf, S. 2010. Batch degradation of phenol in a spouted bed bioreactor system. *Journal of Industrial and Engineering Chemistry*. 16:267–272.

Gallegos, A., Fortunato, M. S., Foglia, J, Rossi, S, Gemini, V., Gomez, L., Gomez, C. E., Higa, L. E., Korol, S. E. 2003. Biodegradation and detoxification of phenolic compounds by pure and mixed indigenous cultures in aerobic reactors. *International Biodeterioration Biodegradation*. 52(4):261–267.

Gonzalez, G., Herrera, M. G., Garcia, M. T., Pena, M. M. 2001a. Biodegradation of phenolic industrial wastewater in a fluidized bed bioreactor with immobilized cells of *Pseudomonas putida*. *Bioresource Technology*. 80:137–142.

Gonzalez, G., Herrera, M. G., Garcia, M. T., Pena, M. M. 2001b. Biodegradation of phenol in a continuous process: comparative study of stirred tank and fluidized-bed bioreactors. *Bioresource Technology*. 76:245–251.

Gonzalez, M. A., Gonzalez, I., Texier, A. C. 2016. Mineralization of 2-chlorophenol by sequential electrochemical reductive dechlorination and biological processes. *Journal of Hazardous Materials*. 314:181–187.

Haritash, A. K., Kaushik, C. P. 2009. Biodegradation aspect of polycyclic aromatic hydrocarbons (PHAs); a review. *Journal of Hazardous Materials*. 169:1–5.

Hasan, S. A., Jabeen, S. 2015. Degradation kinetics and pathway of phenol by *Pseudomonas* and *Bacillus* species. *Biotechnology & Biotechnological Equipment*. 29:45–53.

Heidari, Z., Motevasel, M., Jaafarzadeh, N. 2015. Application of electro Fenton (EF) process to the removal of phenolic compound from aqueous solutions. *Iranian Journal of Oil and Gas Science and Technology*. 4:76–87.

Ho, K. L., Chen, Y. Y., Lin, B., Lee, D. J. 2010. Degrading high- strength phenol using aerobic granular sludge. *Applied Microbiology and Biotechnology*. 85:2009–2015.

Hussain, A., Kumar, P., Mehrotra, I. 2010. Nitrogen biotransformation in anaerobic treatment of phenolic wastewater. *Desalination*. 250:35–41.

Igbinosa, E. O., Odjadjare, E. E., Chigor, V. N., Igbinosa, I. H., Emoghene, A. O., Ekhaise, F. O., Igiehon, N. O., Idemudia, O. G. 2013. Toxicological profile of chlorophenols and their derivatives in the environment: the public health perspective- a review article. *The Scientific World Journal Volume. Article* ID46021:11

Jiang, H. L., Tay, S. T., Maszenan, A. M., Tay, J. H. 2006. Physiological traits of bacterial strains isolated from phenol-degrading aerobic granules. *Journal of the Federation of European Microbiological Societies*. 57:182–191.

Jiang, Y., Ren, N., Cai, X., Wu, D., Qiao, L., Lin, S. 2008. Biodegradation of phenol and 4-chlorophenol by the mutant strain CTM 2. Chin. *Journal of Chemical Engineering*. 16:796–800.

Jiang, Y., Wen, J., Bai, J., Jia, X., Hu, Z. 2007. Biodegradation of phenol at high initial concentration by *Alcaligenes faecalis*. *Journal of Hazardous Materials*. 147:672–676.

Juarez-Jimenez, B., Reboleiro-Rivas, P., Gonzalez-Lopez, J., Pesciaroli, C., Barghini, P., Fenice, M. 2012. Immobilization of *Delftia tsuruhatensis* in macro-porous cellulose and biodegradation of phenolic compounds in repeated batch process. *Journal of Biotechnology* 157(1):148–153.

Juteau, P., Cote, V., Duckett, M. F., Beaudet, R., Lepine, F., Villemur, R., Bisaillon J. G. 2005. *Cryptanaerobacter phenolicus gen. nov., sp nov.*, an anaerobe that transforms phenol into benzoate via 4-hydroxybenzoate. *International Journal of Systematic and Evolutionary Microbiolology*. 55(P1):245–250.

Khemili-Talbi, S., Kebbouche-Gana, S., Akmoussi-Toumi, S., Angar, Y., Gana, M. L. 2015. Isolation of an extremely halophilic arhaeon *Natrialba sp.* C21 able to degrade aromatic compounds and to produce stable biosurfactant at high salinity. *Extremophiles*. 19(6):1109–1120.

Khleifat, K. M. 2006. Biodegradation of phenol by *Ewingella americana*: effect of carbon starvation and some growth conditions. *Process Biochemistry*. 41:2010–2016.

Kılıc, N. K. 2009. Enhancement of phenol biodegradation by *Ochrobactrum sp.* isolated from industrial wastewaters. *International of Biodeterioration & Biodegradation*. 63:778–781.

Kotresha, D., Vidyasagar, G. M. 2017. Phenol degradation in a packed bed reactor by immobilized cells of *pseudomonas aeruginosa* MTCC 4997. *Biocatalyst and Agricultural Biotechnology*. 10:386–389.

Krastanov, A., Alexieva, Z., Yemendziev, H. 2013. Microbial degradation of phenol and phenolic derivatives. *Engineering in Life Sciences*. 13:76–87.

Kumar, A., Kumar, S., Kumar, S. 2005. Biodegradation kinetics of phenol and catechol using *Pseudomonas putida* MTCC 1194. *Biochemical Engineering Journal*. 22:151–159.

Lamichhane, S., Bal Krishna, K. C., Sarukkalige, R. 2016. Polycyclic aromatic hydrocarbons (PHAs) removed by sorption: a review. *Chemosphere*. 148:336–353.

Liu, Y. J., Zhang, A. N., Wang, X. C. 2009. Biodegradation of phenol by using free and immobilized cells of *Acinetobacter sp*. XA05 and *Sphingomonas* sp. FG03. *Biochemical Engineering Journal*. 44:187–192.

Luo, H., Liu, G., Zhang, R., Jin, S. 2009. Phenol degradation in microbial fuel cells. *Chemical Engineering Journal*. 147:259–264.

Maza-Marquez, P., Martinez-Toledo, M. V., Gonzalez-Lopez, J., Rodelas, B., Juarez-Jimenez, B., Fenice, M. 2013. Biodegradation of olive washing wastewater pollutants by highly efficient phenol- degrading strains selected from adapted bacterial community. *Journal of the Federation of European Microbiological Societies of Biodegradation*. 82:192–198.

Monteiro, A. A., Boaventura, R. A., Rodrigues, A. E. 2000. Phenol biodegradation by *Pseudomonas putida* DSM 548 in a batch reactor. *Biochemical Engineering Journal*. 6:45–49.

Mordocco, A., Kuek, C., Jenkins, R. 1999. Continuous degradation of phenol at low concentration using immobilized *Pseudomonas putida*. *Enzyme Microbial Technology*. 25:530–536.

Nair, C. I., Jayachandran, K., Shashidhar, S. 2008. Biodegradation of phenol. *African Journal of Biotechnology*. 7:951–4958.

Nuhoglu, A., Yalcin, B. 2005. Modelling of phenol removal in a batch reactor. *Process Biochemistry*. 40:1233–1239.

Olaniran, A. O., Igbinosa, E. O. 2011. Chlorophenols and other related derivatives of environmental concern: properties, distribution and microbial degradation processes. *Chemosphere*. 83:1297–1306.

Pozo, C., Rodelas, B., Martines-Toledo, M. V., Vilchez, R., Gonzalez-Lopez, J. 2007. Removal of organic load from washing eater by an aerated submerged biofilter and profiling of the bacterial community involved in the process. *Journal of Microbiology and Biotechnology*. 17(5):784–791.

Prasad, S. B. C., Babu, R. S., Chakrapani, R., Ramachandra, R., Rao, C. S. V. 2010. Kinetics of high concentrated phenol biodegradation by *Acinetobacter baumannii*. *International Journal of Biotechnology & Biochemistry*. 6(4):609–615.

Saravanan, P., Pakshirajan, K., Saha, P. 2008. Growth kinetics of an indigenous mixed microbial consortium during phenol degradation in a batch reactor. *Bioresource Technology*. 99:205–209.

Seetharam, G. B., Saville, B. A. 2003. Degradation of phenol using tyrosinase immobilized on siliceous supports. *Water Research*. 37(2):436–440.

Shourian, M., Noghabi, K. A., Zahiri, H. S., Bagheri, T., Karballaei, G., Mollaei, M., Rad, I., Ahadi, S., Raheb, J., Abbasi, H. 2009. Efficient phenol degradation by a newly characterized *Pseudomonas sp*. SA01 isolated from pharmaceutical wastewaters. *Desalination*. 246:577–594.

Singh, A., Kumar V., Srivastava, J. N. 2013. Assessment of bioremediation of oil and phenol contents in refinery waste water via bacterial consortium. *Journal of Petroleum & Environmental Biotechnology*. 4:3.

Stehlickova, L., Svab, M., Wimmerova, L., Kozler, J. 2009. Intensification of phenol biodegradation by humic substances. *International Biodeterioration and Biodegradation*. 63:923–927.

Stoilova, I., Krastanov, A., Yanakieva, I., Kratchanova, M., Yemendjiev, H. 2007. Biodegradation of mixed phenolic compounds by *Aspergillus awamori* NRRL 3112. *International Biodeterioration Biodegradation*. 60:342–346.

Sueoka, K., Satoh, H., Onuki, M., Mino, T. 2009. Microorganisms involved in anaerobic phenol degradation in the treatment of synthetic coke-oven wastewater detected by RNA stable-isotope probing. *Journal of the Federation of European Microbiological Societies*. 291(2):169–174.

Tepe, O., Dursun, A. 2008. Combined effects of external mass transfer and biodegradation rates on removal of phenol by immobilized *Ralstonia eutropha* in a packed bed reactor. *Journal of Hazardous Materials*. 151:9–16.

Tsai, S. Y., Juang, R. S. (2006). Biodegradation of phenol and sodium salicylate mixtures by suspended *Pseudomonas putida* CCRC 14365. *Journal of Hazardous Materials*. B138:125–132.

Van Berkel, W. J., Kamerbeek, N. M., Fraaije, M. W. 2006. Flavoprotein monooxygenases, a diverse class of oxidative biocatalyst. *Journal of Biotechnology*. 124:670–689.

Varma, R. J., Gaikwad, B. G. 2009. Biodegradation and phenol tolerance by recycled cells of *Candida tropicalis* NCIM 3556. *International Biodeterioration and Biodegradation*. 63:539–542.

Viggiani, A., Olivieri, G., Siani, L., Di Donato, A., Marzocchella, A., Salatino, P., Barbieri, P., Galli, E. 2006. An airlift biofilm reactor for the biodegradation of phenol by *Pseudomonas stutzeri OX1*. *Journal of Biotechnology*. 123(4):464–477.

Vinod, A. V., Reddy, G. V. 2006. Mass transfer correlation for phenol biodegradation in a fluidized bed bioreactor. *Journal of Hazardous Materials*. B136:727–734.

Wang, Y., Song, J., Zhao, W., He, X., Chen, J., Xiao, M. 2011. In situ degradation of phenol and promotion of plant growth in contaminated environments by a single *Pseudomonas aeruginosa strain*. *Journal of Hazardous Material*. 192(1):354–360.

Wang, Y., Tian, Y., Han, B., Zhaw, H. B., Bi, J. N., Cai, B. L. 2007. Biodegradation of phenol by free and immobilized *Acinetobacter* sp. Strain PD12. *Journal of Environmental Science*. 19:222–225.

Yingxu, C. S., Kan, W., Liping, L. 2005. Roll of primary active species and TiO2 surface characteristics in UV-illuminated photo degradation of Acid Orange 7. *Journal of Photochemistry and Photobiology A: Chemistry*. 172:47–54.

Zaki, M. 2015. Phenol and phenolic compounds: sources, routes of exposure and mode of action. *Advances in Environmental Biology*. 9:49–55.

Zhang, L., De Schryver, P., De Gusseme, B., De Muynck, W., Boon, N., Verstraete, W. 2008. Chemical and biological technologies for hydrogen sulphide emission control in sewer systems: a review. *Water Research*. 42:1–12.

8 Enhanced Bioavailability and Biodegradation of DDx in an Anoxic Organic Soil

Hiral Gohil and Andrew Ogram[*]

CONTENTS

[*] Corresponding author: Email: aogram@ufl.edu

8.1 INTRODUCTION

Bioremediation is an economically attractive option for the remediation of soils contaminated with hydrophobic organic contaminants (HOCs). One of the major factors that impede bioremediation of HOCs is bioavailability. Bioavailability may be limited because the pollutant may diffuse into the soil with time and become unavailable (Alexander, 1995, 1997). Various theories have been proposed to explain the prolonged persistence of HOCs in soils, including sorption of polycyclic aromatic hydrocarbon (PAH), partitioning of the HOC into nonaqueous phase liquids (NAPLs), or diffusion within micropores or entrapment in the physical matrix of the soil (Alexander, 1997; Luthy et al., 1997).

HOCs such as dichlorodiphenyltrichloroethane (DDT) and its major metabolites DDD (dichlorodiphenyldichloroethane) and DDE (dichlorodiphenyldichloroethylene, collectively known as DDx) undergo an "aging" process with time in soil, such that the DDx molecules have decreased bioavailability over time (Alexander, 1995, 1997). Aging has been shown to decrease rates of loss of DDx by volatilization, leaching, or biodegradation (Boul, 1995). Scribner and coworkers (1992) demonstrated that the longer the compounds remain in soil, the more resistant they become to desorption and biodegradation.

Soil pollution with recalcitrant DDxs can lead to significant ecosystem damage (Ratcliffe, 1967; WHO, 1979; Megharaj et al., 2000; Turusov et al., 2002) and, hence, methods to bioremediate such pollutants are of a high priority. Since bioavailability is an important issue in the degradation of HOCs, a factor that increases accessibility to degrading organisms might increase biodegradation rates. Bioavailability of a compound may be enhanced by dispersion of the soil (Juhasz et al., 1999), leading to release of dissolved organic matter (DOM). Several studies have demonstrated that DOM may increase the solubility and mobility of HOCs (Chiou et al., 1986; Juhasz et al., 1999) and that dissolved organic carbon (DOC) may influence the bioavailability of such compounds (McCarthy and Zachara, 1989; Juhasz et al., 1999).

One means of increasing bioavailability is the use of sodium (Sumner and Naidu, 1998). Sodium is known to disperse soils and increase DOM concentrations (Wood, 1995; White, 1997; Nelson and Oades, 1998; Kantachote et al., 2003; Brady and Weil, 2003;). Both increased DOM and soil dispersion could potentially increase bioavailability of the otherwise inaccessible and aged DDxs (Quensen et al., 1998; Kantachote et al., 2001, 2004). This happens by releasing soil-bound particles or colloids or by making the previously unavailable or protected DDx available to the potential degrading organisms (Kantachote et al., 2004). Kantochote and colleagues (2001, 2004) showed that Na^+ application to DDT-contaminated soil significantly increased the DOM concentrations, anaerobic bacterial numbers, and concentrations of DDT residues in soil solution. This was followed by biodegradation of approximately 95% of the HOC.

The primary objectives of this study are to evaluate the use of Na^+ ions, labile organic carbon and metals to increase biodegradation rates of DDx in microcosms and mesocosms. To explore the potential of Na^+ to increase bioavailability and degradation of DDxs in anoxic organic soils, microcosm experiments were performed. Previous studies on this system demonstrated that anoxic incubations increase DOC (Gohil et al., 2014), which likely increased both bioavailability of DDx and biodegradation rates.

Microcosms were established with a series of different NaCl concentrations, and ground cattail (family *Typhaceae*) was added as a source of carbon and energy for the degrading species. Previous studies on this system demonstrated that the greatest degradation was observed with lactate as an electron donor and DDx as a terminal electron acceptor (TEA) (Gohil et al., 2014). By using cattail in this study, our intention was to increase lactate production because approximately 40% of cattail is cellulose (DeBusk and Reddy, 1998).

Our overarching hypothesis was that the degraders will use DDx as the TEA or cometabolize DDx while using cattail fermentation products for carbon and energy sources and that Na^+ additions will increase degradation rates, potentially due to dispersal of the organic carbon polymers associated with solids, thereby increasing bioavailability of the DDx. Another assumption is that the system would be limited in trace metals that may be required for synthesis of certain enzymes. Therefore, two different types of growth media were compared in microcosm systems. Once these conditions were optimized for biodegradation of DDx in microcosms, the optimum conditions were scaled up to the mesocosm levels.

8.2 MATERIALS AND METHODS

8.2.1 Soils Used

Microcosms were constructed with soil collected from North Shore Restoration Area (NSRA) at Apopka. Lake Apopka is a large (125 km^2 surface area) shallow lake located in the state of Florida, USA. Following collection, the soils were shipped to the University of Florida, Gainesville, and refrigerated at 4°C until use. Soils were passed through a 20-mesh sieve (nominal pore size of 80 μm) and mixed well prior to addition to the microcosms.

8.2.2 NaCl Microcosms

Microcosms were constructed by mixing air dry soils with cattail powder at a rate of 0.04 g cattail powder per gram dry soil. Sieved (20 mesh sieve, nominal diameter of 80 μm) and mixed soils were spiked with 10 μg *p,p′*-DDT per gram dry soil solubilized in methanol. The solvents were allowed to evaporate, and the soils were manually mixed. An aliquot of 10 g dry-weight soil was mixed with cattail (0.04 g cattail powder per gram dry soil) and added to 125-mL serum bottles with 45 mL growth medium. Two different types of growth medium were compared for this experiment. The first set was made using minimal growth medium according to Ou et al. (1978), and a second set contained the more complex growth medium according to Tanner (2007). The Tanner medium is a richer medium that includes a range of growth factors and inorganic nutrients, cofactors, vitamins and trace metals that are not present in the Ou medium.

The Ou medium consisted of (g/L of distilled water) K_2HPO_4, 4.8; KH_2PO_4, 1.2; NH_4NO_3, 1; $CaCl_2 \cdot 2H_2O$, 0.025; $MgSO_4 \cdot 7H_2O$, 0.2; and Fe_2 $(SO_4)_3$, 0.001. The Tanner medium consisted of (amount/L of distilled water) mineral solution 10 mL, vitamin solution 10 mL, trace metal solution 1.5 mL and yeast extract 0.1 g. Ingredients for the individual components starting with mineral solution (g/L of distilled water) were NaCl, 80; NH_4Cl, 100; KCl, 10; KH_2PO_4, 10; $MgSO_4 \cdot 7H_2O$, 20; and $CaCl_2 \cdot 2H_2O$, 4. The vitamin solution consisted of (g/L of distilled water) pyridoxin HCl, 10; thaimine HCl, 5; riboflavin, 5; calcium pantothenate, 5; thioctic acid, 5; *p*-aminobenzoic acid, 5; nicotinic acid, 5; vitamin B_{12}, 5; biotin, 2; and folic acid, 2. Trace metal solution ingredients (g/L of distilled water) were nitrilotriacetic acid, 2; $MnSO_4 \cdot H_2O$, 1; $Fe(NH_4)_2(SO_4)_2 \cdot 6H_2O$, 0.8; $CoCl_2 \cdot 6H_2O$, 0.2; $ZnSO_4 \cdot 7H_2O$, 0.2; $CuCl_2 \cdot 2H_2O$, 0.02; $NiCl_2 \cdot 6H_2O$, 0.02; $Na_2MoO_4 \cdot 2H_2O$, 0.02; Na_2SeO_4, 0.02; and Na_2WO_4, 0.02.

Both media sets were tested for four Na^+ concentrations, including 0, 30, 50 and 80 mg Na^+ per kilogram soil. Sodium was added in the form of NaCl. In the case of autoclaved controls, soils that had been autoclaved for once per day on three consecutive days at 121°C for 30 minutes were included to account for any abiotic losses. A second set of controls referred to as "cattail controls" containing no cattail were used for each set. Cattail controls were included to test for the impacts of cattail as a carbon and energy source by the indigenous populations.

For anaerobic incubations, all sets were purged with N_2, sealed with Teflon-lined butyl rubber stoppers and secured with aluminum crimps. They were then incubated at 27°C in the dark for approximately 2 months. Following incubation, soils were extracted and DDx concentrations determined as described next.

8.2.3 Mesocosm Soil Collection

Soils from NSRA's north boundary at Lake Apopka, Florida, USA site ZNS2017 were collected from a 16-foot-by-16-foot plot on east of Laughlin Road (UTM coordinates: $X = 440603.7$, $Y = 3177660.0$). The site was selected because of high DDx concentration as determined by Pace Analytical Services under contract to the St. Johns River Water Management District. The concentration of *p,p′*-DDE was 5990, for *p,p′*-DDD was 564, and that for *p,p′*-DDT was 15,800 µg/kg soil determined by analytical method EPA 8081 and sample prepared by method EPA 3550. Total organic carbon (TOC) was determined to be 37% by DB Environmental, Inc. Vegetation was removed and soils were mixed on the site using a trackhoe. Following mixing, the soils were transported to the University of Florida Gainesville campus, where the soils were stored in wooden boxes lined with a plastic liner.

8.2.4 Preparation of Anaerobic Mesocosms

Rubbermaid "Farm Tough" (100-gallon) stock tanks made of high-density polyethylene were used for mesocosms. The tank dimensions were as follows: length, 54 in (137.16 cm); width, 35 in (88.9 cm); and height, 23 in (58.42 cm). The tanks had a controllable opening at the bottom of tank for drainage. Prior to filling the tanks with soils, a bed of egg rocks was added to promote drainage.

Four tanks were placed on cement block tables inside a greenhouse. The placement order of the mesocosm tanks was intended to randomize potential differences in temperature within the greenhouse. The order was Control-1, NaCl-1 and Control-2, NaCl-2.

Soil was mixed manually and added to the tanks one wheelbarrow full at a time and thoroughly mixed in the tanks. Water was added to a height of 10 cm above the soil surface to create anaerobic conditions. For all mesocosms, sodium lactate was added to a final concentration of 10 mM in tap water. Lactate solution was added to 10 cm above the soil surface. Ten soil samples were randomly collected at a depth of 5 cm from each mesocosm and mixed to form one composite sample per tank and frozen at −80°C for DDx analysis. The lactic acid included in these additions was added as sodium lactate, which leads to a final Na^+ concentration of 0.015 mg Na^+ per gram dry soil.

Due to the limited availability of NSRA soil, a temporally staggered experimental design was established such that 0.015 mg Na^+ per gram dry soil was added to two experimental tanks and monitored until the redox potential in the tanks stabilized. After 2 weeks stabilization treatment, a second addition of Na^+ was made to bring the final concentration to 80 mg Na^+ per kilogram dry soil. Additions were made in the treatment tanks on the 80th day following anoxic incubation.

Tanner medium (Tanner, 2007) trace metal solution was added to all mesocosms (control and experimental) at the same time as the additional Na^+ described earlier. Trace metal solution was added to both control and NaCl tanks to study the effect of additional NaCl. The additional Na^+ and metals were dissolved in 500 mL distilled water to appropriate concentrations, autoclaved and manually mixed using a shovel on the 80th day and mixed once again on the 81st day to ensure homogeneity.

The ingredients amended (mg/g dry soil) were nitrilotriacetic acid (NTA), 0.0135; $MnSO_4 \cdot H_2O$, 0.00675; $Fe(NH_4)_2(SO_4)_2 \cdot 6H_2O$, 0.0054; $CoCl_2 \cdot 6H_2O$, 0.00135; $ZnSO_4 \cdot 7H_2O$, 0.00135; $CuCl_2 \cdot 2H_2O$, 0.00135; $NiCl_2 \cdot 6H_2O$, 0.00135; $Na_2MoO_4 \cdot 2H_2O$, 0.000135; Na_2SeO_4, 0.000135; and Na_2WO_4, 0.000135.

8.2.5 Sample Collection

Once redox potentials had stabilized, eight samples were randomly taken from each tank and mixed to form a composite sample. Samples were collected every 2 weeks using a 50-mL disposable syringe with the ends cut off. Samples were stored at −80°C until the samples could be analyzed.

8.2.6 Redox Potential (Eh) and Temperature Measurements

Redox potential was determined at a depth of 10 cm below the soil surface every 2 weeks with a portable pH/Eh and temperature meter (HI 9126 Hanna Instruments; Woonsocket, RI). The probe was calibrated before every use. The Eh probe was submerged in each tank to a depth of 10 cm below the soil surface and allowed to equilibrate before taking the reading; temperature readings were taken at the same depth using a temperature probe.

8.2.7 Dissolved Organic Carbon Measurements

One gram soil (dry weight) was diluted to a concentration of 1:10 (soils: distilled water) and incubated for 16 hours at room temperature, followed by centrifugation in a Beckman J2-21 Floor Model Centrifuge (GMI Incorporation Ramsey, MN) (JA-14 rotor) at 4472 × g for 15 minutes. The supernatant was filtered through a 0.22-μm filter (Durapore membrane filters), and the filtered solutions were analyzed on a Shimadzu sample module 5000A Carbon Nitrogen Analyzer (Columbia, MD).

8.2.8 pH Measurements

Soil samples were centrifuged in a Beckman centrifuge Model J2-21 at 7155 × g for 10 minutes. pH was measured from a supernatant using Orion pH meter Model SA720 (Cole-Parmer Instruments, Mount Vernon Hills, IL).

8.2.9 DDx Extraction

DDx extraction could be divided into three stages: soil preparation, accelerated solvent extraction (ASE) and Florisil extraction. The extraction method was based on the United States Environmental Protection Agency method 3545 (U.S. EPA, 1979) with minor modifications developed by Soil Microbial Ecology Laboratory, University of Florida, and chemists at Pace Analytical Services, Ormond Beach, Florida.

8.2.10 Soil Preparation

Soils were separated from liquid by centrifugation in a Beckman J2-21 Floor Model Centrifuge (JA-14 rotor) at 7155 × g for 20 minutes at 4°C. Wet soil samples were allowed to air dry for 2–3 days and then were ground. Soil moisture content was determined in the sample. Moisture was adjusted to 50% moisture on a dry-weight basis, and then the samples were allowed to equilibrate in 4°C refrigerator for 3 days.

8.2.11 Accelerated Solvent Extraction

Soil was mixed with Hydromatrix, a drying and bulking agent, at a ratio of 1:2 and placed into 34-mL stainless steel extraction cells. The remaining headspaces in extraction cells were topped with clean Ottawa sand (Fisher Scientific, Pittsburgh, PA) to decrease the amount of solvent usage. Solvent used for extraction was methylene chloride:acetone (4:1 v/v). Soils were extracted under high pressure about 1200–1400 psi at temperature of 100°C in a Dionex ASE 100 Accelerated Solvent Extractor (Sunnyvale, CA).

The extraction cycle included filling the extraction cell with about 19 mL solvent, followed by the heating cycle where the solvent was heated to 100°C. Static extraction followed the heating cycle, which was performed for 5 minutes, followed by the flushing cycle, which flushed about 19 mL solvent through the cell. The cell was finally purged with N_2 for about 2 minutes, which completed one extraction cycle.

8.2.12 Florisil Extraction

Florisil cleanup used prepacked Florisil "Bond Elute LRC" columns (200-μm particle size) (Varian Inc., Palo Alto, CA), which contained approximately 1 g Florisil packaged into the plastic holder. The column was conditioned using 5 mL hexane:acetone (9:1v/v), loaded with 1 mL of ASE extract, and the column was eluted using 9 mL of solvent.

A volume of 5 mL from the cleanup volume was concentrated to dryness under a gentle stream of air. It was very important to concentrate the samples to complete dryness prior to GC (gas chromatography) analysis to avoid matrix response enhancement (Schmeck and Wenclawiak, 2005). Samples were reconstituted in 1 mL of hexane, and tubes with samples were vortexed and transferred to a 2-mL amber glass GC vials (Fisher Scientific Inc., Atlanta, GA) and crimped with 12 × 32 mm aluminum crimp seals with prefitted PTFE lined septa (Fisher Scientific Inc., Atlanta, GA) for subsequent analysis by GC.

8.2.13 GC Conditions

A Perkin-Elmer autosystem gas chromatograph equipped with an autosampler and an electron capture detector (ECD) was used for analyzing the extracts. Column conditions were He as the carrier gas at a flow rate of 1.0 mL/min, 5% methane in argon as the makeup gas with a flow rate of 50 mL/min, ECD temperature at 350°C, and injector temperature at 205°C in splitless mode. Injection volume was 3 μL in a slow injection mode. GC temperature was programmed at 110°C for 0.5 minutes, followed by a ramp at 20°C/min to 210°C and held for 0 minute, then ramp at 11°C/min to 280°C and held for 6.3 minutes. Under these conditions, the detection limit was calculated at the 90% confidence level to be: 0.007 μg/mL for *o,p′*-DDE; 0.005 μg/mL for *p,p′*-DDE; 0.006 μg/mL for *o,p′*-DDD; 0.009 μg/mL for *p,p′*-DDD; 0.005 μg/mL for *o,p′*-DDT; and 0.010 μg/mL for *p,p′*-DDT (Hubaux and Vos, 1970). These results are similar to the detection limit of 0.010 μg/mL reported by Pace, Inc.

8.3 ANALYSIS OF ORGANIC ACIDS BY HPLC

8.3.1 Sample Preparation

A 1:10 dilution consisting of 1 g soil (dry weight) and 9 mL distilled water was incubated for 16 hours at room temperature. Slurries were then centrifuged in a Beckman J2-21 Floor Model Centrifuge (JA-14 rotor) at 4472 × g for 15 minutes. The supernatant was filtered through a 0.22-μm filter (Durapore PVDF membrane filters, Fisher Scientific, Pittsburgh, PA) under vacuum.

8.3.2 Derivatization

The samples were derivatized using following steps. Pyridine buffer (0.2 mL) was added to the 2-mL sample. The resulting solution was purged with nitrogen to remove CO_2 and O_2 for 4 minutes. The anoxic solution as amended with 0.2 mL of 0.1 M

2-nitrophenyl hydrazine (in 0.25 M HCl) and 0.2 mL of 0.3 M 1-ethyl-3-(3-dimethylaminopropyl) carbodiimide hydrochloride; after mixing the vials were incubated at room temperature for 90 minutes. Following incubation 0.1 mL of 40% KOH was added and the samples were heated in a heating block at 70°C for 10 minutes.

8.3.3 HPLC

Concentrations of fatty acids were determined using high-performance liquid chromatography (HPLC) with UV/VIS detector at 400 nm wavelength and a C8 reverse phase column (22 cm × 1.5 cm). The mobile phases included two solvents. Solvent A was composed of 2.5% *n*-butanol, 50 mM sodium acetate, 2 mM tetrabutylammonium hydroxide, 2 mM tetradecyltrimethylammonium bromide (TDTMABr) with pH adjusted to 4.5 using phosphoric acid. Solvent B differed from solvent A only in containing 50 mM TDTMABr. The injection volume was 100 μL. Retention times for lactate, acetate, propionate and formate were 9.95, 12.62, 14.67 and 13.28 minutes, respectively. Standard fatty acids used were lactate, acetate, formate, propionate, butyrate, succinate, isobutyrate and isovalerate (Albert and Martens, 1997).

8.3.4 Statistical Analysis

Statistical analyses were conducted with JMP software manufactured by SAS (Cary, NC). One-way analysis of variance (ANOVA) was used for microcosm data analysis. Controls were tested against the treatments using Tukey's post hoc test. Mesocosm data sets were compared using Student's *t*-test.

8.4 RESULTS AND DISCUSSION

8.4.1 Microcosms

Although a direct role of Na^+ in DDT degradation is not well documented, its role in soil and organic matter dispersion is well recognized (Wood, 1995; White, 1997; Nelson and Oades, 1998; Kantachote et al., 2003; Brady and Weil, 2003,). Dispersion may lead to the exposure of new surfaces to the solution, thereby increasing diffusion out of the organic matrix and increasing availability to the degrading consortia. Na^+ is not only responsible for dispersing soils but also for increasing the DOC following anoxic incubation.

8.4.2 Comparison of Media

Results using the Ou medium are presented in Figure 8.1, and studies using the Tanner medium, rich in mineral salts, trace metals, and vitamins, are presented in Figure 8.2. Tanner medium microcosms demonstrated that an increase in NaCl concentration resulted in a decrease in DDx concentrations. With respect to the remaining DDx concentrations, 0 mg Na^+ microcosm soils are consistently significantly different from those in 80 mg Na^+ microcosms. For example, 0 mg Na^+ microcosm soils contain 37.5 and 76.9 mmol of *p,p′*-DDE and *p,p′*-DDT, and when compared to that in the case of 80 mg, concentrations are 1.4 and 19.7 mmol, respectively. Similarly, concentrations

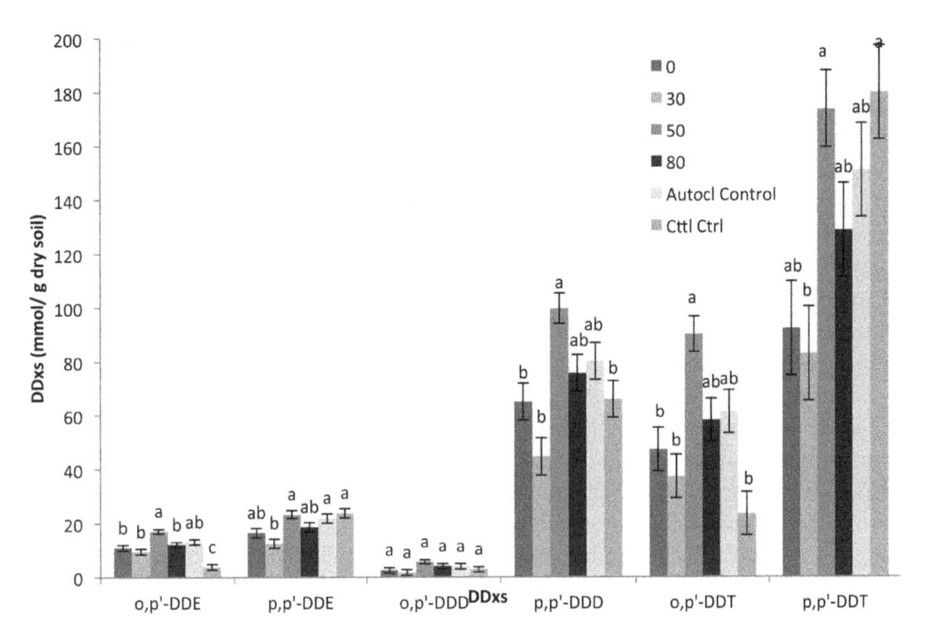

FIGURE 8.1 Concentration of DDxs measured in NSRA soil following incubation in minimal Ou medium (Ou et al., 1978) in microcosms with a range of Na^+ concentrations. Error bars represent +/– one standard deviation. Data labels not connected by the same letter are significantly different $p < 0.05$. "Cttl Ctrl" = cattail control, "Autocl Ctrl" = autoclaved control.

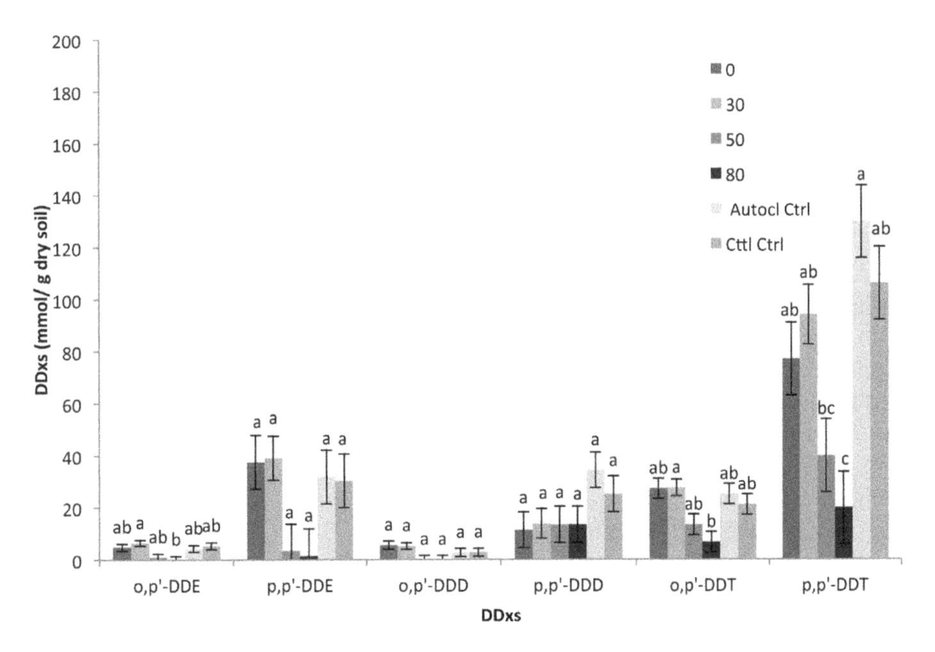

FIGURE 8.2 Concentration of DDx measured in NSRA soil following incubation in microcosms with complex Tanner medium (Tanner, 2007) with a range of Na^+ concentrations. Error bars represent +/– one standard deviation. Data labels not connected by the same letter are significantly different $p < 0.05$. "Cttl Ctrl" = cattail control, "Autocl Ctrl" = autoclaved control.

of *o,p′*-DDE, *o,p′*-DDD and *o,p′*-DDT for 0 mg Na^+ are 4.7, 5.5 and 27.2 mmol per gram dry soil, respectively. When compared to *o,p′*-DDxs in 80 mg microcosms, concentrations are 0 mmol for both *o,p′*-DDE and *o,p′*-DDD and 6.5 mmol for *o,p*-DDT.

When comparing 0 and 30 mg Na^+ microcosms, no significant difference in DDx concentrations was observed, and little difference was observed between the 50 and 80 mg Na^+ microcosms. However, a significant difference was observed when the sets were divided into two groups (0 and 30 mg and 50 and 80 mg). Regardless of Na^+ concentrations, it is evident from Figure 8.2 that the greatest degradation was achieved with 80 mg Na^+ microcosms. Furthermore, Na^+ positively affected degradation, such that the degradation sequence is 80 mg > 50 mg > 30 mg > 0 mg. Findings from this experiment are in accordance with similar studies by Kantachote and colleagues (2001, 2003).

Similar studies (Kantachote et al., 2003) demonstrated that applications of Na^+ at various concentrations to long-term DDT contaminated soils lead to DDT biotransformation under anoxic conditions. They attributed this transformation to the dispersive nature of Na^+, which coincided with increased DOC concentration, anaerobic bacterial numbers, and aqueous solubility of DDxs. The DDxs transformation ranged from 95% to 72% with maximal transformation at 30 mg Na^+ per kilogram dry soil. DDT transformation was repressed at higher Na^+ levels due to flocculation and decreased DOC levels.

Previous microcosm and mesocosm studies on these soils demonstrated that the greatest amount of degradation of DDxs was achievable using lactate as the carbon and electron source, whereas DDxs served as the probable TEAs or were cometabolized (Gohil et al., 2014). These conditions may have promoted growth of a syntrophic community, where lactate fermenting, hydrogen-producing organisms coexisted with a hydrogen-utilizing, DDx dehalorespiring population. An attempt in this study was to simulate lactate production by metabolism of cattail to provide the energy and carbon needs of the degrading consortia.

The cattail control (i.e., no cattail) exhibited lower concentrations of DDx than the autoclaved control, indicating that the organisms capable of degradation are present, but the system is carbon limited. Hence, careful selection of a carbon and electron donor in a TEA-limited anoxic environment can shift the system to dehalorespiration (Suflita et al., 1988). The autoclaved control has the highest concentrations of DDxs, which indicates that changes in DDxs resulted from microbially mediated transformation.

A high variability among replicates was observed for the Ou medium microcosms (Figure 8.1), and no relationship between Na^+ concentration and degradation was observed. Greater degradation of DDx was observed in the microcosms with the Tanner medium than with the Ou medium, likely because the Ou medium is a minimal salts medium, containing no vitamins or trace metals. Tanner medium has higher concentrations and a greater diversity of trace metals and cofactors that may be limiting in the peat soils of NSRA.

8.4.3 Microcosm Conclusions

A positive correlation between the concentrations of Na^+ and DDx degradation was observed in microcosms with the Tanner medium. A likely explanation is that, upon soil dispersion by the added Na^+, the otherwise unavailable DDx were made available to the degrading consortia. Furthermore, with anoxic incubation there is an increase in DOC concentrations (Gohil et al., 2014), which may also increase the solubility, and hence bioavailability of DDx. Both these factors could increase exposure of substrate to the microbes. Another objective in this study was to meet the energy and carbon needs of the degrading consortium via decomposition of cattail.

Higher concentrations of *p,p′*-DDT remaining in the no-cattail controls suggest that the system is carbon limited and that products resulting from the metabolism of the organic carbon in cattail could feed the degrading populations. Another indication is that although organisms capable of degradation are indigenous to the soils, suitable selection of electron donors and carbon source, along with increased bioavailability, results in greater biodegradation of the otherwise inaccessible aged DDx.

Another observation from this study is that additional trace metals, cofactors and vitamins are needed by the degrading population. They may be responsible for triggering the production or activation of degrading enzymes, or may be required as growth factors by the degrading consortia.

8.4.4 Mesocosm Results

As described earlier, Na^+ at a rate of 80 mg/kg dry soil with cattail as the exogenous carbon and electron source gave the greatest degradation of DDx in microcosm studies. Another important amendment in the microcosms exhibiting the greatest loss of DDx was a medium rich in trace metals, vitamins and cofactors, which may have triggered production or activation of degrading enzymes. Mesocosms were established not only to scale up the microcosms but also to check reproducibility of microcosm results.

Within 3 weeks after establishing the mesocosms, redox potentials decreased to those characteristic of anoxic soils (Figure 8.3) and remained below –100 mV for the duration of the experiment. We expected that mixing of the amendments on the

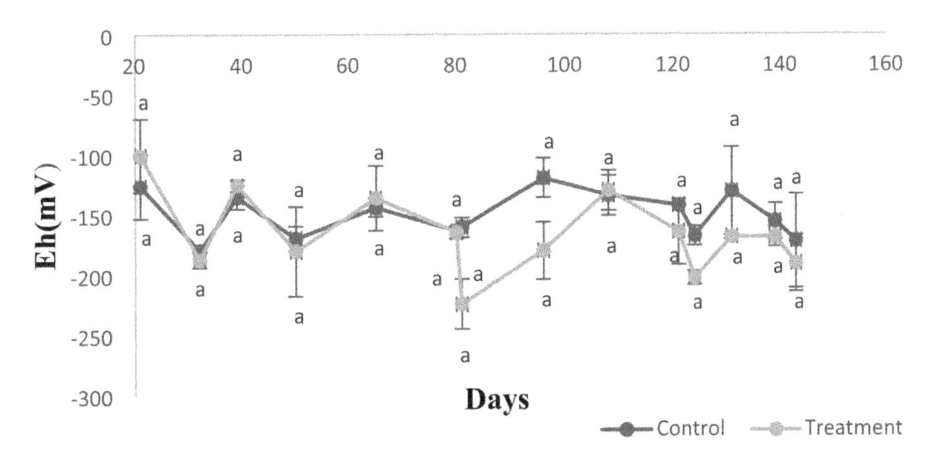

FIGURE 8.3 Redox (Eh) values measured during anaerobic mesocosm incubations. Arrow represents amendments added on the 80th day. Error bars represent +/– one standard deviation based on two replicates. Data labels not connected by the same letter are significantly different $p < 0.05$.

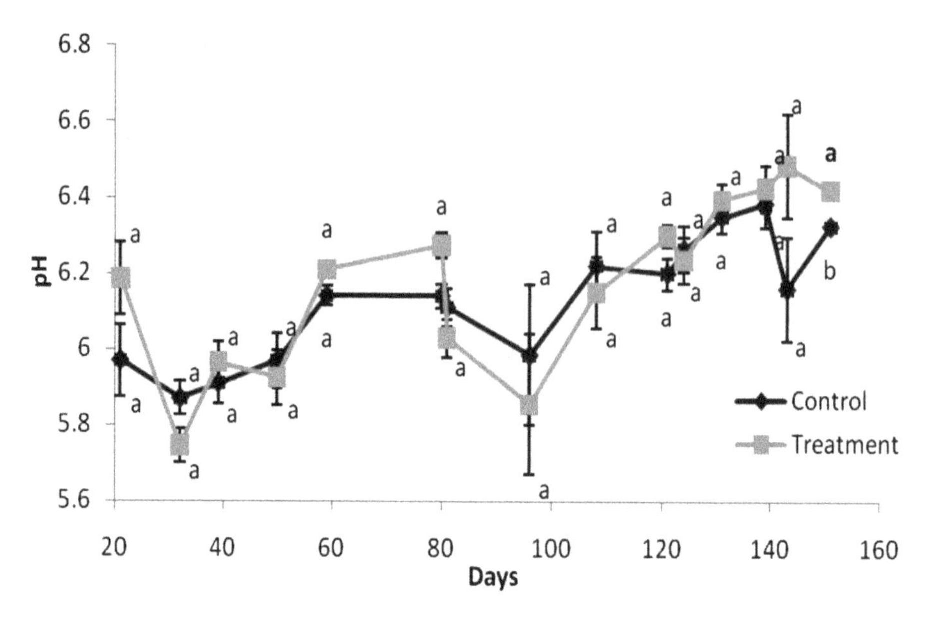

FIGURE 8.4 pH readings during anaerobic mesocosm incubations. Arrow indicates amendments made on the 80th day. Error bars represent +/– one standard deviation. Data labels not connected by the same letter are significantly different $p < 0.05$ based on two replicates.

80th day would incorporate air in the tanks and increase Eh; however, Eh remained low throughout the incubation. A significant decrease in redox potentials was observed in the treatment tank immediately following mixing. This could be because lactate may have distributed into the deeper layers in soils. Increased availability of lactate could have driven the Eh to more negative values. Temperatures fluctuated somewhat throughout the experimental period, although temperatures were similar between control and experimental mesocosms (mean temperatures for control and treatment tanks were 18.1°C and 18.5°C with standard deviation of 2.9 and 3.0, respectively). pH trends (Figure 8.4) were similar between treatment and controls throughout the incubation period, with a significant drop in pH observed following addition of trace elements on the 80th day. Following mixing amendments, the pH gradually increased with a significant difference between control and treatment tanks observed by the end of the experiment (Figure 8.4).

DOC contents were measured in pore waters for each of the samples (Figure 8.5). DOC concentrations increased with incubation time, approximately following the observed decreases in redox potential. DOC for treatment tanks is consistently higher than for control tanks, likely due to the higher concentrations of Na^+ in the treatment tanks. Following a peak in DOC concentrations on the 39th day, DOC concentrations dropped significantly in both control and treatment tanks. This drop in DOC concentrations may be related to a concomitant drop in temperature, which may have decreased production of DOC. DOC concentrations remained relatively constant until the 121st day in incubation, when increases in pH and DOC were observed.

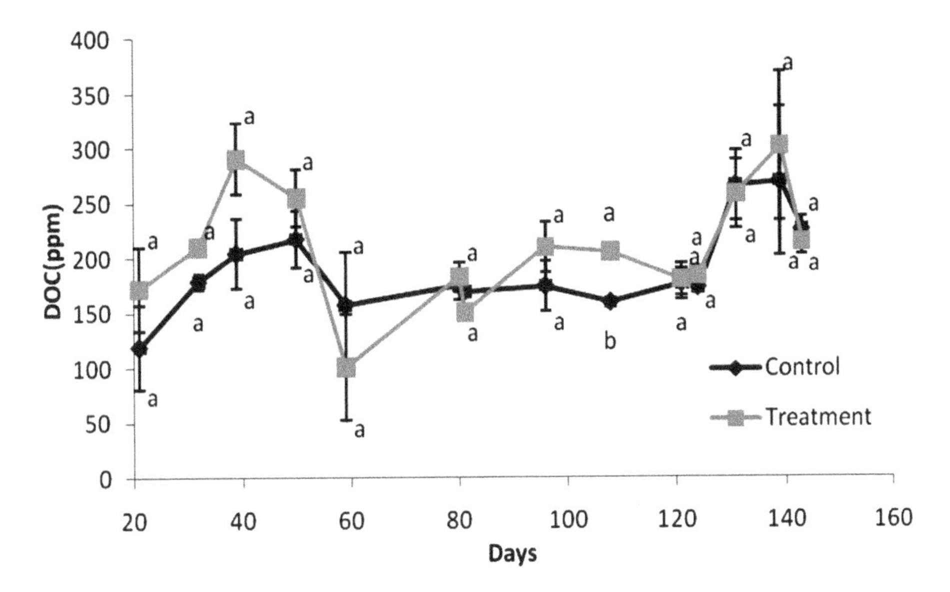

FIGURE 8.5 Dissolved organic carbon (DOC) during anaerobic mesocosm incubations. Arrow indicates amendments made on the 80th day. Error bars represent +/– one standard deviation based on two replicates. Data labels not connected by the same letter are significantly different $p < 0.05$.

The concentrations of potential lactate fermentation products in the measured DOC were determined to distinguish these products from DOC produced from the more complex soil organic matter (SOM) (Figure 8.5). Since there were two peaks with significant release of DOC, samples taken on the 39th and 139th days were chosen for analysis. Results from these days are presented in Figures 8.6 and 8.7,

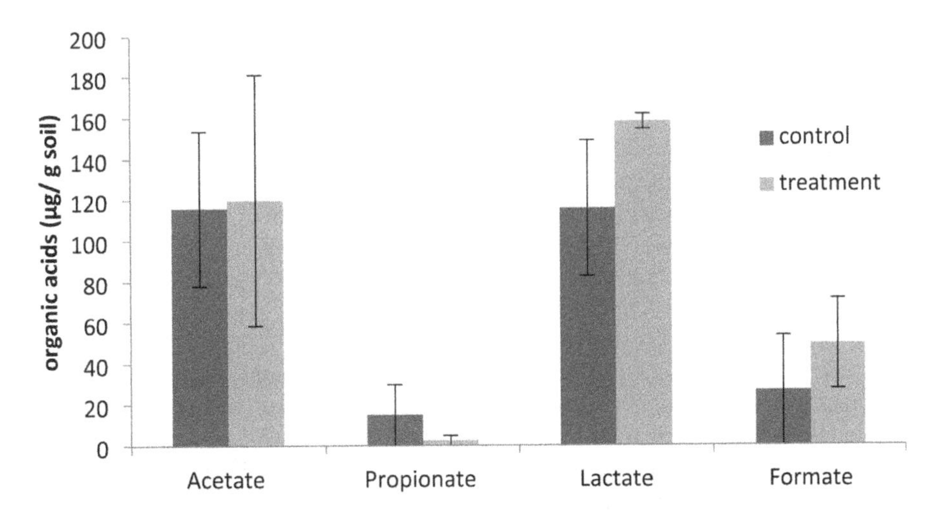

FIGURE 8.6 Fatty acids (µg/g dry soil) in control and NaCl amended (treatment) mesocosm at day 39.

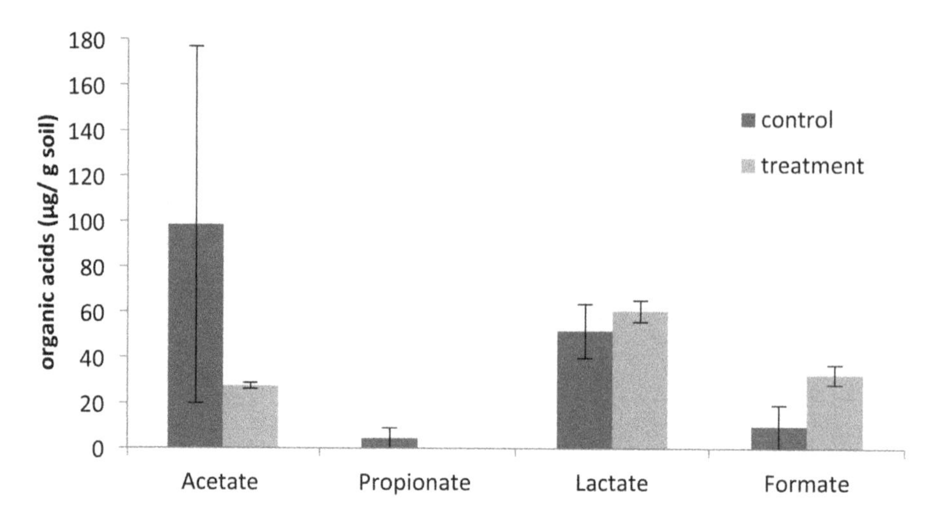

FIGURE 8.7 Fatty acids (µg/g dry soil) in control and NaCl amended (treatment) mesocosms at day 139.

respectively. The dominant short chain fatty acids observed were acetate, lactate and formate, with no butyrate or isobutyrate detected in either treatment or control samples. To check if the dominant DOC was derived from SOM, we calculated the percentage of lactate contributing to DOC, the values for which are as follows: control and treatment tanks on the 39th day accounted for 10% and 8% respectively, while values for 139 days were 6% and 4%, respectively. These values are considered fairly negligible, which suggests that the lactate amended to the tanks was used up very quickly and that the DOM released originated from anoxic incubations in soils.

DDx concentrations in soils followed similar general trends in both controls and Na^+ tanks prior to addition of Na^+ (Figures 8.8–8.11). Prior to Na^+ addition, these tanks may be regarded as replicates because both contained lactate (10 mM) as the carbon and energy source and DDx as potential TEAs. DDx concentrations increased during the initial month approximately for the first 40 days in incubation, which correlates with the "release phase" in the case of lactate mesocosms (Gohil et al., 2014). It should be noted that the peak in DDx concentrations observed on the 39th day corresponds with the peak in DOC concentrations (Figure 8.5), suggesting that release of DOC increases desorption and bioavailability of DDx (Kim et al., 2008). The peak in DDx concentration is followed by a decrease in both the control and treatment tanks. This decrease is in accordance with the "degradation phase" in anoxic lactate mesocosm studies. Furthermore, these studies again support the hypothesis that the system is carbon limited and supplementation with lactate leads to rapid biological degradation.

It is possible that lactate was fermented by syntrophic bacteria to H_2 and organic acids such as formate and acetate, which may have served as electron donors to the dechlorinating population for DDx degradation. The concentrations of all the DDx went to less than 10 mmol/g dry soil by the 50th day and remained

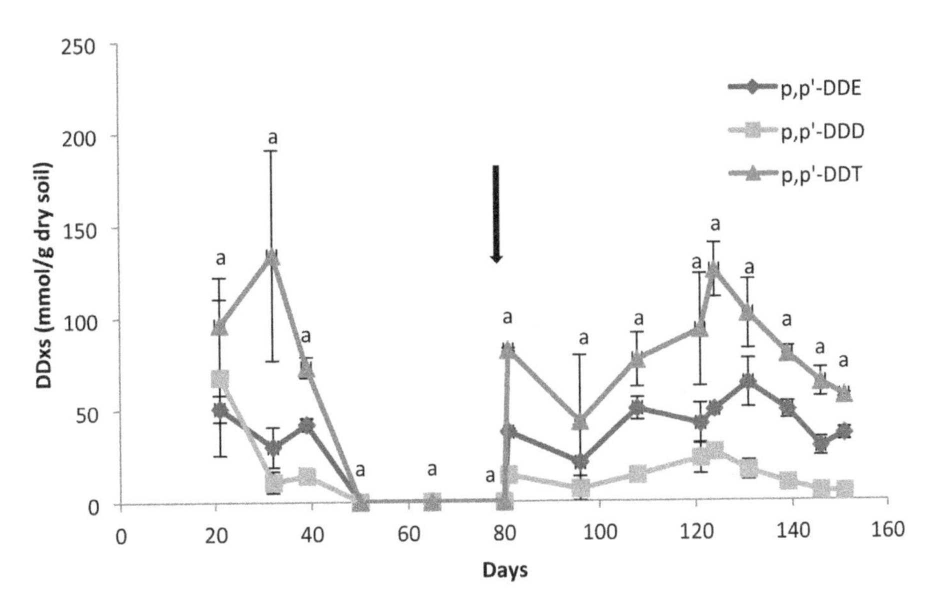

FIGURE 8.8 *p,p′*-DDx concentration in control mesocosms. The arrow at 80 days indicates when trace metals were mixed into the soil. To simplify, the figure data labels are added only in case of *p,p′*-DDT where there was a significant difference between control and treatment. Data labels not connected by the same letter are significantly different $p < 0.05$. Error bars represent +/– one standard deviation based on two replicates.

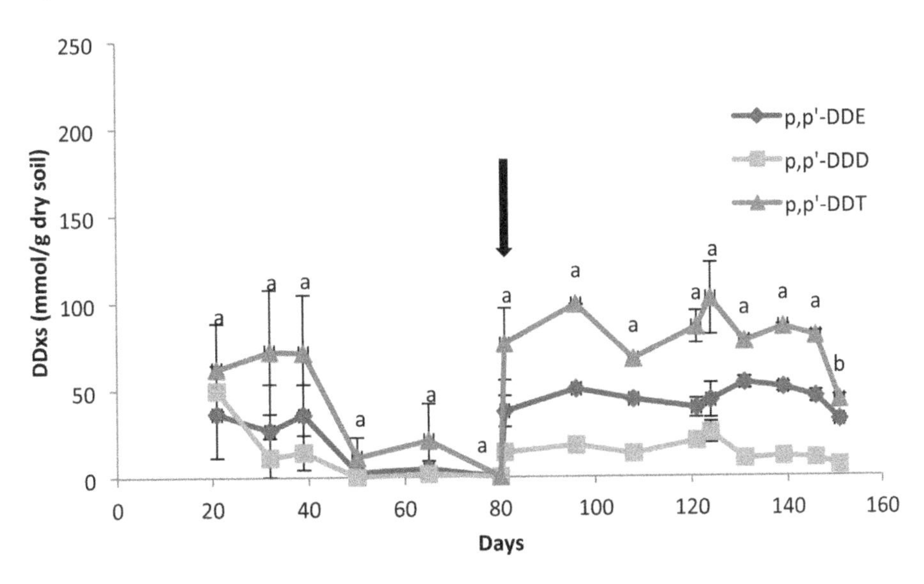

FIGURE 8.9 *p,p′*-DDx concentration in treatment (Na^+) mesocosm soils. The arrow at 80 days indicates when trace metals were mixed into the soil. To simplify, the figure data labels are added only in case of *p,p′*-DDT where there was a significant difference between control and treatment. Error bars represent +/– one standard deviation based on two replicates. Data labels not connected by the same letter are significantly different $p < 0.05$.

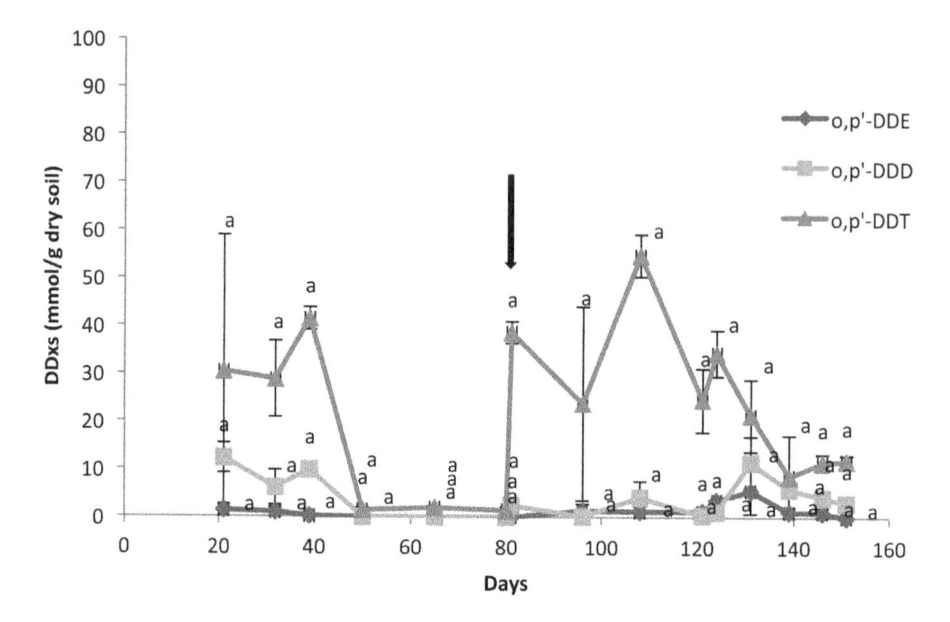

FIGURE 8.10 *o,p′*-DDx concentration in control mesocosm soils. The arrow at 80 days indicates when trace metals were mixed into the soil. Error bars represent +/– one standard deviation based on two replicates.

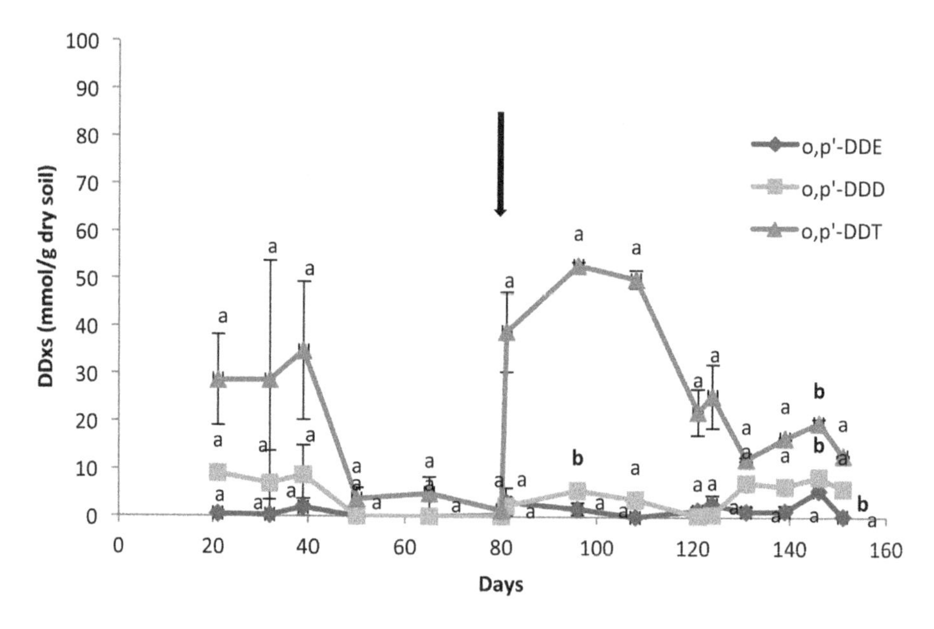

FIGURE 8.11 *o,p′*-DDx concentration in treatment (Na+) mesocosm soils. The arrow at 80 days indicates when trace metals were mixed into the soil. Error bars represent +/– one standard deviation based on two replicates. Data labels not connected by the same letter are significantly different $p < 0.05$.

below that until amendments were made on the 80th day (Figures 8.8–8.11). On the 80th day, Na^+ and trace metal solutions were manually mixed into the treatment tanks, while the control tanks received only trace metal solutions. Following mixing, DDx concentrations increased immediately on the 81st day in both control and treatment tanks.

Mixing may have had several effects on measurable DDx concentrations, including (1) perturbation of soils, which could make the otherwise inaccessible soils available for sampling; and (2) dispersion and homogenization of DOC in the more anoxic or inaccessible zones. Both homogenization of soils and DOC dispersion might have increased the extractability by releasing the bound DDx and made the otherwise inaccessible sites now available for sampling. Since DDx concentration prior to mixing was lower than the detectable limits, it might suggest the presence of DDx degrading consortia. When the DDx concentration increases were observed, trace metals and carbon source (lactate) may have facilitated the degrading consortia.

Analysis of time points toward the end of the experiment shows a significant difference between the control and treatment tanks. Concentrations of all *o,p′*-DDx metabolites (*o,p′*-DDE, *o,p′*-DDD and *o,p′*-DDT) measured in treatment tanks were significantly lower ($p < 0.05$) from the control tanks by the 146th day (Figures 8.9 and 8.10). *p,p′*-DDT was significantly lower in Na^+ amended than in the control tanks by the 151st day (Figures 8.8 and 8.9). Analysis of further time points would have been useful to determine if the decreases in DDx concentrations would continue to be significant or drop below detectable levels, as was observed during the first phase of incubation (prior to mixing amendments 50th to 80th days); however, the experiment was discontinued due to time constraints. Significantly, the observed increases in DDx in both phases match increases in DOC concentrations, both of which precede a significant loss of measurable DDx.

8.5 CONCLUSIONS FOR ANAEROBIC MESOCOSMS

Data presented in Figures 8.8–8.11 indicate loss of the parent and metabolites to less than 10 mmol/g dry soil in all mesocosms within 50 days of incubation. This is in accordance with anoxic lactate mesocosms previously reported (Gohil et al., 2014), strongly suggesting that degradation of DDx is limited by the available carbon. Following amendments, the observed trends indicated a repeated increase followed by a decrease in DDx concentrations. No significant difference was observed between the control and Na^+ treatment tanks, perhaps due to the relatively high variability in some of the data points. Some significant difference was observed between treatment and control tanks with respect to both *o,p-′* and *p,p′*-DDx metabolites.

Greater extraction and degradation of DDx under anaerobic conditions can be explained by earlier studies demonstrating the importance of DOC in increasing the bioavailability of hydrophobic compounds (Kim and Pfaender, 2005; Pravacek, 2005; Kim et al., 2008). These studies showed that anaerobic incubations stimulated the release of bound hydrophobic residues making them more soluble and, hence, more bioavailable. Following the addition of trace metals and Na^+, an increase in DDx concentrations was observed, indicating that as the availability of DDx increased lactate provided energy requirements for the degrading population. Anoxic Eh promotes DOC release (Pravacek, 2005; Kim et al., 2008) and DOC increases the solubility (and hence bioavailability) of DDx. This is in accordance with the DOC data (Figure 8.5), in which higher concentrations of DOC coincide with higher concentrations of DDx, and precede more rapid decreases in DDx concentrations.

8.6 SUMMARY

Na^+ addition may be an inexpensive, safe, sustainable and feasible alternative to the expensive chemical and physical remediation methods. Low concentrations of NaCl increased the biodegradation of aged DDx in microcosms possibly due to the dispersal of soil polymers, resulting in increased bioavailability. Na^+ may have the potential to remove such HOCs that have been present for long periods of time in organic soils. Another advantage is that the technique, being environmentally friendly, may lead to clean up of otherwise contaminated sites. More research is needed to confirm its applicability.

REFERENCES

Albert, D., Martens, C. S. 1997. Determination of low-weight organic acid concentrations in sea water and pore water samples via HPLC. *Mar. Chem.* 56:27–37.

Alexander, M. 1995. How toxic are toxic chemicals in soil? *Environ. Sci. Technol.* 29:2713–2717.

Alexander, M. 1997. Sequestration and bioavailability of organic compounds in soil. In: Linz, D. G., Nakles, D. V. (eds.), *Environmentally Acceptable Endpoints in Soil.* American Academy of Environmental Engineers, Annapolis, MD.

Boul, H. L. 1995. DDT residues in the environment–a review with a New Zealand perspective. *New Zealand J. Agric. Res.* 38:257–277.

Brady, N. C., Weil, R. R. 2003. *The Nature and Properties of Soil.* 13[th]ed., Pearson Prentice Hall Chapter, New York, 10:363–410.

Chiou, C. I., Malcolm, R. L., Brinton, T. I., Kile, D. E. 1986. Water solubility enhancement of some organic pollutants and pesticides by dissolved humic and fulvic acids. *Environ. Sci. Tech.* 20:502–508.

DeBusk, W. F., Reddy, K. R. 1998. Turnover of detrital organic carbon in a nutrient impacted Everglades marsh. *Soil Sci. Soc. Am. J.* 62:1460–1468.

Gohil, H., Thomas, J., Ogram, A. 2014. Stimulation of anaerobic biodegradation of DDT and its metabolites in a muck soil: laboratory microcosm and mesocosm studies. *Biodegradation.* 25:633–642.

Hubaux, A., Vos, G. 1970. Decision and detection limits for calibration curves. *Anal. Chem.* 42:849–855.

Juhasz, A. L., Megharaj, M., Naidu, R. 1999. Bioavailability: the major challenge (constraint) to bioremediation of organically contaminated soils, in remediation of hazardous waste contaminated soils. In: Wise, D. L., Tratolo, D. J., Cichon, E. J., Inyang, H. I.,

Stottmeister, U. (ed.), *Engineering Considerations and Remediation Strategies, section 1–1: Engineering Issues in Waste Remediation* (2nd ed. Vol. 1). Marcel Dekker, New York, pp. 217–241.

Kantachote, D., Naidu, R., Singleton, I., McClure, N., Harch, B. D. 2001. Resistance of microbial population in DDT-contaminated and uncontaminated soils. *Appl. Soil Ecol.* 16:85–90.

Kantachote, D., Singleton, I., McClure, N., Naidu, R., Megharaj, M., Harch, B. D. 2003. DDT resistance and transformation by different microbial strains isolated from DDT-contaminated soils and compost materials. *Compost Sci. Util.* 11:300–310.

Kantachote, D., Singleton, I., Naidu, R., McClure, N. C., Megharaj, M. 2004. Sodium application enhances DDT transformation in a long-term contaminated soil. *Water Air Soil Poll.* 154:1–4.

Kim, H. S., Lindsay, K. S., Pfaender, F. K. 2008. Enhanced mobilization of field contaminated soil-bound PAHs to the aqueous phase under anaerobic conditions. *Water Air Soil Poll.* 189:135–147.

Kim, H. S., Pfaender, F. K. 2005. Effects of microbially mediated redox conditions on PAH-soil interactions. *Environ. Sci. Technol.* 39:9189–9196.

Luthy, R. G., Aiken, G. R., Brusseau, M. L., Cunningham, S. D., Gschwend, P. M., Pignatello, J. J., Reinhard, M., Traina, S. J., Weber, W. J. Jr., Westall, J. C. 1997. Sequestration of hydrophobic organic contaminants by geosorbents. *Environ. Sci. Technol.* 31:3341–3347.

McCarthy, J. F., Zachara, J. M. 1989. Subsurface transport of contaminants. *Environ. Sci. Technol.* 23:496–502.

Megharaj, M., Kantachote, D., Singleton, I., Naidu, R. 2000. Effects of long term contamination of DDT on soil microflora with special reference to soil algae and algal transformation of DDT. *Environ. Pollut.* 109:35–42.

Nelson, P. N., Oades, J. M. 1998. Organic matter, sodicity, and soil structure. In: Sumner, M. E., Naidu R. (eds.), *Sodic Soils: Distribution, Properties, Management, and Environmental Consequences.* Oxford University Press, Oxford, pp. 51–75.

Ou, L.T., Rothwell, D., Wheeler, W., Davidson, J. 1978. The effect of high 2,4-D concentrations on degradation and carbon dioxide evolution in soils. *J. Environ. Qual.* 7:241–246.

Pravacek, T. L., Christman, R. F., Pfaender, F. K. 2005. Impact of imposed aromatic conditions and microbial activity on aqueous- phase solubility of polycyclic aromatic hydrocarbons from soil. *Environ. Toxicol. Chem.* 24:286–293.

Quensen III, J. F., Mueller, S. A., Jain, M. K., Tiedje, J. M. 1998. Reductive dechlorination of DDE to DDMU in marine sediment microcosms. *Science.* 280:722–724.

Ratcliffe, D. A. 1967. Decrease in eggshell weight in certain birds of prey. *Nature.* 215:208–210.

Schmeck, T., Wenclawiak, B. W. 2005. Sediment matrix induced response enhancement in gas chromatograph-mass spectrometric quantification of insecticides in four different solvent extracts from ultrasonic and Soxhlet extraction. *Chromatographia.* 62:159–165.

Scribner, S. L., Benzing, T. R., Sun, S., Boyd, S. A. 1992. Desorption and bioavailability of aged simazine residues in soil from a continuous corn field. *J. Environ. Qual.* 21:115–120.

Suflita, J. M., Gibson, S. A., Beeman, R. E. 1988. Anaerobic biotransformations of pollutant chemicals in aquifers. *J. Indust. Microbiol.* 3:179–194.

Sumner, M. E., Naidu, R. (ed.) 1998. *Sodic Soils.* Oxford University Press, Oxford.

Tanner, R. S. 2007. Cultivation of bacteria and fungi. In: Hurst, C. J., Crawford, R. L., Garland, J., Lipson, D., Mills, A., Stetzenbach, L. (eds.), *Manual of Environmental Microbiology* (Chap. 6, 3rd ed.). ASM Press, Washington, DC.

Turusov, V., Rakitsky V., Tomatis L. 2002. Dichlorodiphenyltrichloroethane (DDT): ubiquity, persistence, and risks. *Environ Health Persp.* 110:125–28.

United States Environmental Protection Agency (U.S. EPA). 1979. Final environmental impact statement for Lake Apopka restoration project. Lake and Orange counties Florida. EPA 904/9-79-43. U.S. EPA. 443.

White, R. E. 1997. *Principles and Practice of Soil Science: The Soil as a Natural Resource*, 3rd ed. Blackwell Science Ltd., Carlton, p. 348.

World Health Organization (WHO). 1979. DDT and its derivatives: Environmental health criteria 9. World Health Organization, United Nations Environment Programme.

Wood, M. 1995. *Environmental Soil Biology*, 2nd ed., Chapman and Hall, Tokyo, p. 150.

9 Use of Corncob Residues (*Zea mays*) and Activated Carbons Obtained from This Material for the Decontamination of Aqueous Media with Ni^{2+}

Rafael Alberto Fonseca-Correa,
Ronal Orlando Serrano-Romero,
Juan Carlos Moreno-Piraján, and*
Liliana Giraldo

CONTENTS

* Corresponding author: Email: jumoreno@uniandes.edu.co

9.1 INTRODUCTION

In recent years, the accelerated industrial growth has generated an increase in pollutants in solid, aqueous and gaseous media. Besides being a serious threat against environmental sustainability, these pollutants directly affect the natural wealth of the world and the quality of life of its inhabitants.

Several processes to remove heavy metals from aqueous solution have been developed and used at an industrial level. These processes include ion exchange, adsorption, reverse osmosis, chemical precipitation/sedimentation, filtration, electrolysis and flotation. Chemical precipitation is the most widely used process, but it produces a large quantity of sludge that then has to be deposited in confinements of toxic waste. On the other hand, the precipitated metals can't be economically recovered from the mud. Therefore, it is necessary that new processes are developed to remove metals from aqueous solutions in a way that the removed metals could be recovered.

Nickel is thought to be essential in animals in trace quantities, although its importance in human metabolism has not been proven. It is considered harmful in high concentrations. Nickel is known for its participation in the activity of some enzymes (ureases), hormone activation and stability of macromolecules [1–3].

The electroplating industry is one of the sources of nickel contamination. Electroplating is an electrochemical process by which a thin layer of metal is generally deposited on a metallic or plastic base. The objects are galvanized to avoid corrosion, to obtain a hard surface or an attractive finish, to purify metals (as in the electrolytic refining of copper), to separate metals for quantitative analysis, or to reproduce a cast (as in the case of electrotyping). The metals normally used in electroplating are cadmium, chromium, copper, gold, nickel, silver and tin. Silver cutlery, chrome car accessories and tinned food containers are typical products of electroplating.

Specifically, nickel is used as a protector and as an ornamental coating on plastics and metal parts, especially those that are susceptible to corrosion such as iron and steel. A nickel plate is deposited by electrolysis of a nickel solution. Another application of nickel is in catalytic hydrogenation processes, where the presence of nickel is 1% with respect to the amount of oil to be hydrogenated [3–5].

Nickel is mainly used in alloys and provides hardness and resistance to corrosion in steel. Some steels contain between 2% and 4% nickel, which are used in automotive parts such as shafts, crankshafts, gears, keys and rods; in spare parts for machinery; and in plates for shielding, given the mechanical properties that the alloy provides. Some of the most important nickel alloys are German silver, invar, monel and permalloy. The nickel coins in use are an alloy of 25% nickel and 75% copper.

The presence of traces of heavy metals is mainly due to the industrial activity. A second factor is the release of metals from minerals caused by their drag or dilution in water. Elements such as arsenic, cadmium, copper, iron, nickel, lead, mercury, chromium and zinc are potentially dangerous to health, and therefore they are subject to control [5].

These metals also appear in traces in fertilizers, and given a low mobility, they accumulate in the superficial horizons of the soil, where the roots of the plants are located.

The absorption of nickel by ingestion of its salts causes severe gastroenteritis, tremors, choreic movements, and paralysis. Continuous contact with the skin or inhalation of carbonyl vapors causes local irritation, asthma pneumoconiosis and sensitizing action. In addition, it causes systemic involvement; induces erythrocytosis, growth retardation and hyperglycemia; and leads to degenerative disorders of the heart, brain, lungs, liver and kidney. Nickel is among the metals that block the entry of calcium.

A major problem with the use of nickel salts in industry is its ability to produce dermatitis. This condition, called scabies or nickel eczema, appears especially in workers engaged in nickel plating [2–5].

An increased risk of cancer of the nasal fossae (ethmoid sinus) and of the lungs has been detected in the workers of the factories, especially those dedicated to the first stages of refining. The carcinogenic risk seems to be associated with exposure to poorly soluble compounds, such as nickel bisulfite and nickel oxide. There may be a higher risk of cancer of the larynx, but the implementation of safety standards and the use of protective implements have considerably reduced the number of cases.

The concentration is proportional to the duration of the exposure—considering its long biological half-life (3.5 years), its concentration remains high even after the exposure has ended.

Exposure of rats for 78 weeks (5 days/week, 6 hours/day, to an aerosol of nickel sulfate: 0.97 mg/m^3, 70% of particles with a diameter less than 1μm) increases the incidence of reactions hyperplastic and neoplastic in the bronchial and alveolar epithelium [1, 2].

In the last 40 years, adsorption on activated carbon has been successfully used for the treatment of municipal and industrial wastewater and is considered the best available technology to eliminate nonbiodegradable and toxic organic compounds present in aqueous solution. Activated carbon is considered the universal adsorbent because it is very versatile due to its large surface area, pore structure, high adsorption capacity and very extensive reactive surface.

It has recently been shown that activated carbon can be successfully applied in the removal of heavy metals present in aqueous solution. Metals that are highly adsorbed on activated carbon are chromium, cadmium, lead, nickel, gold, mercury and copper.

Several studies on the adsorption of nickel from aqueous solution on activated carbon are reported in the literature. In these studies, a variety of activated carbons (powdered, pelletized and granular) synthesized from different carbonaceous sources have been used. It is well known that the adsorption of metal ions depends on the starting material, the treatment to which it is subjected to obtain the activated carbon and finally its acid-base and surface charge properties.

Thus, during the design and optimization of fixed beds or those packed with activated carbon that can be used to adsorb divalent nickel, it is necessary to know the adsorption capacity from the adsorption isotherms and the adsorption speed from the kinetic studies. Further investigation and studies are required in this area.

The objective of the present study is to investigate the adsorption, adsorption capacity and adsorption kinetics of the Ni (II) ions present in aqueous solutions on different types of porous solids synthesized from the corncob (*Zea mays*), as well as to analyze the effect of pH, activation time (for coals) and some thermal effects with immersion calorimetry.

9.2 JUSTIFICATION

There is currently growing worldwide interest in finding efficient methods to treat industrial wastewater pollutants, mainly heavy metals, among many other pollutants present, in which the main objective is to reduce the concentration levels of these pollutants. One of the most important aspects of these investigations is to try to retain these contaminants with materials of high adsorption capacity, with which constant investigations have been carried out with the purpose of determining the adsorption and/or degradation capacities of these contaminants in water. Activated carbons are chosen on the basis of their selectivity, low cost in processing and their compatibility with the environment to retain heavy metals.

As the adsorption capacity of different types of materials depends on the nature and the synthesis process, processes will be carried out to obtain various materials and analyze how these materials are affected in terms of their adsorption capacity.

The corncob (*Zea mays*) is a natural adsorbent, which by its very structure is capable of retaining heavy metals. When treated chemically and/or physically, its structure and surface chemistry change, so that its adsorption capacities in aqueous solution also change according to these parameters. The present work presents the synthesis and modifications of activated carbons to analyze the adsorption capacity of divalent nickel.

9.2.1 Water Contamination

The rational use of water is a fundamental priority. The availability of water in quantity, quality and opportunity is an indispensable requirement for the welfare of the population and economic development. Without this vital resource, the existence of adequate levels of health, agricultural and industrial production and food cannot be conceived. Likewise, water is an essential element to conserve biological diversity, climate stability and quality of life [6].

Technological and industrial development has contributed to the pollution of water. The effluents that the industries generate contain a great variety of pollutants, such as heavy metals (chromium, copper, cadmium, zinc, nickel, mercury, lead and arsenic, among others), in addition to some organic and inorganic compounds that are generally considered toxic substances. These pollutants deserve special attention, since some of them are carcinogenic or may cause other types of damage to the health of living beings, especially human beings. The increase in the levels of toxic heavy metals that are discharged into the aquatic environment from industrial wastewater represents a serious threat to human health, organisms and ecosystems in general. Even when there are different sources of this type of pollutants, the industrial sector is the one that contributes the most with these pollutants, such as nickel.

9.2.2 Heavy Metal Pollution

The environmental problems have acquired global attention with the continuous increase of the contamination of the air, soil and hydric resources. In the list of the most important groups of contaminants are the heavy metals. Due to the limitation of water resources, the removal of heavy metals from wastewater is taking on special importance. Heavy metals are a series of chemical elements that have a relatively high density and some toxicity for humans. Heavy metals are those that have a density at least five times higher than that of water. They have direct application in various industrial processes. The most important are chromium (Cr), cobalt (Co), nickel (Ni), lead (Pb), zinc (Zn), and mercury (Hg).

The term heavy metal refers generally to any metallic chemical element that has a relatively high density and is toxic or poisonous in low concentrations. Heavy metals are natural components of the earth's crust that cannot be degraded or destroyed. To a small degree, they are incorporated into our bodies via food, drinking water and air. In trace amounts, some metals (such as sulfur copper, selenium and copper, among others) are essential for the metabolism of the human body. However, in higher concentrations they can cause poisoning. Heavy metals are considered among the most problematic pollutants in water systems, since they are not normally eliminated through natural processes like organic pollutants. Toxic metals such as Hg, Cd, Cr and Ni, among others, tend to accumulate in organisms causing severe and acute poisoning [7].

The contamination by heavy metals comes mainly from metallurgical processes, battery production, electroplating, extractive metallurgy, tannery industry, dye factories, textile industries, etc.

9.2.3 Nickel Chemistry

9.2.3.1 Nickel Properties

Nickel belongs to the group of transition metals, also called transition elements. This group of chemical elements to which nickel belongs is located in the central part of the periodic table, specifically in block d. One of the characteristics of nickel and the rest of transition metals is that their electronic configuration shows d-orbitals partially filled with electrons. Other properties of these metals include high hardness, high boiling and melting points and high electrical and thermal conductivity. Nickel exists in solid state in nature (ferromagnetic). The magnetic properties of nickel actually make it a very important material for making computer hard drives.

9.2.3.2 Uses of Nickel

Nickel is used in many products. Some examples are electric guitar strings, magnets and rechargeable batteries. The most common application of nickel is its use as an ingredient in steel and other common metal products such as jewelry. Nickel is bonded to iron in an alloy to make stainless steel. Stainless steel has many applications. It is used in kitchen utensils, cutlery, tools, surgical instruments, storage tanks for firearms, car headlights, jewelry and watches. Various types of tin cans are made using nickel alloyed with other metals. Alloys resistant to nickel heat and electricity can also be made. Nickel can be added to other metals, for example with cobalt, to form superalloys. Nickel is also used in a process known as fire testing. This process helps in identifying the types of compounds in a mineral, metal, or alloy. Nickel is able to collect all the elements of the platinum group in this process. It also partially collects gold. In chemistry, nickel is normally used as a catalyst for hydrogenation reactions.

9.2.3.3 Effects of Nickel on Health

Nickel is an element found in the environment only in very small concentrations. Humans can be exposed to nickel through air, water, food, or cigarette smoke. Foods naturally contain small amounts of nickel. Chocolate and fats are known to contain higher amounts of nickel. Its concentration increases in the body when vegetables grown in contaminated soil are consumed in large amounts. It is known that plants accumulate nickel, and as a result nickel intake through vegetables is an important topic of discussion. People who smoke breathe in nickel through their lungs. Nickel is also found in detergents. Skin contact with soil or water contaminated by nickel can also result in exposure to nickel. Nickel is essential in trace quantities, but when it is taken in very high quantities, it can be dangerous for human health. Some of the detrimental effects of high amounts of nickel are as follows:

- High chances of developing lung, nose, larynx and prostate cancer
- Illness and dizziness after exposure to nickel gas
- Lung embolism
- Respiratory failures
- Birth defects
- Asthma and chronic bronchitis
- Allergic reactions such as rashes, mostly with jewelry
- Heart disorders

9.2.3.4 Environmental Effects of Nickel

Nickel is released into the environment by power plants and garbage incinerators. It is deposited on the ground or falls after reacting with raindrops. It usually takes a long period of time for nickel to be removed from the air. Nickel can also end up on the surface of water when it is part of the wastewater. Most of the nickel compounds that are released into the environment are absorbed by sediments or soil particles and become immobilized. In acidic soils, nickel binds to become more mobile and often reaches groundwater.

It is known that high concentrations of nickel in sandy soils can clearly damage plants, and high concentrations of nickel in surface water can decrease the range of growth of algae. Microorganisms can also suffer a decrease in growth due to the presence of nickel, but they usually develop resistance to the metal. Animals and humans are also affected by high concentration of nickel in the environment, as discussed earlier. Those living near refineries are more likely to be affected by nickel contamination. It is not known that nickel accumulates in plants or animals [1–2].

9.2.4 Adsorption of Compounds on Porous Solids

Elimination of compounds from liquids can be performed using physicochemical, chemical, physical and biological processes. Of all these processes, the physicochemical process of adsorption is most commonly used, which utilizes adsorbents such as silica, zeolites and activated carbon, among others, whose demand has increased in recent years; in the United States there is an average annual increase of 6.6% for use in water treatment [8].

Adsorption capacity is the most important property of adsorbents; the value of this parameter determines the amount of water that can be treated per unit mass or volume. The adsorption capacity is related to the surface area, porosity and surface chemistry of the porous solid [9].

The surface chemistry of these solids is often an important factor that controls the adsorption of certain organic compounds in water and depends on the content of heteroatoms of the surface. The oxygen content mainly determines the amphoteric character of these materials that comes from the existence of superficial acidic and basic sites that are separated and therefore are not coincident.

When a solid is immersed in an aqueous solution, an electric charge is generated on the surface by dissociation of surface functional groups, or by the adsorption of ions from the solution. The surface charge depends on the pH of the solution and the characteristics of the solid. It is necessary to know the surface characteristics of the adsorbents to be used, such as surface area, porosity and surface chemistry. The study of the adsorption capacity of the adsorbents is carried out by means of the isotherms of adsorption of the ions under varying conditions of pH, ionic strength and defined temperatures. The adsorption isotherm is generally used as a tool to discriminate between different solids [10].

9.2.5 Activated Carbon

The term "activated carbon" (also active carbon) is applied to a group of porous carbons prepared by the reaction of a carbonized material with oxidizing gases or by carbonization of lignocellulosic materials impregnated with chemical dehydrating agents. All these carbons, which structurally are very disordered solids constituted mainly by carbon, have a high degree of porosity and a high internal surface area, applied, mainly, in adsorption and catalysis processes. More specifically, the main application is the elimination of impurities from gases and liquids through adsorption. The surface of the coal can attract molecules from gas and liquid phases by van der Waals dispersion forces, which cause a higher concentration to be produced on the said surface than in the fluid. Given the fairly inert nature of the carbon surface, the affinity for low-molecular-weight molecules such as nitrogen or oxygen at room temperature or polar molecules such as water is very low. However, the affinity for apolar molecules of a certain volume and molecular weight (for example, hydrocarbons) will be high. This difference in affinity causes the activated carbon to be the preferred adsorbent whenever a separation/purification must be carried out in the presence of moisture or in aqueous solution.

Although activated carbon is considered the universal adsorbent, due to the variety of applications to which it can be applied, the number of applications has been growing continuously in recent years, as progress has been made in controlling its properties. Traditionally, it was considered that porosity and surface area were the parameters that defined quality, but today it is known that the surface chemistry of activated carbon plays a very important role in its adsorption capacity [6–10]. In fact, porosity is the necessary condition for activated carbon to perform its adsorbing mission, but it is not enough, and in many cases adsorption is not possible if its surface is not modified properly to be able to interact specifically with the compound to be adsorbed. In addition, the application of activated carbons to new technological processes requires much more sophisticated materials than the classic coals in powder or granular form. This has led to the development of new physical forms such as fibers, fabrics, felts, monoliths, etc. Moreover, the need has arisen to prepare coals that have the ability to separate molecules according to the size or molecular form, the so-called carbon molecular sieves, which play an important role in purification and catalysis processes.

9.2.5.1 Structure

All activated carbons, regardless of the physical form in which they are manufactured, are made up of carbon, usually with a relatively low number of heteroatoms (mainly hydrogen and oxygen) and inorganic components that constitute what are called ashes (which depend on the material precursor). However, the main characteristic of activated carbon is the porous structure.

The activated carbon structure, which is based on completely disordered graphene layers, corresponds to that of a non-graphitizable material (i.e., a material that has not passed through a fluid state during the carbonization process). The macromolecular or polymeric structure of the precursor remains during the thermal treatment, because a crosslinking occurs which prevents fusion, with only the loss of small molecules by pyrolysis. This crosslinking is the one that leads to a rigid structure, with very little mobility, which avoids the ordering that occurs during the thermal treatment process (e.g., graphitization), and giving rise to a porous structure [11]. The pyrolysis that takes place throughout the carbonization process produces the loss of heteroatoms such as hydrogen, oxygen and nitrogen in the form of volatile matter; the residual carbon atoms are grouped in stacks of graphene sheets that are crisscrossed at random. Since these layers of graphene are grouped irregularly, they leave gaps or spaces between them, which may be partially blocked by the tars that are produced in the carbonization process; the heat treatment leads to the conversion of these tars into disorganized coal. This disorganized carbon found in the carbonized material will be the first to react with the oxidizing gases during the activation process, so that the pores are free for the adsorption process. The degree to which activation takes place will determine the final porosity of activated carbon, but it is normal for activated carbon to have a wide range of pore sizes, hence its consideration as a universal adsorbent.

Most authors consider that the structure of activated carbon can be represented approximately as shown in Figure 9.1 [11]. This structure resembles a piece of

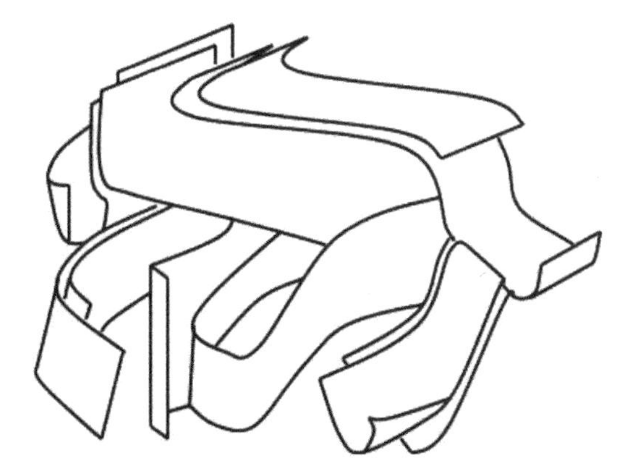

FIGURE 9.1 A schematic model of activated carbon structure [11].

crumpled paper, with spaces between the sheets, which correspond to the smallest pores and which, consequently, are considered to have a slit shape. According to the recommendation of the IUPAC [12], the pores are classified into three main groups: (1) micropores, with dimensions less than 2 nm (a more rigorous classification would further differentiate these micropores into narrow micropores up to 0.7 nm in size, and those between 0.7 and 2.0 nm in size); (2) mesopores, with dimensions between 2 and 50 nm; and (3) macropores, with dimensions greater than 50 nm. Although most of the adsorption that takes place in an activated carbon occurs in the micropores, as a consequence of the adsorption potential created by the proximity of the pore walls, the meso- and macropores play a very important role in the adsorption processes because they serve as access routes for the molecules to the micropores located inside the particles.

The porous structure of an activated carbon is a function of the precursor used in the preparation, the method of activation, as well as the degree of activation achieved. This is the reason why the surface area and the porosity (pore volume) can vary so widely from one activated carbon to another, and the use in a given application is conditioned by the pore size distribution. Thus, while an essentially microporous carbon may be suitable for the adsorption of gases and vapors and, to some extent, for the separation of molecules of different dimensions, a coal with a well-developed additional mesoporosity may be necessary to accelerate the kinetics of adsorption of molecules of greater dimensions like those that are usually found in aqueous solution.

In addition to all the foregoing regarding the porous structure of the activated carbons, the surface chemistry is at least as important with respect to its effect on the adsorbing properties. In a pure carbon such as graphite, the adsorption takes place mainly through the van der Waals forces, and this indicates that the apolar organic molecules will adsorb preferably with respect to polar molecules such as ethanol or water. As stated above, this is an important property of the activated carbon for the adsorption of gases in the presence of moisture. However, the presence of oxygen, hydrogen and nitrogen in the form of functional groups on the surface of the carbon exerts a great effect on the adsorbent properties, especially against polar or polarizable molecules [13].

The presence of these heteroatoms on the surface of the coal can be due to the precursor used in the production, the activation method, or the artificial introduction after the manufacturing process by means of the so-called posttreatments. For example, the unsaturated free radicals on the surface of the carbon (mainly at the edges of the pores) can chemisorb oxygen, giving rise to superficial oxygen groups that are by far the most important from the point of view of the effect on the adsorbent properties.

How are the superficial oxygen groups formed? At the edges of the graphene planes, where there are no more carbon-carbon bonds, unsaturated carbon atoms play an important role in chemisorption. In the case of graphite, the area of the edges is very small in relation to the basal plane (walls of the pores), and therefore the graphite is not characterized by chemically adsorbing oxygen. However, non-graphitizable microcrystalline carbons such as activated carbon have a much more disordered structure and their edges have much more surface area. This leads to a greater possibility of oxygen chemisorption.

Oxidation of the activated carbon surface, which is inherent to production by physical activation (controlled reaction of char with an oxidizing agent), leads to the formation of different types of oxygen surface groups [11]. On the other hand, these surface groups are not formed exclusively by reaction with oxygen, but also by reaction with other oxidizing gases in the gas phase (ozone, carbon dioxide, nitric oxide, etc.) or in an aqueous solution (nitric acid, hydrogen peroxide, hydrogen, sodium hypochlorite, etc.). The chemical nature and the amount of oxygen surface groups of a given carbon are a function of the surface area, particle size, ash content and the experimental conditions used in its manufacture. It is known that there are a variety of surface oxygen groups, ranging from carboxyl to carbonyl groups, phenol, ether, chromene, etc. (Figure 9.2), which gives the charcoal an amphoteric character, so that it can be an acid or a base in aqueous solution.

The ash content and its composition also vary greatly from one coal to another and are functions of the precursor used. They can range from less than 1% for activated carbons prepared from relatively pure precursors to more than 10% when prepared from a mineral coal. Thus, the ash content of an activated carbon prepared from a lignocellulosic material (wood, bone and fruit peel, etc.) usually ranges between 1% and 3%. The main constituents of the ash in this case are usually oxides of silicon, magnesium, calcium, iron, aluminum, sodium, etc. When the precursor is a lignite

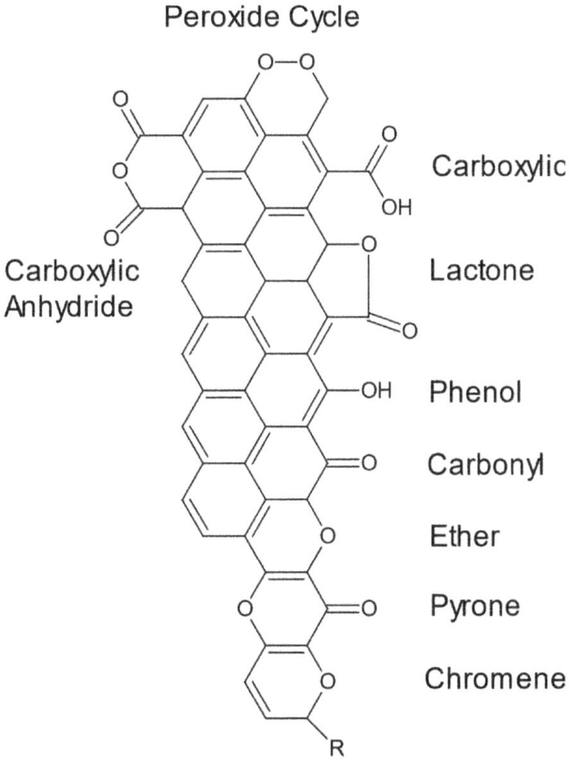

FIGURE 9.2 Types of surface oxygen groups in an activated carbon [11].

or any other mineral coal, the main components of the ash are silicates and alumino-silicates, with smaller amounts of calcium, magnesium, potassium and sodium and traces of zinc, lead, tin, etc. The components of the ashes play a very important role in the adsorption processes since they modify the interactions of the carbon surface with the molecule that is intended to be adsorbed.

9.2.5.2 Characterization

The characterization of an activated carbon is of high importance for both the producer and the user and must cover adsorbent, chemical, physical, mechanical, etc., properties. For this reason, a large number of characterization methods have been developed, which can be found in the specialized literature. Here is an introduction to this characterization, with emphasis on the adsorbent and chemical properties, indicating additionally some of the most common tests used in quality control.

9.2.5.2.1 Porosity

The first stage in the characterization of the adsorbent properties of an activated carbon is, frequently, the determination of the surface area, followed by an analysis of the porosity and the determination of the applicability to a given case. The surface area is usually determined by the adsorption of a gas or vapor under isothermal conditions. The adsorption isotherm is analyzed by applying a given equation to determine the so-called monolayer volume (the volume corresponding to covering the surface of the solid with a layer of adsorbed molecules), from which the specific area of the coal is calculated if the area occupied by each molecule is known. The most commonly used equation to calculate the specific surface area is the BET equation [12], although it is a model that does not adapt exactly to the porosity typical of activated carbons, fundamentally constituted by micropores. The BET equation is usually expressed as:

$$\frac{p/p^0}{n(1p/p^0)} = \frac{1}{n_m C} + \frac{C-1}{n_m C}(p/p^0) \tag{9.1}$$

where n_m is the monolayer capacity, p is the equilibrium pressure of each experimental point, p^0 the saturation pressure of the gas at the adsorption temperature, and C is a constant that is a measure of the enthalpy of adsorption.

The representation of $(p/p^0)/[n(1-p/p^0)]$ in function of p/p^0 should be a straight line with a slope equal to $(C-1)/n_m$, from which the value of n_m is calculated. In the case of activated carbons, the linearity range of the representation of this equation is limited to the relative pressures 0.05–0.15 [13, 14].

The concept of surface area should be handled with care in the case of activated carbons because the application of any of the traditional equations (e.g., BET) can give unreal values (in the bibliography, values higher than 4000 m^2/g are given, when the maximum value that corresponds to a layer of graphite that adsorbs on both sides is 2600 m^2/g), because not only the molecules adsorbed on the walls of the pores are accounted for, but also all the layers of molecules adsorbed inside the said micropores [15]. For this reason, it is advisable to use the term "apparent" or "equivalent" surface.

The characterization of the porosity is very complex in the case of activated carbons because good results are not obtained with techniques such as x-ray diffraction, which is usual in the characterization of inorganic adsorbents such as zeolites. The adsorption of gases and vapors is the most conventional technique for the characterization of activated carbons, although there are others that have been used in recent years such as immersion calorimetry and electron microscopy. The complex and disordered nature of the activated carbon, with pores that are close in dimensions to those of the molecules that are adsorbed, makes the interpretation of the adsorption data difficult. For example, it is known that the adsorption of nitrogen at–196°C, the usual technique for determining the BET surface, can be problematic in coarse with narrow micropores because the low temperature used in the adsorption can hinder the access of nitrogen molecules to pores with dimensions close to 0.4 nm at the normal time used in the adsorption measurement [16–18].

The theory of micropore filling proposed by Dubinin [19], which has been developed in successive stages since 1947, has been successfully applied in the characterization of the microporosity of activated carbons. This theory starts from the principle of the characteristic curve of Polanyi's potential theory. The most usual way to represent this equation is the one that corresponds to that proposed by Dubinin and Radushkevich (DR).

$$Log V = Log V_0 - D(Log(p/p^0))^2 \tag{9.2}$$

where D is a constant related to the free energy characteristic of adsorption and V_0 is the volume of micropores. By representing log V according to (log $p/p^0)^2$ you must obtain a straight line that, when extrapolated, will give the value of V_0.

In many cases the adsorption of several gases and vapors in an activated carbon can be described by a single characteristic curve, which is a demonstration of the validity of the equation. However, there are many activated carbons in which this does not occur, because exponent 2 is used, which is a particular case of Dubinin's theory. Considering the three most common types of adsorption isotherms that can be found in activated carbons, as the isotherm changes in type, the increase in the degree of activation of a common char decreases, thus deviations from the DR equation occur.

Experimental studies of gas adsorption on solids allowed Brunauer, Deming, Deming and Teller (BDDT) [15] to classify the corresponding isotherms into six large groups. Type I isotherms occur when adsorption is limited to some molecular layers. This situation occurs in the chemisorption, where the asymptotic approximation to a limit quantity indicates that all the surface sites are practically occupied. In the case of physical adsorption, type I isotherms are found in microporous solids, whose pore diameters do not exceed a few angstroms, and these constitute approximately the total volume of the pores. Type II isotherms are more frequently found in nonporous solids or solids with pore diameters greater than micropores. The point of inflection of the isotherm occurs when it is close to completing the formation of the first monolayer, and with the increase of the relative pressure, a second layer and still others are formed until the surface saturates.

Type III and V isotherms are characteristic of systems in which adsorbate-adsorbent interactions are weaker than adsorbate-adsorbate interactions. Type IV isotherms are characteristic of mesoporous solids.

To each of these isotherms various theoretical models have been associated that are used in order to provide information regarding the porous structure and the surface area of the solid.

9.2.5.2.2 Surface Chemistry

Although the adsorption capacity of an activated carbon depends on the porous texture and the volume of pores, there are other factors that condition the behavior of a carbon in an adsorption process. Since carbon is essentially composed of carbon, it will selectively adsorb apolar compounds, but the presence of defects in the structure, the inorganic matter of the ash and, especially, of the surface groups has a direct effect on the adsorbent properties, especially toward molecules polar or polarizable. This means that the behavior of an activated carbon cannot be interpreted only as a function of the surface area and the pore size distribution. Therefore, carbons with the same surface area but prepared by different methods can show completely different adsorbent characteristics. It is therefore essential to characterize the surface chemistry of the activated carbons in addition to the porous texture.

The nature and quantity of the inorganic components of the ash vary widely depending on the type of the precursor used in the manufacture and exert a certain effect on the adsorption as a consequence of which it can produce specific interactions with the molecule to be adsorbed. However, the presence of surface groups (especially oxygen, hydrogen and nitrogen) has a much greater effect. The origin of these heteroatoms linked to the carbon structure is in the precursor itself, the activation method used, or in the posttreatment. In the case of oxygen, which is the most important and usual type of group, it is formed fundamentally (although it can partly be in the precursor itself) when the carbon is exposed to the air after its manufacture, by a chemisorption process, or it is introduced artificially, by treatment of the coal with an oxidizing agent. Figure 9.2 shows a diagram of the oxygen functional groups found on the surface of activated carbon, which determine the acidic and basic character. The acidic character is normally associated with surface groups such as carboxyl, lactone and phenol, while the most basic character is assigned to groups such as pyrone, chromene and ether.

Many attempts have been made to unambiguously characterize the oxygen surface groups using a variety of experimental techniques such as acid-base selective titration, infrared spectroscopy, x-ray photoelectron spectroscopy and programmed thermal desorption. Boehm proposed the selective evaluation as a method to characterize the superficial groups of oxygen, taking four bases of different strengths, $NaHCO_3$, Na_2CO_3, NaOH and $NaOC_2H_5$ and hydrochloric acid [20, 21]. Strong acidic surface groups such as carboxylic groups are neutralized with $NaHCO_3$, the weaker base, while NaOH will neutralize the carboxylic and phenolic groups. The presence of intermediate acidity groups can be detected with Na_2CO_3, which can neutralize lactoses and carboxylic acids. Sodium ethoxide, the strongest base is used to obtain quantitative information of all acidic groups, including those of low acidity. The amount of each group in particular can be deducted by difference.

The use of IR spectroscopy is difficult in the case of carbons activated by the high absorption of carbon, which is why it is usually used with a diffuse reflectance cell. In spite of this, the interpretation of the spectra is not easy because each group produces bands at different frequencies that can overlap with those of the other groups. X-ray photoelectron spectroscopy is a surface technique that measures the bond energies of the internal electrons of surface atoms and has been used lately in the characterization of surface groups.

With regard to the basic groups of the activated carbon surface, its characterization is much more confusing than that of the relatively acidic groups. The adsorption of bases on the surface of the coals is normally irreversible, while the ratio of acids is reversible and much lower. Even Boehm went on to show that a graphitized carbon black could adsorb relatively large amounts of acid, even if it did not have superficial oxygen groups. This has led several authors to conclude that the basic character of the activated carbons is due mainly to the π electrons on the surface of the graphene layers:

$$C\pi + 2\,H_2O \rightarrow C\pi H_3O^+ + OH^- \tag{9.3}$$

where $C\pi$ specifies a carbon surface with electrons π [22]. A confirmation is the fact that the amount of acid used in the titration decreases when the oxygen content increases, which is explained by the location of the electrons caused by the high electronegativity of oxygen, which decreases the basicity of the basal plane.

9.2.5.2.3 Manufacturing

Activated carbons are manufactured industrially from carbon-rich precursors, mainly lignocellulosic materials (wood, bone and fruit peel, sawdust, etc.), peat, lignite, sub-bituminous coals, petroleum pitch, coke, etc. The criteria that are usually used to select the precursor are, among others, the following:

- Low content of inorganic components
- Availability and low cost
- Resistance to degradation during storage
- Ease of activation (calcined coke is a very difficult material to activate, while wood is easily activated)
- Good adsorbent quality

The lignocellulosic materials correspond to 47% of the total precursors used worldwide in the manufacture of activated carbon and are followed by mineral coal (around 30%) and peat. Controlled world production is approximately 400000 tons/year, an amount that must be added to China and other countries for which reliable data are not available [8]. The production data are always lower than the consumption data because of the fact that a part of the used coal (especially if it is granular) is regenerated and reused, without the data appearing as production.

Within a type of precursor, e.g., lignocellulosic materials, there may be large differences between the different materials that are used. Thus, a high density is important because it contributes to the performance and mechanical properties of the final carbon, while a high volatile content is important because it can save energy in manufacturing and allows a better control of the activation process and the final performance. Materials with low density such as pinewood, which also contains a high percentage of volatiles, lead to activated carbons with high pore volume and low density, but if the process is modified to decrease the carbon loss during carbonization or if coal is densified, the quality of activated carbon can be increased considerably. On the other hand, coconut husk and fruit bones have a high density and adequate amount of volatiles, so they produce hard granular activated carbons with high pore volume, very suitable for a large number of applications.

There are two industrial methods of producing activated carbon, one called physical activation (also called thermal activation) and another called chemical activation. Although different types of furnaces can be used for the carbonization and activation processes (rotary, multiple hearth and fluidized bed), the most used is the rotary kiln [29].

9.2.5.2.4 Physical Activation

This process involves two well-differentiated stages: carbonization (or pyrolysis) of the precursor and controlled gasification of the resulting char. In the first stage, the elimination of the atomic species other than carbon takes place and a fixed coal mass with a very rudimentary porous structure is produced. The experimental conditions used, such as the heating rate, residence time, final temperature, etc., control the performance of the process, but do not have a marked effect on the porous texture of the char. The carbonization process is usually carried out in rotary kilns or multiple hearth furnaces. Since pyrolysis produces tars, which can remain in the porous texture of the carbonized material, the temperature used (a minimum of 500–600°C) produces a carbonization of the same giving rise to a coal more disorganized than the solid resulting from the carbonization of the skeleton. As a consequence, the adsorbent capacity of the carbonized material is very low and has to be increased by the activation process, by controlled reaction with water vapor, carbon dioxide, or a mixture of both.

In the case where granular activated carbon is desired, the precursor is chopped and sieved to the desired size before carbonization. If the precursor is a mineral coal, it is necessary to include additional stages such as preparing briquettes before chopping and oxidation with air (to prevent the coal from passing through a fluid stage). In the case of pellets, the finely divided precursor (or char) is mixed with a binder and then heated to produce a mass that can be extruded. The resulting pellets are carbonized and activated in the usual way. Although the binder that is added is usually the best-kept secret of each manufacturer, it is known that some of those used are coal tar or wood, lignosulfonic acids, etc.

The carbonization must be followed by a process of activation with the oxidizing gases at temperatures ranging between 800°C and 1000°C, which first eliminate the disorganized carbon coming from the carbonization of the tars and then develop the incipient char porosity. Since the two activating gases commonly used are carbon dioxide and water vapor, the reactions involved are:

$$C + CO_2 \rightarrow 2\,CO \rightarrow \Delta H = +159\ \text{kJ/mol} \quad (9.4)$$

$$C + H_2O \rightarrow CO + H_2 \rightarrow \Delta H = +117\ \text{kJ/mol} \quad (9.5)$$

Both reactions are important, even in the case of the reaction with water vapor, because at high temperatures the displacement reaction that is catalyzed by the carbon surface is in equilibrium:

$$CO + H_2O \rightarrow CO_2 + H_2 \qquad \Delta H = -41 \text{ kJ/mol} \tag{9.6}$$

Since the two activation reactions are endothermic, an external heat input is necessary to maintain the reaction temperature, which is achieved by introducing controlled amounts of air into the furnace to burn the necessary amount of natural gas (when needed) and the gases produced in the activation reactions (CO and H_2) that, on the other hand, are inhibitors of the activation reactions:

$$CO + \tfrac{1}{2}\, O_2 \rightarrow CO_2 \quad \Delta H = -285 \text{ kJ/mol} \tag{9.7}$$

$$H_2 + \tfrac{1}{2} O_2 \rightarrow H_2O \rightarrow \Delta H = -238 \text{ kJ/mol} \tag{9.8}$$

Thus, the combustion of the two inhibitors not only decreases the concentration in the furnace but also increases the partial pressure of the activating agents. On the other hand, the small amount of air introduced into the furnace produces even small combustion of the coal according to the reaction

$$C + O_2 \rightarrow CO_2 \rightarrow \Delta H = -406 \text{ kJ/mol} \tag{9.9}$$

helping to maintain the oven temperature.

In some industries, a mixed activation of carbon dioxide and water vapor is carried out by the introduction of a flue gas stream and steam [8]. Air is not usually used as an activating agent because the reaction with the carbon is too exothermic (the reaction is 100 times faster than with carbon dioxide), producing excessive external burning of the particles and hindering the control of the reaction.

The development of porosity in a carbonized material subjected to activation is a function of the activating agent used. In the early stages of the process, the elimination of disorganized carbon occurs, which is more reactive than the skeleton, which leads to the opening of the partially blocked charred pores and the corresponding increase in the volume of micropores. Then the pore volume increases with the degree of activation.

9.2.5.2.5 *Applications in Liquid Phase*

The adsorption of organic and inorganic compounds in aqueous solution by means of activated carbon is one of the most important applications. Coals used in the liquid phase usually have a wider porosity than those used in the gas phase, to facilitate the faster diffusion of liquids to adsorption centers. At the same time, the larger pore size allows the adsorption of molecules of larger dimensions, more common in solution. The fact that the adsorption is in the liquid phase also affects the particle size that must be used, since it must be considered that the smaller the size, the easier it will be to access the centers where the adsorption takes place. However, a small particle size produces a pressure drop that makes the real process difficult if working with a fixed bed. In addition, the use of activated carbon powder requires a filtration process at the end of the adsorption. Finally, powdered activated carbon is usually used in batch operations, while granular or pellet-shaped carbons are used in continuous operations.

Although the selection of an activated carbon for a given application in the liquid phase is done following the same pattern as in the case of gas phase, the description of the stages is made here, because it is somewhat more complex. The first thing that must be stated is that not only is it necessary to know the adsorption capacity, but also the adsorption speed, so it is necessary to carry out dynamic tests. The adsorption capacity for each solute is studied by determining the isotherm under equilibrium conditions. Once the coals that exhibit the highest adsorption capacity have been selected, dynamic tests are made by passing a flow through a bed of activated carbon and determining the fracture curve. The activated carbon that allows to treat the largest volume of solution before the breakage of the column occurs, will be the most appropriate one for practical application.

Among the applications of activated carbon in the liquid phase the most outstanding is the treatment of water, both drinking and wastewater (in fact, this application constitutes more than 40% of consumption in liquid phase). Natural waters are contaminated with natural and artificial organic compounds, halogenated compounds and bacteria and viruses. Natural organic matter is composed mainly of residues of the metabolism of living beings, which produce bad taste and smell. Artificial organic compounds include benzene, toluene, phenols, halomethanes, detergents, dyes, etc. Trihalomethanes are the most important artificial organic compounds present in drinking water, as a result of chlorine disinfection treatment. These compounds are strongly adsorbed on activated carbon, and this is the main reason why the number of drinking water plants that use activated carbon in the last stage of polishing the water before passing to the consumer grows.

The adsorption of organic and inorganic compounds in solution can be carried out in both, powdered and granular coals, but for the treatment of drinking water, the tendency is to preferably use granular carbon because it is easier to regenerate and to produce a lower pressure drop at the passage of water. Coal dust is often used to solve specific pollution problems that must be solved quickly and normally is put in suspension to proceed later to its sedimentation. The granular or pellet carbon is the one used in the polishing phase of water, helping to eliminate halomethanes and other substances that have resisted the processes prior to the said operation. The treatment of wastewater is more complex, and for better understanding it is advisable to use one of the many treatises published on the subject [15].

An important and traditional application in liquid phase is related to the food industry, in the process of discoloration or purification, mainly of sugars and sweeteners. The sugar industry has used many adsorbents over the last two centuries, including the so-called bone coal, but today activated carbon is preferably used in the last stage of purification, adsorbing substances that not only color but also hinder crystallization. Activated carbon powder is preferably used in this application and is operated batchwise or continuously. In the first case, the syrup is mixed with the activated charcoal and, after the necessary contact time, the filtration is carried out; in the continuous process the syrup is passed through the carbon bed, to which a quantity of diatoms is usually added to facilitate the passage of the liquid.

Some industries prefer to use granular activated carbon for continuous processes, due to the greater ease of regeneration. The use of activated carbon is also very normal in the beverage industry, alcoholic or not. For example, it is used to eliminate components that impart bad taste to beer, bad taste and smell to vodka, etc. It is used in the production of brandy for the elimination of acids, furfurals and tannins, without other important components such as acetaldehyde and alcohols being adsorbed. Activated charcoal was also used in the manufacture of decaffeinated coffee: caffeine was extracted with organic solvents and activated carbon adsorbed caffeine from the solution. Nowadays, the use of extraction with supercritical carbon dioxide, which does not involve organic solvents, is preferred.

Other applications in liquid phase include gold and silver mining (adsorption of the metal in the form of a cyanide complex for gold), dry-cleaning (to control the color of the solvents and remove oils, dyes and other compounds that cause the spots), aquariums, dialysis, etc.

9.2.6 Adsorption Process

It is the phenomenon by which a liquid, gas, or vapor on contact with a solid surface adheres to it and forms a layer of adsorbate. The substance that exerts the retention action is generally called the adsorbent and the liquid, gas, or vapor that is retained on the surface is called the adsorbate.

Three phenomena can intervene in physical adsorption: monomolecular adsorption, multimolecular adsorption and condensation in pores or capillaries. When there is adsorption there is a balance between the adsorbed and non-adsorbed molecules so that in an adsorbent-adsorbate system under temperature and pressure conditions it is defined and reproducible.

The types of adsorption that are presented are physical or Van der Waals adsorption and chemical adsorption or chemisorption. The first is characterized by small heats of adsorption and is one where the equilibrium of adsorption is reversible and is reached at low temperatures, is not very specific, and can occur in all condensed systems. The second is highly specific in its nature and depends on the chemical properties of the adsorbate and the surface of the adsorbent, among which the bond is much stronger, being able to generate chemical bonds of ionic or covalent type [23].

The adsorption of metal ions from aqueous solution is usually a physical processes, but the nature of the sample must be considered to determine if there are other compounds that can generate chemical-type adsorption. The selectivity of different ions to a specific type of solid adsorbent is given by their sizes and by the sizes of the pores of the material, achieving a specific selectivity.

An adsorption isotherm is the relationship that exists at a given temperature between the amount of substance adsorbed at equilibrium and the equilibrium pressure or concentration. In the specialized literature, there are several mathematical models, with different conceptions about the adsorption process and that allow their description. The most used ones are briefly discussed in the following sections.

9.2.6.1 The Freundlich Adsorption Isotherm

The isotherm can be written for the adsorption of a solute in solution as:

$$N = K \cdot C^{1/n} \tag{9.10}$$

where

N = mass of solute (grams or moles) adsorbed per unit mass of adsorbent
C = equilibrium concentration of the solute in solution (mol/L)
K and n = empirical constants

With logarithms:

$$\log N = \log K + \frac{1}{n}\log C \tag{9.11}$$

The graph of log N as a function of log C gives a straight line where the slope equals $1/n$ and the intercept is log K [23–25].

9.2.6.2 The Adsorption Isotherm of Langmuir

Its initial definition was made for a solid-gas system, and assumes that during the process of adsorption of a gas on a solid, a dynamic equilibrium occurs where the evaporation rate is equal to that of condensation. It is based on the following assumptions: (a) only a monomolecular adsorption takes place, (b) the adsorption is localized and (c) the adsorption heat is independent of the surface coating. Where the adsorption consists of two opposite reactions: one the condensation of the gas molecules on the surface of the solid and the other the evaporation of the gas molecules adsorbed from the surface toward the body of the gas phase. As the adsorption progresses, only the molecules that hit surface sites not covered by adsorbed molecules are adsorbed, so at the beginning of the adsorption the condensation rate of the molecules is high and decreases with time as the free area decreases on the surface. Desorption is carried out, i.e., the molecules are released from the surface by thermal agitation. Their speed depends on the amount of surface covered and increases each time the surface is occupied [23, 26].

For the adsorption of a solute in solution the equation is:

$$\frac{C}{N} = \frac{1}{N_m} \cdot C + \frac{1}{KN_m} \tag{9.12}$$

where

C = molar equilibrium concentration of the solute in the solution
N = number of moles adsorbed per gram of sorbent
N_m = number of moles of solute required per gram of adsorbent, to have a monolayer of solute on the surface of the adsorbent
K = constant

The graphical representation of C/N as a function of C should give a straight line whose slope is $1/N_m$ and the intercept is $1/KN_m$ [27, 28].

At different temperatures different values of K are obtained, which is why the lines are called isotherms. K is an equilibrium constant for the distribution of the material on the surface and in the solution.

Another relationship between the volume retained by a porous material and the equilibrium pressure or concentration is deduced by Stephen Brunauer, Paul Emmett and Edward Teller equation, on the basis of a model that assumes the initial formation of a monolayer adsorbed at low pressure followed by a succession of other layers superimposed as the pressure increases. They are obtained by applying the concept of Langmuir adsorption in a generalized form, the following expression for the volume retained per gram of adsorbent:

$$v = \frac{v_m cf}{1-f} * \frac{1}{(1+(c-1)f)} \tag{9.13}$$

where

v = volume adsorbed per unit mass of adsorbent

v_m = volume adsorbed per unit mass of adsorbent, when a monolayer has formed on the surface

f = p/p_o where p_o is the saturation pressure and p the equilibrium pressure

c is about the same as $e^{(E1-EL)/RT}$, where E_1 is the heat of gas adsorption in the initial monolayer and E_L is the heat of gas liquefaction.

Several methods have been developed for the determination of the surface area of porous materials, using some of the relationships indicated earlier. All are based on the balance between molecules adsorbed on the surface and molecules in the phase surrounding the surface. Generally, an inert gas is used in front of the surface and experimental conditions are used to allow the use of the perfect gas equation to calculate the balanced moles. The final calculation of the area is made by comparison with a reference area that in a large number of cases is the cross-sectional area of a nitrogen molecule.

9.2.7 Immersion Calorimetry

In immersion calorimetry, the thermal effects resulting from submerging a solid in a solvent, generally of nonpolar type, with which the solid does not present chemical interactions, are measured. These heats of immersion can be related to the surface area of the solid considered, through the models developed by Dubinin and Stoeckli [29–33].

The technique of immersion calorimetry has been used for several years to make accurate assessments of the surface interactions of porous solids.

The following section gives a brief summary of the theoretical arguments developed mainly by Dubinin and Stoeckli, which allow to quantitatively relate the heat of immersion with the surface area in the case of activated carbons.

9.2.7.1 Immersion of Porous Solids

It has been shown that in the case of microporous solids [34] without external surface, the enthalpy of immersion ΔH_i is related to the isosteric heat of adsorption q^{isost} by the expression:

$$-\Delta H_i(T) = \int_0^1 q^{isost.}(T;\Theta)d\Theta - \Delta H_{vap.}(T) \tag{9.14}$$

where Θ is the degree of micropore filling, W/W_0. The negative sign is due to q^{isost} and due to the fact that ΔH_i is negative.

Dubinin defines [35] the net heat of adsorption (q^{net}) as:

$$q^{net} = q^{isost.} - \Delta H_{vap.} \tag{9.15}$$

where q^{isost} is the isosteric heat and ΔH_{vap} is the enthalpy of vaporization of the adsorbate.

For the general case of n moles adsorbed, the previous equation is transformed:

$$q^{net} = E\left[\left(\ln\frac{n_a^0}{n_a}\right)^{\frac{1}{n}} + \left(\frac{\alpha}{n}T\right)\left(\ln\frac{n_a^0}{n_a}\right)^{\left(\frac{1}{n}\right)-1}\right] \tag{9.16}$$

where α is the coefficient of thermal expansion of the adsorbate, n_a represents the amount adsorbed to the relative pressure p/p_0, n_a^0 corresponds to the adsorption limit value, E_o is the characteristic energy for the adsorption of steam and Θ is degree of micropore filling (n_a/n_a^0).

In the case of activated carbons, with $n = 2$, we obtain the following equation:

$$q^{net} = \beta E_o\left[\left(\ln\frac{1}{\Theta}\right)^{\frac{1}{2}} + \left(\frac{\alpha T}{2}\right)\left(\ln\frac{1}{\Theta}\right)^{-\left(\frac{1}{2}\right)}\right] \tag{9.17}$$

so that

$$-\Delta H_i = \int_0^1 q^{net}\, d\Theta \tag{9.18}$$

By replacing this equation in Eq. (9.16) we get

$$-\Delta H_i = \beta E_o \int_0^1 \left(\ln\frac{1}{\Theta}\right)^{\frac{1}{2}} d\Theta + \frac{\alpha T \beta E_o}{2}\int_0^1 \left(\ln\frac{1}{\Theta}\right)^{-\frac{1}{2}} d\Theta \tag{9.19}$$

So, Eq. (9.19) can be written in an integrated way:

$$\Delta H_i (J/mol) = \frac{\beta E_o \sqrt{\pi}(1+\alpha T)}{2} \tag{9.20}$$

which is valid for filling micropores in carbons activated at temperature T.

For a carbon with a micropore volume W_o and an adsorbate with molar volume V_m, Stoeckli and Kraehenbüehl propose the equation:

$$\Delta H_i = -\frac{\beta E_o W_o \sqrt{\pi}(1+\alpha T)}{2V_m} \tag{9.21}$$

which expresses the immersion enthalpy ΔH_i in J/g.

This equation shows the relationship that exists between the enthalpy of immersion and the volume of the micropores, which indicates that the said enthalpy corresponds to the process of filling them and is different to the enthalpy of wetting of open or nonporous surfaces. Equation (9.21) is known as the Stoeckli-Kraehenbüehl equation [33].

The experimental enthalpy of immersion (ΔH_{exp}) of the activated carbons has two types of contribution: one due to the micropores and the other due to the wetting of the external surface (S_{ext}), as proposed by Stoeckli, Bansal and Donnet [36]. This can be expressed by means of Eq. (9.21).

$$\Delta H_{exp.}\left(J/g\right) = \Delta H_i + h_i S_{ext.} \tag{9.22}$$

where h_i represents the specific enthalpy (J/m^2) determined with different wetting liquids on solids that do not have porosity, usually a graphite [37].

Equation (9.22) can be expressed as:

$$S_{ext.}(J/m^2) = \frac{\Delta H_i(J/g)_{exp.}}{h_i(J/m^2)} - \frac{\Delta H_i(J/g)}{h_i(J/m^2)} \tag{9.23}$$

That is

$$S_{ext.}(m^2/g) = A_{TOTAL} - A_{MICROPO.} \tag{9.24}$$

The latter shows that by determining the immersion heat of an activated carbon in a nonpolar liquid solvent, the total area thereof can be determined.

For the general case of a solid that also has a contribution of mesopores and macropores, the total area is expressed as:

$$A_{TOTAL} = A_{MICROP.} + A_{MESOP.} + A_{MACROP.} + S_{abierta.} \tag{9.25}$$

where A_{TOTAL} is the total area, A_{MICROP} is the micropores area, A_{MESOP} is the mesopores area, A_{MACROP} is the macropores area and $S_{abierta}$ is the area corresponding to the open nonporous surface.

When the activated carbon has no external surface or meso- and macropore development, the experimental enthalpy (ΔH_{exp}) is equal to the enthalpy (ΔH_i) calculated with the Stoeckli-Kraehenbüehl equation, Eq. (9.21).

9.2.8 Adsorption Kinetics

The study of adsorption kinetics in wastewater is important because it allows a better understanding of the reaction route and the mechanism of the reaction. In addition, it is important to predict the time in which the adsorbate will be removed from the aqueous solution and thus be able to design a suitable treatment plant. Any adsorption process is controlled by the following stages.

External transport: The mass transport by diffusion of the adsorbate molecules from the fluid phase to the surface of the solid.

Internal transport: The transport of the adsorbate inside the particle by migration of the molecules from the external surface of the adsorbent to the surface within the pores and/or by diffusion of the adsorbate molecules through the pores of the particle.

Adsorption process: The molecules in the pores are adsorbed from the solution to the solid phase. This stage is relatively quick, compared to the first two steps; therefore, the local equilibrium between the two phases is assumed.

In kinetic modeling these three steps are grouped. The application of the kinetic model only depends on the initial and final concentration of the solution at different time intervals. Currently, there are certain empirical mathematical models that are used in obtaining kinetic parameters that may be useful in later calculations.

9.2.8.1 Pseudo-First-Order Model

Lagergren (1898) [38] proposed this speed kinetic equation for adsorption in liquid-solid systems, which is derived from the adsorption capacity of the solid. It is one of the most used equations of speed for the adsorption of a solute in an aqueous solution. According to several authors, the adsorption speed is directly proportional to the equilibrium potential, the difference between the initial and equilibrium concentration of the adsorbate ($q_e - q$). Therefore, the kinetic equation of pseudo-first-order can be expressed as:

$$\frac{dq}{dt} = k_1(q_e - q) \tag{9.26}$$

where q_e is the amount of solute adsorbed in the equilibrium per unit mass of the adsorbent (mg/g), q is the amount of solute adsorbed at one time t, k_1 is the pseudo-first-order rate constant (h–1) and t is the contact time (h). When $q = 0$ to $t = 0$, Eq. (9.26) is integrated and we obtain:

$$Log(q_e - q) = Log q_e - \frac{k_1 t}{2.303} \tag{9.27}$$

The speed constant k_1 can be calculated from the graphic log (q_e–q) against t for different adsorption parameters such as pH, temperature, adsorbate concentration, adsorbent dose, particle size and agitation speed.

9.2.8.2 Pseudo-Second-Order Kinetic Model

Ho and McKay (1999) [38] used a model based on a pseudo-second-order equation to describe the kinetics of heavy metal removal in an adsorbent in particular. To derive the equation from this model, the following reactions were considered:

$$2P^- + M^{2+} \leftrightarrow MP_2 \tag{9.28}$$

$$2P^- + M^{2+} \leftrightarrow MP_2 \tag{9.29}$$

where M is a metal ion, P and HP are polar sites in the adsorbent. The mathematical model obtained is represented by Eq. (9.30), and it is assumed that the adsorption capacity is proportional to the number of active sites occupied in the adsorbent, i.e., this model is based on the capacity of adsorption to equilibrium. This model is represented by:

$$\frac{dq}{dt} = k_2(q_e - q)^2 \tag{9.30}$$

Equation (9.30) is integrated when $q = 0$ at $t = 0$ and we obtain the equation in a linear form:

$$\frac{t}{q} = \frac{1}{k_2 q_e^2} + \frac{t}{q_e} \tag{9.31}$$

where q_e is the maximum adsorption capacity for second-order kinetics (mg/g), q is the amount of solute adsorbed at a time t, k_2 is the pseudo second rate constant (g/mg/h) and t is the time

$$h = k_2 {q_e}^2 \tag{9.32}$$

the graph $t = q$ against t at different adsorption parameters will give us a linear relationship that will allow the calculation of q_e, k and h.

9.3 ADSORBENTS

9.3.1 Preparation of Activated Carbons from Corncob

Corn is one of the basic foods for the population in Latin America, its scientific name is *Zea mays*. The fruit of the corn is harvested in four months, starting from the moment the seed is placed in the ground. Corn plants are subject to various precautions so that the product can be used properly either in its first phase of growth, when the grain is tender corn, or in the second phase of maturation, when the fruit is the corncob. The best-known types of maize in Central America and part of South America are white, yellow and black; currently other varieties are also available. The cob from different kinds of corn constitute the wrapping that protects the corn from weather conditions, including direct rays of the sun, wind and rain, as well as from birds and various other pests.

The cob as a corn wrap comes from a cylindrical trunk, which is attached to the plant at the bottom, this trunk is approximately 3 cm long and about a centimeter in diameter from where superimposed and overlapping leaves emerge, forming a protective layer on the corncob. The position of the corncob on the plant is diagonal and downward, i.e., the upper part of the ear is placed downward so that it can resist its own weight and water can drain properly. An important part of the plant necessary for the development of the ear and growth of the grain is the hair of the corn, which is a set of thin fibers whitish in color, which later turns brownish, and finally dark brown when the grain is almost ripe. The function of the hairs is to internally transport oxygen to the fruit. These hairs also serve as a kind of packing that helps in sealing the spaces between each row of grains in the ear, thus

FIGURE 9.3 Cultivation of *Zea mays.* (From: maizedoctor.cimmyt.org/es/component/content/8/8?task=view)

preventing direct penetration of water and insects. In the outer part of the ear, the hair of the corn leaves between the cobs and falls; by placing the pipe at the top where the water flows (rain or artificial irrigation) it deflects and does not penetrate directly into the grain, the ear wrap is sealed with corn hair, forming a hermetic protective layer for the corn grain. Cob is a by-product of corn and has various domestic and artisanal uses. It is used for making scrubbing pads and also serves as food for some animals. Dry cob is abundant in different regions of Colombia, and it is used in large quantities as fodder for livestock. In the field, together with corncobs, they are used as fuel to light fire for cooking. Peasants use the dry corn leaves to wrap tobacco leaves and roll them into cylindrical shapes to make cigarettes or cigars of cob.

Figures 9.3 and 9.4 show the cultivation of maize and a corncob. The corncob is used as a starting material to obtain different porous solids used in the present study.

The raw sample (corncob) is ground to a particle size of 3 mm, then it is dried at 110°C for 6 hours, referenced as TM0. The corncob is ground to a particle size of 3 mm, then it is dried at 110°C for 6 hours, it is carbonized in a stainless steel tubular oven with a progressive heating of 5 K/min to 450°C for 4 hours in a nitrogen atmosphere. Activation of the char is then carried out by activation with CO_2 at 850°C for 2, 4 and 8 hours, to obtain four solids with different textural characteristics referenced as TM2, TM4 and TM8, respectively.

The samples of corncob are ground to a particle size of 3 mm, then dried at 110°C for 6 hours, impregnated with concentrated H_2SO_4 (98.5%) in a ratio of 1:1.8 p/p (dry material:acid) for 72 hours, and then it is washed until the end of acidity. The sample is referenced as TMSA

FIGURE 9.4 Corncob is the fibrous part or the heart that remains of the cob after shelling (i.e., removal of the grains or kernels). (From: http://phenobarbital.wordpress.com/2010/01/21/maiz-y-tuza-o-lo-que-estamos-comiendo/).

The corncob is ground to a particle size of 3 mm, then dried at 110°C for 6 hours, impregnated with H_3PO_4 at 50% in ratio 1:2 p/p (dry material:acid) for 72 hours, and then it is washed until the end of acidity. The sample is referenced as TMPA.

9.3.2 Adsorbates

To study the porous solids synthesized in this work and analyze their adsorption capacity, Ni (II) and Cr (III) ions were used as adsorbates.

Solutions were prepared from their corresponding salts using analytical grade reagents to have solutions of concentration ranging from 10 to 450 ppm. All the solutions were prepared by dilution using deionized water and their concentrations were determined using the atomic absorption technique, previously preparing their corresponding calibration curve.

9.3.3 Instrumental and Analytical Techniques Used for the Characterization

9.3.3.1 Thermogravimetric-Differential Thermal Analysis

Thermogravimetric-differential thermal analysis (TG-DTA) evaluates the content of light volatiles of the starting material, the temperature at which cracking occurs, and the decomposition of the most unstable molecules [39–41]. For the thermogravimetric

FIGURE 9.5 Equipment TG-DSC Netzsch STA 409 PC.

analysis of the starting material, 20 mg of scale was taken, which was subjected to a nitrogen flow of 100 mL/min, a linear heating rate of 5 K/min and was brought to a final temperature of 1273 K [42], the analysis was performed on a TG-DSC Netzsch STA 409 PC, shown in Figure 9.5.

9.3.3.2 Next Analysis

The next analysis of the samples was carried out based on standard methods (such as American Society for Testing and Materials – **ASTM** International), stipulated for each process. In this way, moisture determination was carried out in accordance with ASTM 2867 revision 2004 [43]. The total ash content was determined in accordance with ASTM 2866 revision 2004 [44]. The determination of the volatile matter content of the samples was made in accordance with ASTM 5832 revision 2003 [45]. The fixed carbon content was calculated by difference with respect to the other tests of the next analysis.

9.3.3.3 Textural Characterization: Adsorption Isotherms N_2 at 77 K

The textural characterization of carbonaceous materials was performed by physical adsorption of N_2 at 77 K. To perform this analysis, the samples were previously degassed at 573 K for 24 hours. The apparent surface areas were calculated from the BET equation, the micropore volume V_0 (N_2), was obtained by applying the Dubinin-Radushkevich equation to the nitrogen adsorption data (density of the N_2 liquid = 0.808 g/cm^3). The total volume of pores V_t was calculated from the volume adsorbed at the relative pressure of 0.99, and the volume of mesopores by difference [46–48]. An automatic IQ2 sortometer from Quantachrome Instruments (Boynton

Beach, Miami, FL, USA) was used for these determinations. The carbonaceous samples TM2, TM4 and TM8, of around 0.100 g, were analyzed by this method. TM0, TMSA and TMPA were degassed at 60°C to prevent their decomposition.

9.3.3.4 Functional Groups

In order to determine the quantity and types of oxygenated groups located on the surface of the prepared carbonaceous materials, samples of these were immersed in solutions of NaOH, Na_2CO_3 and $NaHCO_3$ 0.1 M. The most used bases are $NaHCO_3$ (p*Ka* = 6.37), Na_2CO_3 (p*Ka* = 10.25), NaOH (p*Ka* = 15.74). According to Boehm, carboxylic groups are only valued by $NaHCO_3$, the difference between the acidity valued by $NaHCO_3$ and Na_2CO_3 corresponds to the content of lactones, and phenolic groups and carbonyls are obtained from the difference between the acidity registered with NaOH and Na_2CO_3. Finally, the hydrochloric acid gives an estimate of the total basicity of the material [49, 50].

In the proposed methodology, approximately 100 mg of the adsorbent was immersed in 25 mL of the solution, in 50 mL plastic containers; the solutions were left at 298 K, shaken manually twice a day for 48 hours; occasionally N_2 was bubbled over the solutions with the objective of removing atmospheric CO_2. Finally, 10 mL aliquots of the previously standardized acid or base solutions, were titrated and the milliequivalents gram were determined by difference. [49, 50].

9.3.3.4.1 Boehm Titrations

Approximately 1 g of sample was weighed, 50 mL of the selected acid or base was added, heated at 298 K for 48 hours with shaking, and a filtered 10 mL aliquot was taken and titrated potentiometrically with the base or acid, as needed. (The titration was done in triplicates.)

9.3.3.5 X-ray Diffraction

X-ray diffraction is a nondestructive technique used for obtaining structural information of the materials. This technique was important to clarify the structure of coal, thanks to the work of Hirsh, Diamond and Franklin [51–53]. Yoshizawa et al. [53] have contributed to the study of the carbon activation process. They used the methodology devised by Hirsch and Diamond to apply a Fourier analysis to the diffractograms of the coals. By means of this technique of Fourier analysis, structural parameters of carbon were studied, such as the number of layers that compose the stacks of aromatic planes and the fraction of stacked structure, which is equivalent to the fraction of crystalline structure. Yoshizawa et al. called this technique stacking X Ray Diffraction (STAC-XRD), an acronym from the English Generalized Analysis of Coal by XRD (standard analysis of coal by XRD) and applied it to the study of the physical activation of lignocellulosic residues, of mineral coals and to the chemical activation of carbons of different ranges [53].

The x-ray diffraction of the prepared activated carbon materials was performed on an x-ray diffractometer using CuKα radiation operated at 40 kV and 14 mA. The activated carbons were macerated to a fine powder before making the measurements with this technique. The powder was placed on a support to measure the x-ray spectrum, the data is taken from $5° < 2\theta < 80°$ at a speed of 0.1° (2θ) per minute.

9.3.3.6 Fourier-Transform-Infrared Spectroscopy

The samples were subjected to a Fourier-transform-infrared spectroscopy (FT-IR) analysis, for which 0.1 g of the activated carbon was crushed and mixed with potassium bromide (to remove dispersing effects of the large crystals). This powder mixture was compressed in a mechanical die press to form a translucent pellet through which the light beam of the spectrometer can pass. Then it was read on a Thermo-Nicolet 6700 FT-IR [54].

9.3.3.7 Quantification of Nickel (Atomic Absorption)

To determine the content of nickel present in the aqueous solutions after reaching equilibrium, an atomic absorption analysis was carried out with the corresponding lamp. Initially, calibration curves with primary standard of nickel nitrate II between 1 and 30 ppm were obtained; the pH was adjusted to 4.5 and the concentration of the solutions was measured using atomic absorption spectrometry (AAS). For this, the standard test method E-841-04 (ASTM 2004) was followed. The data for the adsorption isotherms were obtained by using 0.1 g of each solid with a 25 mL of solutions of the metal of known initial concentrations in a range of 20–500 mg/L; the pH was adjusted to 4.5 at 298 K for 96 hours and the concentration of the ions was measured under the same conditions in which the calibration curve was prepared. For the absorbance readings, an Analyst 300-Perking Elmer atomic absorption device was used, ASTM-D 3682-78 standard [55].

9.3.3.8 Scanning Electron Microscopy

The Scanning Electron Microscope (SEM) consists of an optical column and an electronic console. The optical column has a camera that is at high vacuum (approximately 2×10^{-6} torr) where the sample to be analyzed is placed. The image of the microscope is formed by an electron beam directed toward the sample. The beam is generated from an electronic gun, which has a tungsten filament as a cathode that is heated by a thermionic emission system at a temperature that exceeds 2700 K. This filament emits electrons in abundance that produce a high negative potential with respect to the anode, which are rapidly accelerated toward it through the electronic column. The electron beam passes through two or three electromagnetic capacitor lenses whose function is to decrease the diameter of the beam. Any radiation from the sample can be used to provide a signal and each of these is the result of some interaction between the incident electrons and the sample, providing different information [56–57].

A microscope is an optical system that magnifies small objects to be examined with natural light or light emitted from an artificial source. In an electron microscope, an electronic beam, instead of light, is used to form the image. Scanning electron microscopy (SEM) has a range of magnifications of 10–180000x and, together with x-ray detectors, makes it possible to identify very small minerals and examine their shape and distribution [56].

The SEM allows one to observe directly the morphology and porosity of materials. In this technique, a beam of electrons in the vacuum causes the excitation of secondary electrons in the sample, which generate signals that are captured as an image, whose details depend on the magnification that is taken.

9.3.3.9 Determination of the Adsorption Capacity of Contaminating Metal Ions Nickel (II) from Aqueous Solution

The adsorption of each ion was carried out separately with the samples of porous solids obtained using 0.5 g of the adsorbent material; the pH was adjusted by dropwise addition of diluted HNO_3 or NaOH [58]. Solutions of 100 mL at 35 ppm were prepared. The evaluation of the kinetics and follow-up of the concentration drop were carried out by taking measurements at intervals of 15 minutes at the beginning and prolonging them at the end until 92 hours.

This experiment was carried out in a specially designed assembly to determine the changes in concentrations. This assembly is called the batch adsorber (or batch type). The adsorber consists of a 500- or 1000-mL Erlenmeyer flask in which the solution of the solute or adsorbate is placed with the porous solid. The latter is placed in a bag made of nylon mesh so that there is no friction between the particles themselves and the wall of the container, thus avoiding the formation of fine powders of carbon or ceramic that may interfere in the analysis of the ions in aqueous solution.

The adsorber or flask is partially introduced in a bath of constant temperature to keep the temperature of the solution constant. The constant temperature bath consists of an acrylic container and a water recirculator. The solution was kept in constant agitation by means of a Teflon-coated magnetic stirring bar that was acted upon by means of a magnetic plate placed under the constant temperature bath.

9.3.3.10 Adsorption Kinetics

The study of the adsorption kinetics was carried out with the optimal adsorption parameters obtained at different initial concentrations, with a mass ratio of carbon/volume of solution of 50 mg/50 mL. During the adsorption process at different time intervals, small aliquots were separated and then the nickel content was determined, respectively.

9.3.3.11 Immersion Calorimetry

The immersion enthalpies of the materials prepared in benzene were determined in a Calvet-type heat conduction microcalorimeter with a stainless steel calorimeter cell (Figure 9.6). Around 150–200 mg of the solids were weighed in a glass ampoule and degassed for 3 hours at 250°C and the vial was sealed. The calorimeter was assembled whose cell could contain 10 mL of the solvent. When the equipment reaches thermal equilibrium, the bulb breaks, the solid gets wet by the liquid and the heat generated is recorded as a function of time. Finally it is electrically calibrated [58–62].

9.4 RESULTS

9.4.1 Analysis of Corncob

During this process, cellulose units, the main component of lignocellulosic materials such as corncob, undergo depolymerization transforming into smaller units, volatile

FIGURE 9.6 Calvet-type calorimeter.

matter is released and a more carbon-rich solid is produced [63]. Table 9.1 shows the next and last analysis of the corncob and the porous solids synthesized in this study, which as mentioned in the experimental section are labeled as follows: TM0—raw material, TM2—activated carbon obtained at the activation time of 2 hours, TM4—activated carbon obtained at the activation time of 4 hours, TM8—activated carbon obtained at the activation time of 8 hours, TMPA—porous solid treated with phosphoric acid and TMSA—porous solid treated with sulfuric acid. The content of volatile matter decreased as the corn material was treated, while the carbon content increased, which is beneficial for the synthesis of the activated carbons.

On the other hand, the oxygen content decreased in the samples and the fixed carbon content increased. The percentage of fixed carbon was between 8.5% and 78.5%, the percentages of nitrogen and hydrogen decreased, while the sulfur content was minimal. In summary, it was clearly observed that the processes of activation and oxidation of the starting material significantly increased the carbon content and decreased the percentages of nitrogen, oxygen, hydrogen and volatile matter and increased the fixed carbon and the ash content for both carbons (Table 9.1).

9.4.2 Activation of Corncob

In activated carbons, the carbon content increased with a proportional decrease in other chemical constituents (Table 9.1). This increase in the carbon content was mainly due to two variables that were modified during this study: (a) the time,

TABLE 9.1
Next and Last Analysis of the Starting Materials, Carbonized and Activated Carbons

		Last Analysis (Dry Basis)			
Material	**C (%)**	**H (%)**	**N (%)**	**S (%)**	**O[a] (%)**
TM0	44.5 ± 0.2	5.3 ± 0.2	1.09 ± 0.01	0.03 ± 0.00	49.08 ± 0.01
TM2	77.5 ± 0.2	4.1 ± 0.2	1.02 ± 0.01	0.02 ± 0.00	17.32 ± 0.01
TM4	84.8 ± 0.2	2.7 ± 0.2	0.92 ± 0.01	0.01 ± 0.00	11.57 ± 0.01
TM8	88.5 ± 0.2	2.3 ± 0.2	0.89 ± 0.01	0.00 ± 0.00	8.31 ± 0.01
TMPA	89.6 ± 0.2	2.1 ± 0.2	0.82 ± 0.01	0.00 ± 0.00	7.48 ± 0.01
TMSA	91.2 ± 0.2	1.8 ± 0.2	0.77 ± 0.01	0.00 ± 0.00	6.23 ± 0.01

		Last analysis (dry basis)		
Material	**Residual Humidity (%)**	**Volatile Material (%)**	**Ashes (%)**	**Fixed Carbon (%)**
TM0	12.4 ± 0.1	75.6 ± 0.1	3.5 ± 0.1	8.5 ± 0.1
TM2	6.7 ± 0.1	17.4 ± 0.1	8.4 ± 0.1	67.5 ± 0.1
TM4	1.2 ± 0.1	14.5 ± 0.1	9.8 ± 0.1	76.9 ± 0.1
TMPA	1.5 ± 0.1	10.4 ± 0.1	9.6 ± 0.1	78.5 ± 0.1
TMSA	1.3 ± 0.1	9.1 ± 0.1	8.4 ± 0.1	81.5 ± 0.1

Note: [a] Calculated by difference.
TM0: raw corncob; TM2 and TM4: activated charcoal obtained from the corncob at 2 and 4 hours; TMSA: corncob treated with sulfuric acid; TMPA: corncob treated with phosphoric acid.

which causes the texture of the activated carbons to vary and (b) the oxidizing and/or dehydrating character of the used acids (phosphoric and sulfuric), which facilitated the loss of hydrogen and oxygen and converted large carbonated units into smaller units with greater carbon richness [63]. These acids tend to generate an increase in the aromaticity of the carbon, loss of the aliphatic character and formation of a rigid, crosslinked solid [64, 65]. The corncob is basically composed of lignin, cellulose and hemicellulose. When carbonizing at 450°C, only cellulose and hemicellulose are depolymerized in smaller units to give rise to a large mass of fixed carbon, since lignin decomposes at higher temperatures [66, 67]. When the mat is impregnated with the aforementioned acids, the hydrolysis of the lignin, present in the material, is facilitated, so that the surface chemistry of the solid changes and generates a particular porosity [63]. If, for example, the case of impregnation with phosphoric acid is analyzed, after the loss of water, the phosphoric acid is composed of a mixture of polyphosphoric acids, including species such as H_3PO_4, $H_4P_2O_7$ and $H_5P_3O_{10}$ and other species of general formula $H_{n+2}P_n O_{3n+1}$ [64]. During the activation process, with the increase in temperature there is a slight increase in the ash content that can be attributed to small impurities of the inorganic material [64].

9.4.3 Thermal Analysis of Corncob

The profile of the carbonization of the raw corncob sample (Figure 9.7) indicates that there is no significant change in the weight of the material up to 200°C. Subsequently, a rapid change occurred between 200°C and 430°C. This corresponds to a typical change of lignocellulosic materials. This profile has been observed in the carbonization of various carbonaceous materials such as coconut shell [6], bamboo [68, 69] and palm peel [69] that are not mixed with inorganic activating agents. Lignocellulosic materials such as corncobs consist of three main components: cellulose, hemicellulose and lignin. Cellulose is a partially crystalline linear polysaccharide consisting of up to 3000 glucose units in long chains. Hemicellulose is similar, but it is composed of an order of magnitude fewer units. Lignin is more complex. It is a three-dimensional polymer of phenylpropane units bound by C—O—C or C—C bonds. In addition to linking the units of phenylpropane, ether and the C—C bond is connected with cellulose and hemicellulose. Therefore, lignin acts as a "cement" for lignocellulosic structures. This composition and structure of the lignocellulosic materials determine their decomposition profile. As the reaction temperature increases, the adsorbed species are released. This is followed by the decomposition of the hemicellulose (between 200°C and 260°C), cellulose (between 240°C and 350°C) and lignin (between 280°C and 500°C). Up to 430°C, decomposition is dominated by depolymerization and a chain break through ether groups and carbon-carbon bonds. As shown in the one obtained with Termo Gravimetric Analysis (TGA) (Figure 9.7), a rapid change was observed at the beginning of the hemicellulose decomposition, which continued in the decomposition of lignin.

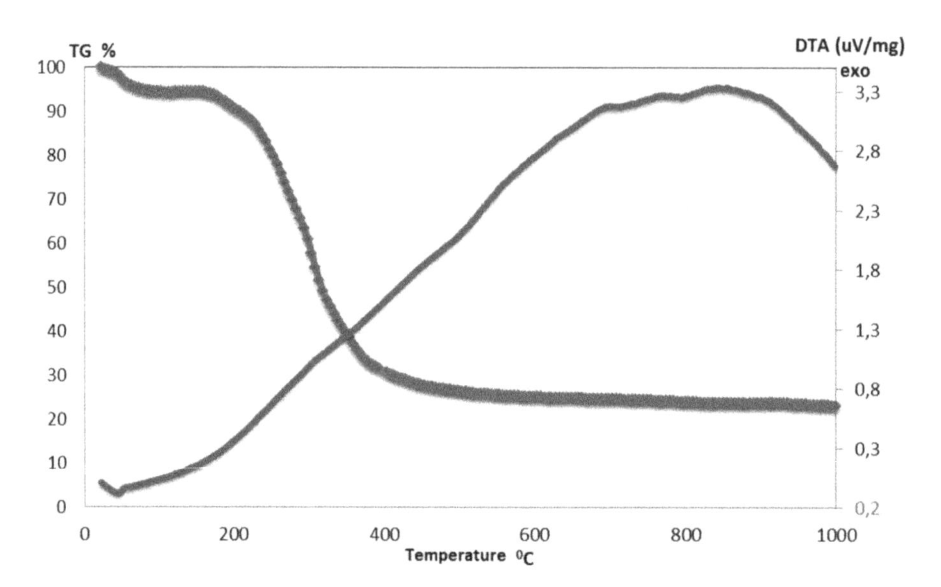

FIGURE 9.7 TGA and differential thermal analysis (DTA) profiles for corncobs as a precursor to the activated carbons obtained at a heating rate of 2°C/min in a nitrogen atmosphere.

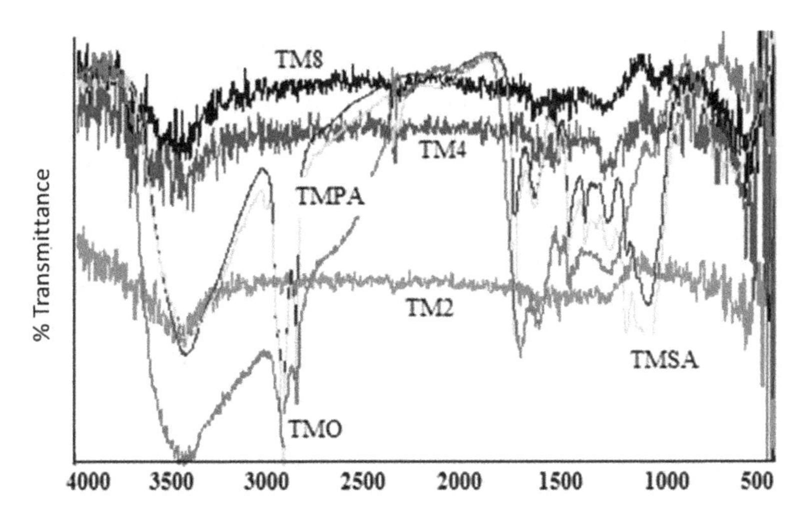

FIGURE 9.8 Infrared spectrum with diffuse reflectance cell corresponding to the activated carbons obtained from the corncob: Samples: TM0, TM2, TM4, TM8, TMSA and TMPA.

9.4.4 Infrared Analysis of Porous Solids Obtained from Corncob

The interpretation of the infrared spectra with Fourier transform with diffuse reflectance (DRIFTS) for solids such as activated carbons must be carried out carefully because each group originates several bands, so that each of these could include contributions from several groups [70]. Figure 9.8 shows the diffuse reflectance infrared Fourier transform spectroscopy (DRIFTS) spectrum, wave numbers (ν, cm^{-1}) versus percentage of transmittance, of C—C bands recorded at 1200 cm^{-1} (C—O in ethers); 1049–1276 cm^{-1} (alcohols); 1100 cm^{-1} and 3400 cm^{-1} (C—OH phenolic); 1585–1600 cm^{-1} (C=C aromatic); 2900 cm^{-1} (C—H aliphatic); and 1150 cm^{-1}, 1700 cm^{-1} and 3400 cm^{-1} (carboxylic acids) for the different samples synthesized in this study. In the regions ~3400 and ~1700 cm^{-1} the carbons obtained at higher temperatures had less pronounced bands compared to the original sample, showing the difference between each of the thermal treatments.

For samples TM0, TMPA and TMSA, corncob spectra have bands at 1650, 1060 and 1000 cm^{-1}, which are the characteristic peaks of cellulose and hemicellulose adsorption. In addition, the bands at 1700, 1570, 1460 and 855 cm^{-1} are characteristic of the aromatic skeleton of lignin macromolecules such as *p*-coumaryl, coniferaldehyde, and synapillic acid units, which is typical of herbaceous angiosperms (HGS lignins) [71]. After carbonization and activation (TM2, TM4 and TM8), the intensity of bands decreased and even disappear in some cases. This is due to the pyrolysis of the corncob during pre-carbonization and later to the activation process. The oxidized precursor (corncob) shows a broad band at about 3488–3100 cm^{-1}, which is assigned to the O—H mode of stretching of the hexagonal groups and the water adsorbed. The band at 2920 cm^{-1} is assigned to the C–H stretching and vibration, band representing the alkyl groups such as methyl and methylene groups, and is observed in all the carbons with different intensities. The variation in intensity in each activated carbon is due to the different activation processes, which leads to a different rate at which the CH_3 groups are removed from the substituted aromatic rings during the different times of the pyrolysis. The band at around 1610 cm^{-1} is attributed to the stretching and vibrations of C=O of carbonyl groups and is observed in all carbons with different intensities. All activated carbons also show the band around 1180 cm^{-1}, which is attributed to stretching and vibration C–O of the carbonyl groups. The result of FT-IR spectroscopy indicates that the carbons produced are rich in different functional groups on the surface.

9.4.5 Study of the Acidity of Activated Carbons

Table 9.2 shows the content of acidic and basic groups of the starting material, the prepared and modified carbons of the carbons in aqueous suspension and pH.

It is established that the different materials obtained had an acidic pH (4.7–6.2), which means that these materials showed a predominance of acid functional groups on their surface, which is very important considering the process it was desired to study, since this characteristic is important for the process of adsorption of metal ions and tends to favor this. Activated carbons at different times show a slight change in carbon acidity as the basic groups are formed at temperatures above 1073 K [72], suggesting that the basic sites on the surface are essentially of the Lewis type, associated with regions rich in π electrons [73], in general, oxidation with sulfuric acid and phosphoric acid (TMSA, TMPA) increased its acid content, especially oxidation with sulfuric acid.

TABLE 9.2
Textural Parameters of the Activated Carbons Obtained from the Corncob Determined from the Isotherms of N_2 at 77 K and Titrations in an Acidic Medium

Carbon	S_{BET} (m^2/g)	V_{micro} (cm^3/g)	V_{meso} (cm^3/g)	Total Acidity (meq NaOH/g)	pH
TMO	327	0.17	0.56	1.44 ± 0.02	6.2 ± 0.2
TM2	532	0.23	0.63	1.40 ± 0.2	5.8 ± 0.2
TM4	874	0.37	0.55	1.47 ± 0.2	5.5 ± 0.2
TM8	1120	0.46	0.43	1.36 ± 0.2	5.3 ± 0.2
TMPA	154	–	1.56	1.59 ± 0.2	5.0 ± 0.2
TMSA	86	–	1.88	1.62 ± 0.2	4.7 ± 0.2

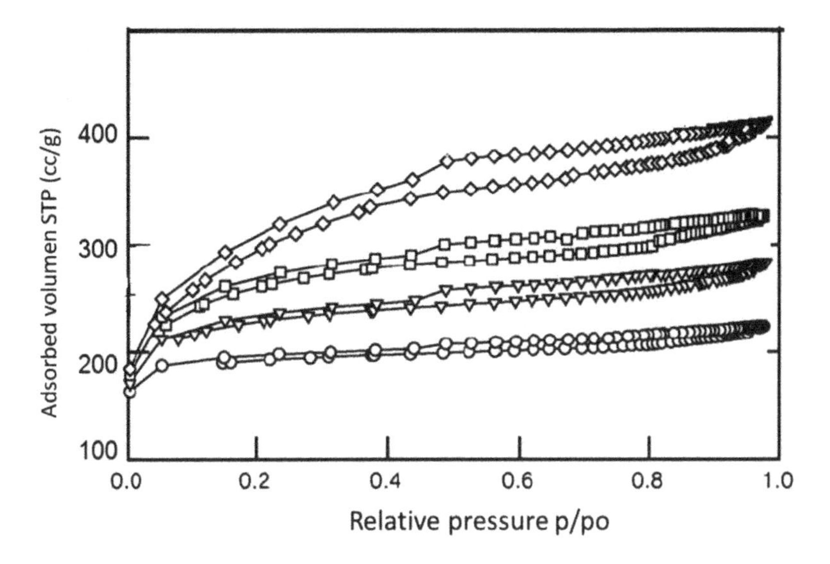

FIGURE 9.9 Isotherms of adsorption-desorption of N_2 at 77 K for coals activated from corncob (O) TM0 (▽) TM2 (□) TM4 and (◊) TM8.

9.4.6 Textural Analysis of the Different Carbons from the N_2 Isotherms and Scanning Electron Microscopy Analysis

The nitrogen adsorption isotherms for the raw materials treated with phosphoric and sulfuric acid (TMPA and TMSA, respectively) showed very low areas, as shown in Table 9.2. The TM0 shows together with the samples TMSA and TMPA the lowest values of BET areas.

The isotherms of the activated carbons obtained from corncobs (TM2, TM4 and TM8) were adjusted to type IV isotherms according to the IUPAC classification. Figure 9.9 shows the adsorption-desorption isotherms of N_2 at 77 K for activated carbons synthesized from the corncob. The adsorption isotherms of N_2 for the TMPA and TMSA samples are not shown in the figure because, as reported in Table 9.2, these materials were those with the lowest surface area developed.

The three analyzed materials show a hysteresis cycle that has the following sequence: TM8 < TM4 < TM2, which indicates an increasing volume of mesopores. It is also observed that the greater the volume of mesopores (as seen by the height of the hysteresis curve), the smaller the volume of micropores (as seen by the amount of N_2 adsorbed at low relative pressures). Table 9.2 shows that the development of the total surface area developed by the TM8 coal (1120 m^2/g) was great for this type of material synthesized under our experimental conditions. The results show that the textural properties of the synthesized solids are strongly dependent on the treatment of the material. When the tile is subjected to a longer activation time, a greater surface area and a more heterogeneous pore

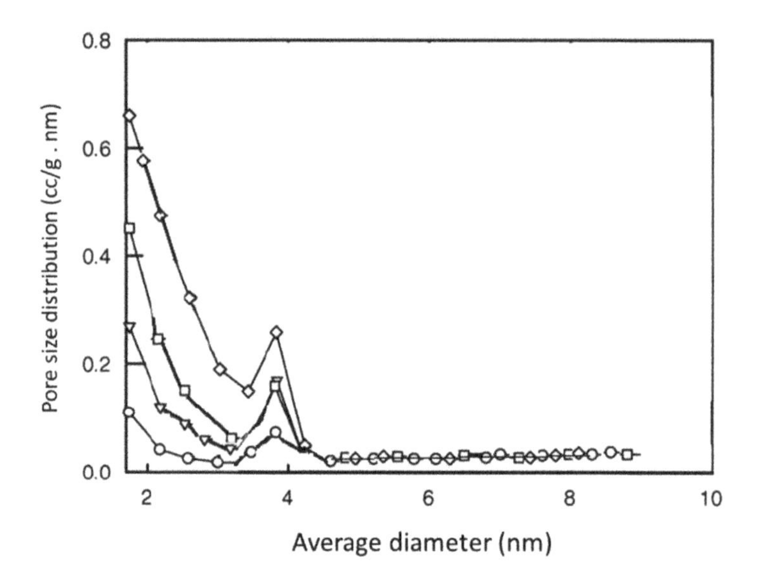

FIGURE 9.10 DFT pore distribution of activated carbons synthesized from the corncob. (O) TM0 (▽) TM2 (□) TM4 and (◊) TM8.

distribution are generated, as shown in Figure 9.10. The area generated for the raw material (TM0) and the corresponding samples of the material treated with the acids, showed the development of areas that are interesting if one thinks about the application of these materials in environmental chemistry. The total basic number (meq of HCl/g) was zero for all the materials synthesized in this work, which indicates that the carbons were acids, a value that is presented in Table 9.2.

The MEB microphotographs of the synthesized materials are shown in Figure 9.11. In these microphotographs it can be seen that the morphology of the raw material has a basic porous structure, which makes it a natural adsorbent; this is in good agreement with the results of the analysis of the isotherms of N_2 at 77 K. When this material is treated with acids (samples TMPA and TMAS) the structure is modified developing a porous structure with a larger pore distribution, and the distribution of pores at the micro level is also altered. These microphotographs are not shown here.

This suggests that a blockage is being generated inside the pores during the acid treatment of the starting sample that modifies its structure and that is reflected in the decrease in surface area, as shown in the results reported in the Table 9.2.

In Figures 9.11. (b), (c) and (d) are shown the MEB corresponding to the activated carbons (TM2, TM4 and TM8) which show a more heterogeneous porous system product of the activation treatment, with a greater development of pores.

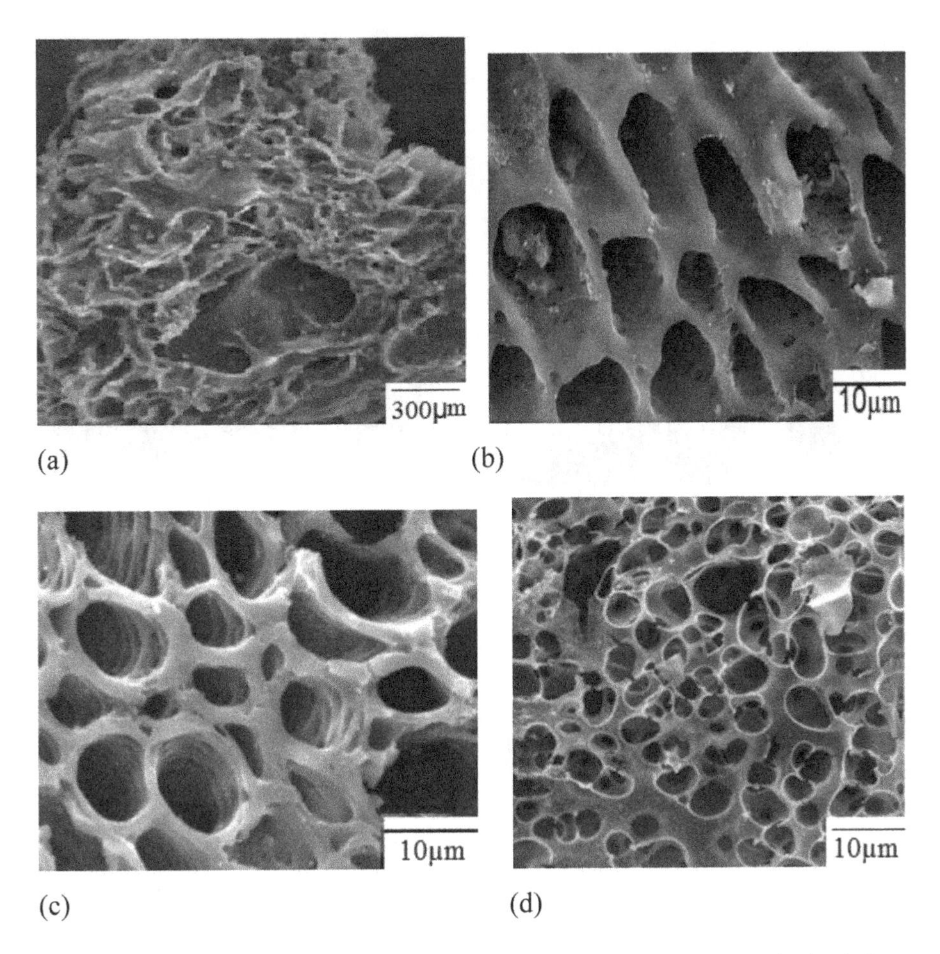

FIGURE 9.11 Scanning electron microscopy (SEM) of materials synthesized from corncob: (a) TM0 (b) TM2 (c) TM4 (d) TM8 (taken at 20 kV, magnification: 1000x)

9.4.7 Analysis of the DRX

Activated carbons produced from corncobs at different times of activation can be characterized crystallographically by x-ray diffraction.

From Figure 9.12, it can be seen that most of the carbons obtained have broad, small peaks, centered around 26° and 44°, respectively. For these materials the peaks can be assigned to flexures of the graphite plates (0 0 2) and (1 0) plans, respectively according to the calculations made with Bragg's law and the Scherrer equation.

It is very interesting to note that the XRD spectra for the samples TM2, TM4 and TM8 show very broad and poorly defined peaks, which is associated with the poor organization of the activated carbons and their heterogeneity in the pores. Otherwise, it happens with the other materials, whose pore distribution is more homogeneous and its DRX peaks are more defined.

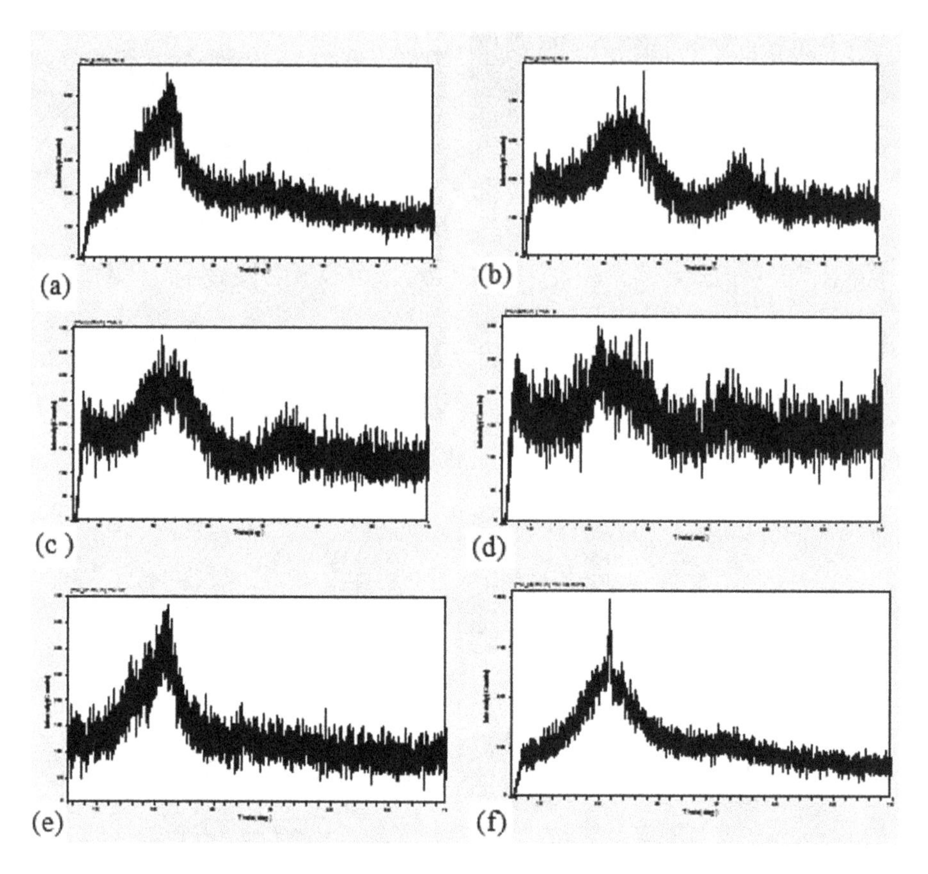

FIGURE 9.12 Diffractograms of the different synthesized coals. (a) TM0 (b) TM2 (c) TM4 (d) TM8 (e) TMPA (f) TMSA.

9.4.8 Distribution of Species for Nickel (II)

In aqueous solution, several hydro complexes of nickel are formed by successive acid dissociation reactions. Nickel in solution can form different species according to the following reactions:

$$\mathbf{Ni^{2+} + H_2O \rightleftharpoons Ni(OH)^{+} + H^{+}} \qquad \mathbf{pK_1 = 9.86} \tag{9.33}$$

$$\mathbf{Ni(OH)^{+} + H_2O \rightleftharpoons Ni(OH)_2 + H^{+}} \qquad \mathbf{pK_2 = 10.14} \tag{9.34}$$

$$\mathbf{Ni(OH)_2 + H_2O \rightleftharpoons Ni(OH)_3^{-} + H^{+}} \qquad \mathbf{pK_3 = 11.0} \tag{9.35}$$

Using the values of these equilibrium constants, the species diagram is made, which is illustrated in Figure 9.13 in which the different species that make up

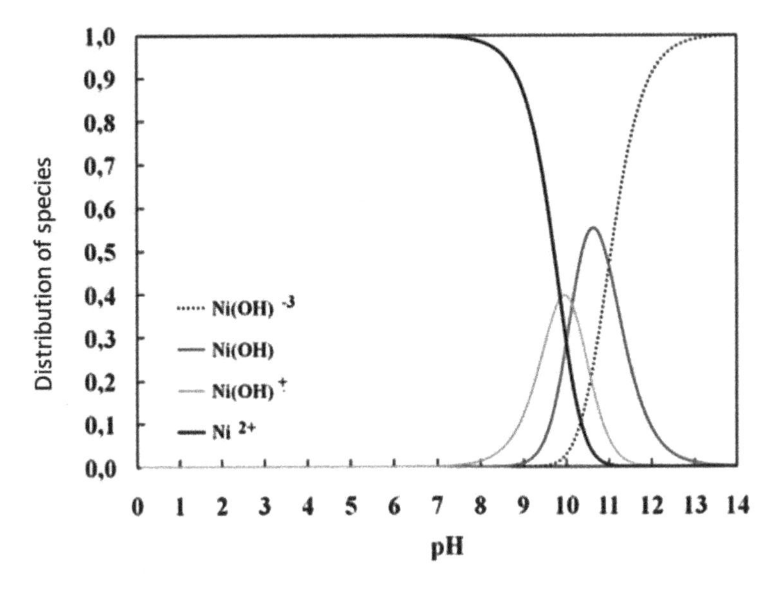

FIGURE 9.13 Diagram of distribution of nickel species.

nickel in the aqueous solution are observed, as well as their percentage distribution. From the speciation diagram can be established which are the predominant complexes within the pH range; from the diagram it is established that the predominant form as Ni^{2+} is at pH around 6.0.

In this work, the adsorption experiments were also carried out using the synthesized materials to analyze the variation of the adsorption capacity with respect to pH. For this, measurements were made within the range 2, 3, 4, 5, 6 and 7, finding that the maximum adsorption capacity was 5.0, so at this pH the adsorption experiments from aqueous solution were carried out.

9.4.9 Analysis of the Results of the Isotherms from Aqueous Solution for Nickel (II)

As presented in Section 9.4.8, it is necessary and important to determine the isotherms for each of the synthesized materials in aqueous solution in order to establish their adsorption capacity, in this case for Ni (II). Again, this study provides a general perspective of how adsorption occurs in the system under study, in this case, carbon solutions of Ni (II), which will also allow to establish the efficiency of activated carbons obtained from tusa of corn with respect to its capacity of adsorption with respect to Ni (II). Langmuir and Freundlich models were used in order to adjust the experimental results to a suitable model that allows to reproduce the obtained experimental results. The isotherms corresponding to the adsorption of Ni (II) from aqueous solution at controlled pH (pH = 5.0 in this study) on the activated carbons synthesized from the corncob and on the raw material are shown in the

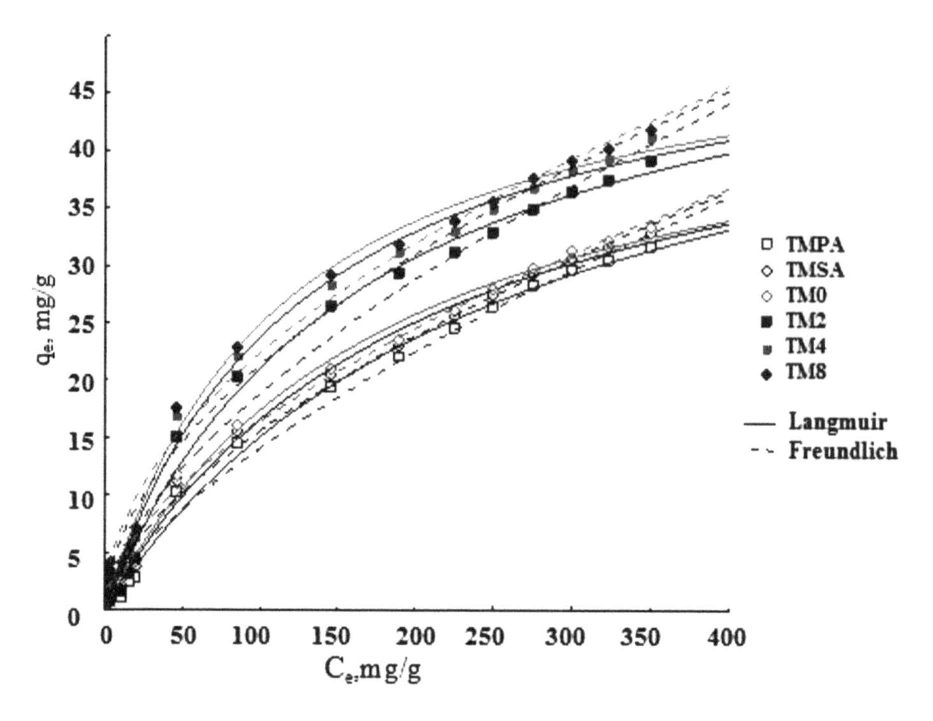

FIGURE 9.14 Adjustment of Ni (II) adsorption from aqueous solution: Freundlich and Langmuir models. Conditions: pH: 5.0. and temperature: 298 K.

Figure 9.14; the experimental results and the adjustment to the models of Freundlich and Langmuir are shown in the figure.

In Table 9.3 the correlation and constant coefficients corresponding to these models are presented. In comparison with the raw material, the activated carbons obtained at different activation times show a greater capacity of adsorption of Ni (II) ions with this material (TM2, TM4 and TM8).

TABLE 9.3
Constants and Correlation Coefficients of the Langmuir and Freundlich Models, Corresponding to the Ni (II) Adsorption Isotherms on Activated Carbons Synthesized from a Corncob

	Langmuir			Freundlich		
	q_e	K	R^2	K_F	n	R^2
TMPA	30.2134	0.023844	0.99954	4.34652	1.436543	0.97534
TMSA	32.4312	0.025984	0.99976	5.34562	1.346786	0.96534
TM0	32.8754	0.026873	0.99943	6.08763	1.895432	0.97543
TM2	38.2219	0.032872	0.99876	10.87654	1.994321	0.97432
TM4	40.7644	0.068645	0.99887	20.98626	2.540838	0.96432
TM8	43.6453	0.231655	0.99856	38.75430	2.946353	0.96789

The results presented in Table 9.3 show that the regression coefficient (R^2) of all synthesized materials are between 0.999976 and 0.99887 for the Langmuir model, suggesting that the Ni (II) adsorption on activated coals from corn may be well described by this model.

From Figure 9.14, the maximum amount of Ni (II) (q_e) adsorbed on the different samples of activated carbon is synthesized with 43.6453 mg/g with sample TM8 determined at pH 5.0. This value is very interesting because it is a value comparable to that reported in the literature using other types of materials, and in some cases it is superior, showing the feasibility of using this material to retain Ni (II). In order to establish the possible mechanism of Ni (II) adsorption on the respective adsorbates obtained in this study, the experimental data were adjusted, as mentioned before, to the Freundlich and Langmuir models. The constants and parameters of the equations for this system were calculated by linear regression using the shape of the isothermal equations. The results are given in Table 9.3.

Table 9.3 shows that, based on the correlation coefficient, the best fit for the studied isotherms corresponding to the elimination by Ni (II) adsorption at the pH studied, is the Langmuir model. The values of R^2 are close to unity, which allows this deduction to be made, so the adsorption must be a monolayer adsorption [74–79].

For the study of Ni (II), the modified samples were also used to analyze the effect of each variable on the adsorption capacity of the metal study carried to pH = 5.0 and T = 298 K. The data of adsorption equilibrium and their corresponding isotherms of adsorption are shown in Figure 9.14. It can be observed that the adsorption capacity of the CAs was dependent on the degree of thermal modification. The adsorption capacity increased progressively with the increase of the activation time. Here, two trends were observed: the materials TMPA, TMSA and TM0 in the first group tend to present a similar adsorption capacity in equilibrium, showing that in the case of the Ni (II) adsorption on the crude material and those modified with acids, the treatment does not generate a change with respect to its final adsorption capacity. The adsorption capacities of the materials TM2, TM4 and TM8 in the second group differ from those of the materials in the previous group, but among them the difference of adsorption toward the nickel is not very significant.

The maximum adsorption capacity is the constant q_e of the Langmuir isotherm, and these constants are given in Table 9.3. The values of q_e varied between 30.2134 and 43.6453 mg/g, with TM8 being the activated carbon with the highest adsorption capacity. The lines (___) represent the adjustment isotherms to the Langmuir model and the lines (- - - -) represent the adjustment corresponding to the isotherms of the Freundlich model. From the above, the adsorption capacities of the CAs were deduced in the following order in terms of the adsorption capacity of divalent nickel:

$$TM8 > TM4 > TM2 > TM0 > TMSA > TMPA$$

This decreasing order is very similar to the concentrations of carboxylic sites in these CAs determined by Boehm titration. Therefore, the adsorption capacity of the CAs was dependent on the carboxylic sites, and the adsorption capacity is proportional to the concentration of carboxylic sites, as well as the concentration of the total

TABLE 9.4
Parameters of the Kinetics of Adsorption of Ni (II) on Activated Carbons at 298.15 K

		Pseudo-First-Order			Pseudo-Second-Order		
Sample	C_o (mg/L)	q_e (mg/g)	k_f (mg/g·min)	R^2	q_e (mg/g)	k_s (mg/g·min)	R^2
TM0	450	27.876	0.034	0.87	276.56	0.0012	0.99
TM2	450	32.965	0.025	0.88	288.49	0.0011	0.99
TM4	450	36.763	0.021	0.89	289.54	0.0015	0.99
TM8	450	42.977	0.022	0.88	294.44	0.0012	0.99

acid sites. It was assumed that the capacity of adsorption was more dependent on the concentration of carboxylic sites than that of hydroxyl sites (the presence of this is very low). This assumption is based on the facts that, in the modification by thermal treatment, the concentration of carboxylic sites was increased more significantly than that of the hydroxyl sites, and most of the carboxylic sites are deprotonated at the experimental pH, pH = 5.0.

9.4.10 Nickel Adsorption Kinetics

In order to investigate the mechanisms that control adsorption processes, such as mass transfer and chemical reaction, pseudo-first-order kinetics was studied, as well as pseudo- second-order kinetics that, as mentioned, was widely used to perform the mathematical interpretation of the adsorption speed of solid adsorbents in liquid solution [74, 78, 79]. The pseudo-first-order kinetics (Lagergren equation) and the pseudo-second-order kinetics follow Eqs. (9.26) and (9.31).

The calculated parameters of the pseudo-first-order kinetics, as well as the pseudo-second-order model are shown in Table 9.4, using the materials with greater adsorption capacity: TM2, TM4, TM8 and the raw material. The obtained values of q_e and corresponding correlation coefficient (R^2) from the pseudo-first-order equations did not show to be the appropriate ones to illustrate the kinetic relation of Ni (II) adsorption on activated charcoal from the corncob (Figure 9.15).

On the other hand, the calculated values of q_e for the pseudo-second-order kinetics are adjusted according to the results presented in Table 9.4. The calculated values of R^2 were between 0.99 and 1. Therefore, experimental kinetic adsorption data correlated well for pseudo-second-order kinetics, as shown in Figure 9.16. The result obtained in this kinetic study also coincided with the results of the previous studies [76–78].

9.4.11 Results of Immersion Calorimetry with Nickel (II)

For nickel, immersion calorimetry was performed for all the activated carbons and solids synthesized in solutions of NaOH and 0.1 M HCl. The results showed that

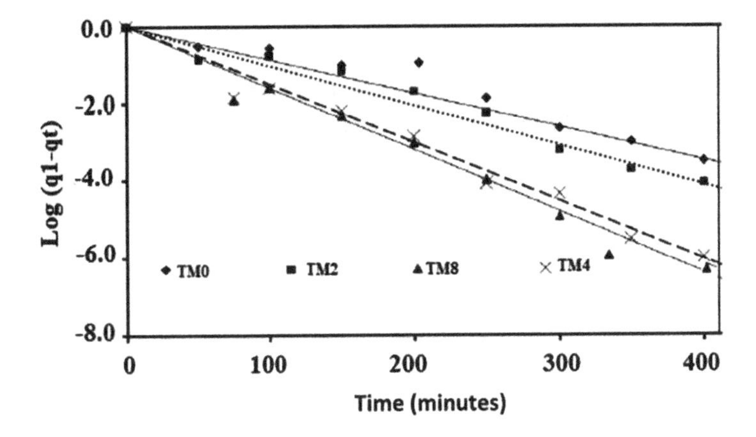

FIGURE 9.15 Kinetics of pseudo-first-order nickel on coals activated from the corncob ($T = 298$ K, C0 = 450 mg/L, $m = 5.0$ g/L).

immersion in both NaOH and HCl solutions generates negative immersion enthalpy values, indicating that such a process is exothermic. The values obtained for the heats of immersion in sodium hydroxide and hydrochloric acid in the synthesized materials at 298 K are presented in Table 9.5. Higher values were obtained for immersion in sodium hydroxide solutions, this is because the synthesized materials possess a pH with acid character, and therefore the heat generated comes from the interaction of the groups with this character and NaOH.

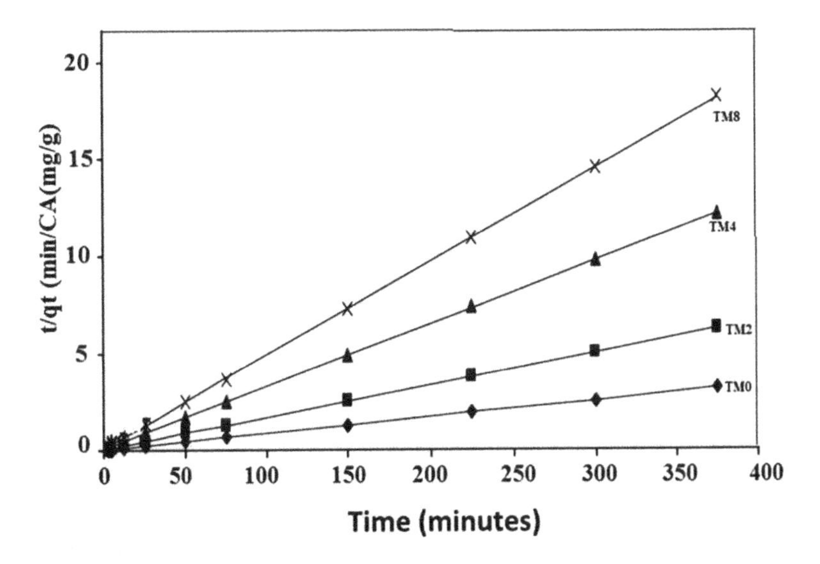

FIGURE 9.16 Pseudo-second-order kinetics of Ni (II) on activated carbons synthesized from the corncob ($T = 298$ K, C0 = 450 mg/L, $m = 5.0$ g/L).

TABLE 9.5
Surface Chemical Characteristics of Activated Carbons Synthesized from the Corncob by Immersion Calorimetry

Activated Carbon	$-\Delta H_{inm}$ NaOH 0.1 M (J/g)	$-\Delta H_{inm}$ HCl 0.1 M (J/g)
TMSA	12.5	2.3
TMPA	19.4	4.7
TM0	22.4	7.5
TM2	26.4	6.4
TM4	28.6	10.5

The results for the immersion enthalpies reported in Table 9.5 correspond to the average of ten determinations, and the standard deviations are between 0.20 and 1.0 J/g. In this system, enthalpic values registered were also detected when carrying out the immersion calorimetry experiments in HCl since, in spite of the basicity of the same, that when evaluating them by titrations was zero, small effects are observed here. These calorimetric results are very interesting since they show that the calorimetric technique is highly sensitive, which means that there are some functional groups that allow you to establish the presence of acidic groups [78–80].

9.5 CONCLUSION

The capacity of nickel adsorption on the CAs synthesized was determined, presenting a range between 30.2134 and 43.6453 mg/g under the experimental conditions of this investigation, a range that is comparable and in some samples exceeds those reported in the literature. The adsorption of Ni (II) on the CAs conforms to the Langmuir model. The kinetic process of adsorption of activated charcoal nickel corn cobs is adjusted to the pseudo-second-order. Immersion calorimetry has been shown to be a versatile technique for monitoring and interpreting the adsorbate-adsorbent adsorption process.

ACKNOWLEDGEMENTS

The authors thank the Framework Agreement between Universidad de los Andes, Universidad Nacional de Colombia and the act of agreement established between the Chemistry Departments of both universities. The authors also appreciate the grant for funding *research programs for Associate Professors, Full Professors, and Emeritus Professors* announced by the Faculty of Sciences of the University of the Andes, 20-12-2019–2020, 2019, according to the project "Enthalpy, free energy and adsorption energy of activated carbon interaction and solutions of emerging organic compounds".

REFERENCES

1. Samarghandi, M. R., Azizian, S., Shirzad, M. S., Jafari, S.J., Rahimi S. Removal of divalent nickel from aqueous solutions by adsorption onto modified holly sawdust: equilibrium and kinetics. Iranian Journal of Environmental Health, Science and Engineering Iranian Association of Environmental Health (IAEH). 8(2) (2011):181–188.
2. Kalyani, S., Srinivasa, P. R., Krishnaiah, A. Removal of nickel (II) from aqueous solutions using marine macroalgae as the sorbing biomass. Chemosphere. 57 (2004):1225–1229.
3. Bulut, Y., Tez, Z. Removal of heavy metals from aqueous solution by sawdust adsorption. Journal of Environmental Sciences. 19 (2007):160–166.
4. Ramezankhani, R., Sharifi, A. A. M., Sadatipour, T., Abdolahzadeh, R. A mathematical model to predict nickel concentration in Karaj river sediments. Iranian Journal of Environmental Health, Science and Engineering. 5 (2008):91–94.
5. Park, Y. J., Suh, M. Y., Park, K. K., Choi, K. S., Lee, K. Y., Kim, W. H. Adsorption studies of nickel (II) ions onto amorphous alumina. Analytical Science & Technology. 13 (2000):433–439.
6. CNA. Boletín de la Red Nacional de Monitoreo de la Calidad del Agua Mexicana. (2000): 4–15.
7. Wang, S., Zhang, M., Li, B., Xing, D., Wang, X., Wei, C., Jia, Y. Comparison of mercury speciation and distribution in the water column and sediments between the algal type zone and the macrophytic type zone in a hypereutrophic lake (Dianchi Lake) in Southwestern China. Science of the Total Environment. 417–418 (2012, Feb 15): 204–213.
8. Rodriguez-Reinoso, F. Production and Applications of Activated Carbons: Handbook of Porous Solids. In: Schüth, F., Sing, K. S. W., Weitkamp, J. (eds.), Wiley-VCH, 2002.
9. Marsh, H., Heintz, E. A., Rodriguez-Reinoso, F. (eds.). Introduction to Carbon Technologies. Universidad de Alicante, Spain, Secretariado de Publicaciones, 1997.
10. Jankowska, H., Active Carbon. Ellis Horwood limited, Chichester, England, 1991.
11. Rodriguez-Reinoso, F., Sepulveda-Escribano, A. Porous carbons in adsorption and catalysis. In: Nalga, H. S. (ed.), Handbook of Surfaces and Interfaces of Materials, Vol 5. Academic Press. New York, 2001, pp. 309–355.
12. Rouquerol, F., Rouquerol, J., Sing, K. Adsorption by Powders and Porous Solids. Academic Press, San Diego, 1999.
13. Rodriguez-Reinoso, F., Molina-Sabio, M., Muñecas, M. A. Effect of microporosity and oxygen surface groups of activated carbon in the adsorption of molecules of different polarity. The Journal of Physical Chemistry. 96 (1992):2707–2713.
14. Rodriguez-Reinoso, F., Linares-Solano, A. Microporous structure of activated carbons as revealed by adsorption methods. In: Thrower, P. A. (ed.), Chemistry and Physics of Carbon, Vol 21. Marcel Dekker Inc. New York, 1989, pp. 1–146.
15. Marsh, H., Rodriguez-Reinoso, F. Activated Carbon. Elsevier. Amsterdam, 2006.
16. Rodriguez-Reinoso, F., Lopez, G. J. D., Berenguer, C. Activated carbons from almond shells—I: Preparation and characterization by nitrogen adsorption. Carbon. 20 (1982):513–518.
17. Rodriguez-Reinoso, F., Garrido, J., Martin-Martinez, J. M., Molina, M., Torregrosa, R. The combined use of different approaches in the characterisation of microporous carbón. Carbon. 27 (1989):23–32.
18. Rodriguez-Reinoso, F. An overview of methods for the characterisation of activated carbons. Pure and Applied Chemistry. 61 (1989):1859–1866.
19. Dubinin, M. M., Radushkevich, L. V. Equation of the characteristic curve of activated charcoal. Proceedings of the Academy of Sciences of the USSR. 55 (1947):331.

20. Boehm, H. P., Voll, M. Basische Oberflächenoxide auf Kohlenstoff—I. Adsorption von säuren. Carbon. 8(2) (1970):227–240.
21. Boehm, H. P. Surface oxides on carbon and their analysis: A critical assessment. Carbon. 40(2) (2002):145–149.
22. Fabish, T. J., Schleifer, D. E. Surface chemistry and the carbon black work function. Carbon. 22 (1984):19–38.
23. Laidler, K., Meiser, J. Fisicoquímica, 2nd ed. Cecsa, México, 2002.
24. Rodriguez-Reinoso, F. El Carbon Activado como solvente universal De: Sólidos porosos, preparación, caracterización y aplicaciones. Universidad de los Andes. 2007, p. 21
25. Maron, S. H., Prutton, C. F. In: Limusa, S. A., De C. V. (eds.), Fundamentos de Fisicoquímica. México, 1994.
26. Shaw, D. In: Alambra, S. A. (ed.), Chap. 6. Introducción a la Química de Superficies y Coloides. España, 1970.
27. Giraldo, L., Moreno, J. C., Gómez, A., Polanía, A. Caracterización de carbón activado por calorimetría de inmersión. Revista Colombiana de Química. 23(1) (1994):27.
28. Mcclellan, A. L., Hansberger, H. F. Cross-sectional areas of molecules adsorbed of solids surfaces. Journal of Colloid and Interface Science. 23 (1967):577.
29. Dubinin, M. M., Polyakov, N. S., Kataeva, L. I. Basic properties of equations for physical vapor adsorption in micropores of carbon adsorbents assuming a normal micropore distribution. Carbon. 29 (1991):481.
30. Dubinin, M. M. Microporous structures of carbonaceous adsorbents. Carbon. 20 (1982):195.
31. Donnet, J. B., Papirer, E., Wang, W., Stoeckli, H. F. The observation of active carbon by scanning tunneling microscopy. Carbon. 31 (1993):182.
32. Stoeckli, H. F. Microporous carbons and their characterization: the present state of the art., Carbon. 28 (1990):1.
33. Stoeckli, H. F., Kraehenbüehl, F. The enthalpies of immersion of active carbons in relation to the Dubinin theory for the volume filling of micropores. Carbon. 19 (1981):353.
34. Stoeckli, H. F., Centeno, T. A. On the characterization of microporous carbons by immersion calorimetry alone. Carbon. 35 (1997):1097.
35. Silvestre–Albero, J., Gómez, C., Sepúlveda-Escribano, A., Rodríguez–Reinoso, F. Characterization of microporous solids by inmersion calorimetry. Colloids and Surfaces A. 187 (2001):151–165.
36. Roque-Malherbe, R. "Adsorción física de gases", Enpes (Ed.). La Habana. 1897
37. Bansal, R. C., Donnet, J. B., Stoeckli, H. F. Active Carbon. Marcel Dekker. New York, 1988.
38. Ho, Y. S., Mckay, G. The sorption of lead (II) ions on Peat. Water Research. 33(2) (1999):578–584.
39. Diaz, C. M., Briceño, N., Baquero, M. C., Giraldo, L., Moreno, J. C. Influence of temperature in the processes of carbonization and activation with CO_2 in the obtainment of activated carbon from African palm pit. study of the modification of characterization parameters. Internet Journal of Chemistry. 6 (2003).
40. Moreno, J. C., Giraldo, L., Huertemendía, M., de las Pozas del Río, C., Ladino, Y. Immersion Calorimetric in the characterization of carbonaceous materials. Marsella, Francia. Mayo 25-28 de 2005.
41. Giraldo, L., Moreno, J. C., Gómez, A., Polanía, A. Caracterización de carbón activado por calorimetría de inmersión. Rev. Col. Quím. 23(1), (1994): 27.
42. Giraldo, L., Moreno, J. C. Microcalorimety: Application to study of the heats of activated carbon into phenol aqueous solutions. CARBON´04. An International Conference on Carbon. Providence, Rhode Island, USA. 2004.
43. Lowell, S. Introduction to Powder Surface Area. Ed John Wiley & sons. New York, 1979, p. 15.

44. Estándar Test Method for Total ash Content of Activated Carbon. ASTM D2866 – 94, 2004.
45. Landau, M. V. Transition metal oxides. In: Schuth, F., Sing, K. S., Weitkamp, J. (eds.), Handbook of Porous Solids. Vol 3. Wiley VCH. Weinheim, 2002, p. 1766.
46. Laidler, K., Meiser, J., Fisicoquímica. Ed. Cecsa. ed 2th. México, 2002.
47. Glasstone, S., Lewis, D. Elements of Physical Chemistry, Ed. Van Nostrand Company, Inc., Segunda Edición, 1960., pp. 558–570.
48. Maron, S. H., Prutton, C. F., Fundamentos de Fisicoquímica. Ed. Limusa S.A. De C.V., 1994.
49. Dubinin, M. M., Polyakov, N. S., Kataeva, L. I. Basic properties of equations for physical vapor adsorption in micropores of carbon adsorbents assuming a normal micropore distribution. Carbon, 29, (1991), 481.
50. Dubinin, M. M. Microporous structures of carbonaceous adsorbents. Carbon, 20, (1982), 195.
51. Stoeckli, H. F., Microporous carbons and their characterization: the present state of the art., Carbon, 28, (1990), 1.
52. Stoeckli, H. F., Kraehenbüehl, F. The enthalpies of immersion of active carbons in relation to the Dubinin theory for the volume filling of micropores. Carbon, 19, (1981), 353.
53. Stoeckli, H. F., Centeno, T. A. On the characterization of microporous carbons by immersion calorimetry alone. Carbon, 35, (1997), 1097.
54. Silvestre–Albero, J., Gómez, C., Sepúlveda-Escribano, A., Rodríguez–Reinoso, F. Characterization of microporous solids by inmersion calorimetry. Colloids and Surfaces A, 187, (2001), 151–165.
55. Roque-Malherbe, R., "Adsorción física de gases", Enpes (Ed.). La Habana. 1897.
56. Reyes, S. A. M. Microscopía electrónica y microanálisis a la solución de problemas geoquímicos. VII Congreso Nacional de Geoquímica. Instituto de Geología, UNAM. 1997.
57. Shaw, D., Introducción a la Química de Superficies y Coloides. (Eeditorial.), Alambra, S. A. España., 1970. Cap. 6.
58. Ruiz, C. Trabajo de Grado. Departamento de Química. Universidad Nacional de Colombia. Bogotá, 1990.
59. Fonseca, R. A. Trabajo de Grado. Departamento de Química. Universidad Nacional de Colombia. Bogotá, 1993.
60. Ruiz, C., Giraldo, L., Gómez, A. Construcción de celdas calorimétricas metálicas con escudos adiabáticos. Revista Colombiana de Química. 23(1) (1994):15.
61. Moreno, J. C., Giraldo, L., Gómez, A. A batch type heat conduction microcalorimeter for immersion heat determinations: design and calibration. Thermochimica Acta. 290(1) (1996).
62. Moreno, J. C., Giraldo, L. Determination of the immersion enthalpy of activated carbon by microcalorimetry of the heat conduction. Instrumentation Science & Technology. 28(2) (2000):171.
63. Rodríguez-Reinoso, F., Molina-Sabio, M. Carbones activados a partir de materiales lignocelulósicos. Quibal. Química e Industria. 45(9) (1998):563–571.
64. Hsisheng, T., Yeh, T-S., Hsu, L-Y. Preparation of activated carbon from bituminous coal with phosphoric acid activation. Carbon. 36(9) (1998):1387–1395.
65. Huidobro, A., Pastor, A. C., Rodríguez-Reinoso, F. Preparation of activated carbon cloth from viscous Rayon. Part IV. Chemical Activation. Carbon. 39(3) (2000):389–398.
66. Berkowitz, N. On the differential thermal analysis of coal. Fuel. 36 (1957):355–373.
67. Kudo, K., Yoshida, E. On the decomposition process of wood constituents in the course of carbonization. Journal of the Japan Wood Research Society. 3(4) (1957):125–127.

68. Zhang, J., Shi, Q., Zhang, C., Xu, J., Zhai, B., Zhang, B. Adsorption of neutral red onto Mn-impregnated activated carbons prepared from *Typha orientalis*. Bioresource Technology. 99 (2008):8974–8985.
69. Hameed, B. H., Chin, L. H., Rengaraj, S. Adsorption of 4-chlorophenol onto activated carbon prepared from rattan sawdust. Desalination. 225 (2008):185–198.
70. Figueiredo, J. L., Pereira, M. F. R., Freitas, M. M. A., Orfao, J. J. M. Modification of the surface chemistry of activated carbons. Carbon. 37(9) (1999):1379–1389.
71. Rai, D. I., Sass, B. M., Moore, D. A. Chromium(III) hydrolysis constants and solubility of chromium(III) hydroxide. Inorganic Chemistry. 26 (1987):345–355.
72. Studebaker, M. L. The chemistry of carbon black and reinforcement. Rubber Chemistry and Technology. 30 (1957):1400–1484.
73. Lopez-Ramon, M. V., Stoeckli, F., Moreno-Castilla, C., Carrasco-Marin, F. On the characterization acidic and basic surface sites on carbons by various techniques. Carbon. 37(8) (1999):1215–1221.
74. Latimer, W. M. The Oxidation States of the Elements and Their Potentials in Aqueous Solution. Prentice-Hall, Englewood Cliffs. NJ, USA, 1952, p. 281.
75. Bernhard, M., Brinckman, E. E., Sadler, P. J. The Importance of Chemical Speciation in Environmental Processes. Springer. New York, USA, 1986.
76. Fierro, V., Torne-Fernandez, V., Montane, D., Celzard, A. Adsorption of phenol onto activated carbons having different textural and surface properties. Microporous and Mesoporous Materials. 111 (2008):276–287.
77. Hameed, B. H., Rahman, A. A. Removal of phenol from aqueous solutions by adsorption onto activated carbon prepared from biomass material. Journal of Hazardous Materials. 160 (2008):576.
78. Giraldo, L., Moreno, J. C., Huertas, J. I. Heats conduction micro-calorimeter with metallic reaction cells. Instrumentation Science & Technology. 30(2) (2002):177–186.
79. Giraldo, L., Moreno, J. C. Determinación de la entalpía de inmersión de carbón activado en soluciones acuosas de fenol y su relación con la capacidad de adsorción. Revista Colombiana de Química. 32(1) (2003):45–54.
80. Giraldo, L., Moreno, J. C. Determinación de la entalpía de inmersión y capacidad de adsorción de un carbón activado en soluciones acuosas de plomo. Revista Colombiana de Química. 33(2) (2004):87–97.

10 Bioremediation—With Special Reference to Hydrocarbon Degradation

Harish Chandra and Ramesh Chandra Dubey[*]

CONTENTS

10.1 INTRODUCTION

Bioremediation is a process that uses microorganisms to convert environmental deleterious substances into nonhazardous forms or bring them to a tolerable level. It is one of the best technologies known for combating water pollution, or in a broader sense, the wastewater problem. Due to the advancements in the field of biotechnology, there is a possibility to modify the genetic makeup of bacteria, which in turn

[*] Corresponding author: profrcdubey@gmail.com

are capable of degrading two or more environmental pollutants. The degradation of biodegradable waste is a natural phenomenon and does not require much effort for degradation and assimilation. It can be easily managed by burying the waste in a large pit or leaving it as is. However, waste containing pesticides, hydrocarbons and chemicals is recalcitrant and not easily degraded. Managing such waste requires a lot of effort and may involve physical methods like burning in incinerators. However, these processes release harmful fumes, which can be a reason for air pollution and a leading cause of respiratory diseases.

Anthropogenic activities and inappropriate solid waste management are responsible for the release of a huge amount of wastes in the environment each year. Of these wastes, some industrial wastes are well regulated through the laws of government. However, certain wastes, such as oil spills, are accidentally released into the environment. A number of pollutants are known to cause soil, water and air pollution, of which hydrocarbon or petroleum pollutants are most common. They contaminate land and water bodies by accidental spills, like the Alaska oil spill in 1989 and oil spills during the Gulf War, leakage from pipelines and other human activities. Different well-developed chemical and physical methods are available to detoxify the sites where contaminants persist, but the major disadvantages of these methods are that they are not economical and the cost of the process is quite high. Different strategies have been developed so far to combat these recalcitrant but most promising technique is through Bioremediation (Figure 10.1).

Bioremediation consists of using naturally occurring or laboratory-cultivated microorganisms to reduce or eliminate toxic pollutants. Petroleum products are a rich source of energy, and some organisms are able to take advantage of this and use hydrocarbons as a source of food and energy. This results in the breakdown of these complex compounds into simpler forms such as carbon dioxide and water. Bioremediation thus involves detoxifying hazardous substances instead of merely transferring them from one medium to another. This process is less disruptive and can be carried out at the site, which reduces the need for transporting these toxic materials to separate treatment sites. Crude oil can be accidentally or deliberately released into the environment leading to pollution (Thouand et al., 1999). Different countries have regulatory bodies that monitor the level of release of hydrocarbons in the environment. The permissible limit of hydrocarbon is well-documented by most of the countries; however, the release of these hydrocarbons in the environment, such as in water, may cross the permissible limits set by the regulatory bodies (Spence et al., 2005).

The presence of these pollutants may cause adverse effects on living as well as nonliving components of the ecosystem (Mueller et al., 1992). To overcome or dispose of these contaminants, use of large, electrically operated incinerators is an available option, or the other option is to bury them in the fields (U.S. EPA 2001; ITOPF 2006). One of the disadvantages of these methods, particularly with incinerators, is the emission of dangerous fumes. In the case of burying, there may be the possibility of contaminating the groundwater (Pye and Patrick, 1983). Some technologies are being developed that markedly enhance the microbial destruction or degradation of organic pollutants that otherwise would have

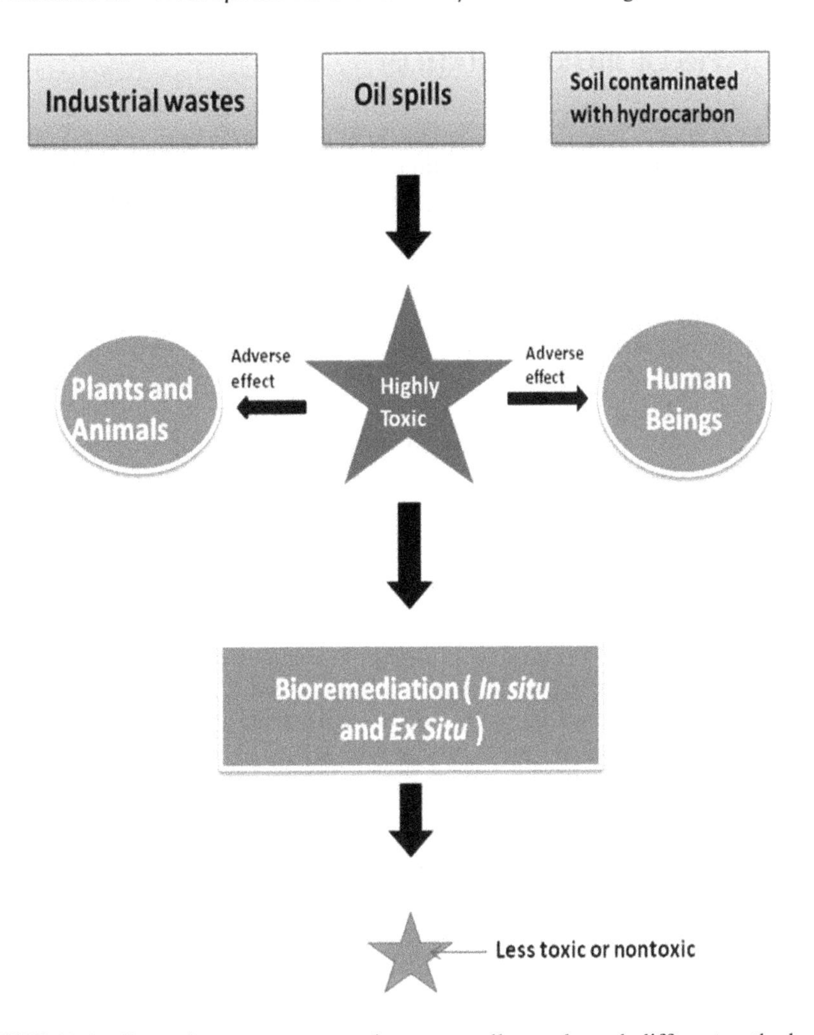

FIGURE 10.1 Strategies to overcome environment pollutant through different methods.

persisted even after the cleanup of polluted groundwater or soil using physical and chemical methods (Alexander, 1994). The success of the bioremediation efforts in the cleanup of the 1989 oil spill caused by the tanker *Exxon Valdez* (Atlas and Bartha, 1998) in Prince William Sound, in the Gulf of Alaska, has drawn the attention of scientific communities to solve the problem by using economic and nonhazardous methods to clean up oil spills in the aquatic environment. A number of techniques are available—physical and chemical—to clean up the aquatic environment, but all these techniques have many demerits. The bioremediation process involving microorganisms and plants, however, can be effective as well as economical. These techniques have now become more popular due to the use of new genetically engineered bacteria that have the ability to combat these orthodox pollutants.

10.2 TYPES OF BIOREMEDIATION

10.2.1 Biostimulation

In this process, nutrients and oxygen (liquid or gas form) are added to contaminated water, which increase or support the growth of bacteria, which in turn cause the remediation of contaminated water.

10.2.2 Bioaugmentation

Microorganisms having the capability to clean a particular contaminant are added to the contaminated water or site. Bioaugmentation is the process in which genetically engineered microorganisms have a gene that functions to degrade a particular pollutant and has catalytic capabilities to degrade aromatic hydrocarbons (Mrozika and Segetb, 2009).

10.3 BIOREMEDIATION APPLICATION

Bioremediation can take place under both aerobic and anaerobic conditions. In aerobic conditions, microorganisms use the available atmospheric oxygen. In the presence of sufficient oxygen, microorganisms convert organic contaminants to carbon dioxide and water. Under anaerobic conditions with a lack of oxygen, the microorganism breaks down chemical compounds in the soil to release the energy. Bioremediation applications are divided into two categories: in situ bioremediation and ex situ bioremediation.

10.3.1 In Situ Bioremediation

In situ bioremediation is treatment of the contaminated site, whether it is groundwater or contaminated soil. This kind of bioremediation process occurs on the site. It has certain advantages, i.e., it is cost-effective, excavation is not required and it does not create much dust as compared to ex situ techniques. Contaminated soil or groundwater is treated at the place or location at which it is found. It is also possible to treat a large volume of soil at once. It is slower than ex situ techniques, is difficult to manage and is most effective at sites with sandy soil (having greater pore size or permeability). In situ bioremediation includes bioventing and biosparging.

10.3.1.1 Bioventing

It is also known as soil venting or soil vacuum extraction. It is used for removing oily contaminants above the water table. It is a very common and important process. The in situ technique involves supplying air and nutrients through wells to contaminated soil or site to stimulate the native bacteria. A well is bored near the point of contamination but above the water table, a vacuum is applied and volatile emissions are vented. This causes oxygenated air to come into contact with the undissolved contaminated subsurface material, which gets biodegraded. Appearance of CO_2 in the extraction well indicates the biodegradation activity. It is used for removing simple hydrocarbons and can be used where the contamination is deep under the surface. Bioventing techniques have been successfully used to remediate soils contaminated by petroleum hydrocarbons, nonchlorinated solvents, some pesticides, wood preservatives and other organic chemicals.

10.3.1.2 Biosparging

In situ sparging was first implemented in Germany in 1985. The procedure includes injection of air under pressure below the water table to increase groundwater oxygen concentration and enhance the rate of biological degradation of contaminants by naturally occurring bacteria. The purpose of biosparging is to increase the contact between the soil and groundwater. Biosparging can be used to reduce petroleum constituents that are adsorbed to soil within the capillary fringe, below the water table, or dissolved in groundwater.

The injection of peroxide is used to deliver oxygen to stimulate the native microorganisms by circulating hydrogen peroxide to enhance the bioremediation of organic contaminants.

10.3.2 Ex Situ Bioremediation

Ex situ bioremediation requires excavation of contaminated soil or pumping of groundwater before they can be treated. It can be faster and can be used to treat a wider range of contaminants.

Ex situ techniques include slurry-phase bioremediation and solid-phase bioremediation.

10.3.2.1 Slurry-Phase Bioremediation

This is a relatively rapid process compared to the biological treatment process. This process is similar to the conventional activated sludge process. Waste is taken in a reactor in the form of slurry and mixed with the microorganisms and organic and inorganic nutrients. Nutrients and oxygen are added, and conditions in the bioreactor are controlled to create an optimum environment for the microorganisms to degrade the contaminants. A neutralizing agent, a surfactant and a dispersant are added to hasten degradation through pH adjustment, toxicity reduction and area of action.

After treatment, solids are allowed to settle, which are then disposed of or treated further if they still contain pollutants. The success of the process is highly dependent on the chemical properties of the contaminated material.

10.3.2.2 Solid-Phase Bioremediation

The solid-phase bioremediation process includes the following techniques:

Land farming: This is a simple technique in which contaminated soil is excavated and spread over a prepared bed. It is then tilled time to time until pollutants are degraded. The target of this method is to stimulate normal inhabitant biodegradative microorganisms and facilitate their aerobic degradation of contaminants (Maila and Cloete, 2004). In general, the practice is limited to the treatment of superficial 10–35 cm of soil. Since land farming has the potential to reduce monitoring and maintenance costs, as well as cleanup liabilities, it has received much attention as a disposal alternative.

Composting: This involves mixing contaminated soil with nonhazardous organic substances such as manure or agricultural wastes. The rich content of this mixture enhances the microbial population and elevates the temperature. Microorganisms present in the mixture consume contaminants in soils, ground and surface waters and air. The contaminants are digested, metabolized and transformed into humus and inert by-products, such as carbon dioxide, water and salts (Khan and Anjaneyulu, 2006). It is effective in degrading or altering many types of contaminants, such as chlorinated and nonchlorinated hydrocarbons, solvents, heavy metals, pesticides, petroleum products and explosives. Compost used in bioremediation is referred to as "tailored" or "designed" compost in that it is specially made to treat specific contaminants at specific sites.

Biopiles: This is an ex situ bioremediation and is a combination of two technologies, i.e., land farming and composting, in which modified cells are constructed as aerated composted piles (Hazen et al., 2000). It is generally applied for the treatment of surface contamination with petroleum hydrocarbons. It is a modified version of land farming to minimize losses of the contaminants by leaching and volatilization. It also provides appropriate conditions for aerobic and anaerobic microorganisms.

Bioreactors: Slurry reactors or aqueous reactors are used for ex situ treatment of contaminated soil and water pumped up from a contaminated plume. Bioremediation in reactors involves the processing of contaminated solid material (soil, sediment and sludge) or water through an engineered containment system. A slurry bioreactor may be defined as a containment vessel and apparatus used to create a three-phase (solid, liquid and gas) mixing condition to increase the bioremediation rate of soil bound and water-soluble pollutants as a water slurry of the contaminated soil and biomass (usually indigenous microorganisms) capable of degrading target contaminants.

10.4 SOURCE OF HYDROCARBON POLLUTANTS IN ENVIRONMENT

Among all the pollutants known so far, hydrocarbon pollutants have a great importance because they affect all forms of life. Petroleum compound is a mixture of low- and high-molecular-weight hydrocarbons, and its crude form consists of saturated alkanes, branched alkanes, alkenes, naphthenes, naphthene aromatics, large aromatic molecules like resins, asphaltenes and hydrocarbons containing different functional groups such as carboxylic acids and ethers (Shukla and Singh, 2012). These hydrocarbons my reach the environment not only by accidents on oil platforms and through ships used for hydrocarbon transport but also by discharging water used to wash the tanks of tanker vessels into the sea (Arulazhagan et al., 2010). Hydrocarbons can also get access to the environment through garages, petrol stations, oil storage sites and timber yards Most of the hydrocarbons such as ethylene, methane, propylene, benzene, toluene, xylene, cumene and styrene are released into the environment during their industrial production and use.

10.5 MICROORGANISMS KNOWN FOR BIOREMEDIATION OF HYDROCARBONS

The utilization of microorganisms for the production of antibiotic and other useful products such as in the food industry is well known. However, it has now been shown that some microorganisms or genetically modified microorganisms can

TABLE 10.1
Bacteria and Fungi Known for Degrading Hydrocarbons

Bacteria	Fungi
Achromobacter, Acinetobacter, Actinomyces, Aeromonas, Alcaligenes, Arthrobacter, bacillus, Brevibacterium, Erwinia, Flavobacterium, Klebsiella, Lactobacillus, Leucothrix, Moraxella, Nocardia, Peptococcus, pseudomonas, etc.	*Aspergillus, Aureobasidium, Botrytis, Candida, Cephalosporium, Cladosporium, Debaryomyces, Fusarium, Hansenula, Helminthosporium, Mucor, Oidiodendron, Paecilomyces, Phialophora, Penicillium, Rhodosporidium*

effectively degrade waste or be utilized as growth factors or for metabolic activities. Some bacterial and fungal species are known to have hydrocarbon-degrading abilities (Table 10.1).

10.6 MECHANISM OF MICROBIAL DEGRADATION OF HYDROCARBONS

The mechanism of petroleum or hydrocarbon degradation involves the catabolic pathway used by the microorganisms. Theses catabolic pathways are as follows:

1. *alk* pathways (for degradation of alkanes)
2. *nah* pathway (for degradation of naphthalene)
3. *Xyl* pathway (for degradation of toluene)

Biodegradation of petroleum products can be enhanced by elevating the temperature of the medium and use of thermophilic bacteria (Nzila, 2018). An increase in the temperature of the medium, i.e., petroleum product, increases the affinity of bacteria toward petroleum hydrocarbon. This is possible due to the decreased viscosity and increased diffusion coefficient (Tse and Ma, 2016). However, at low temperatures, the rate of degradation is lower. The dioxygenases—enzymes produced by the microorganisms, particularly by thermophilic bacteria—act on polyaromatic hydrocarbons resulting in the formation of monohydroxylated and then dihydroxylated aromatic compounds. Other enzymes that are important in biodegradation are cytochrome P450 hydroxylase, integral membrane di-iron alkane hydroxylases, membrane-bound copper containing methane monooxygenases and soluble di-iron methane monooxygenases (Van Beilen and Funhoff, 2007).

10.7 BIOREMEDIATION OF HYDROCARBONS

Petroleum hydrocarbons are well known to possess health-hazardous chemicals that cause negative effects on human beings as well as on plants and animals. Hydrocarbon pollution has increased due to the emission of vehicles, oil spills and manufacturing of these hydrocarbons. So, there is a need for some ecofriendly methods that can effectively minimize the level of contaminants to below permissible

TABLE 10.2
Microorganism Used in the Bioremediation of Hydrocarbons

S. No.	Name of Microorganisms	Pollutant/Contaminant	Reduction	References
1	*Serratia marcescens* BC-3 + *Glomus intraradices*	Petroleum-contaminated site	74.0%	Dong et al. (2014)
2	*Pseudomonas*	Kerosene	46.30%	Nduka et al. (2012)
3	*Candida catenulata*	Petroleum hydrocarbon	84.0%	Joo et al. (2008)
4	*Pseudomonas aeruginosa* strains S4.1, S5 3, *Bacillus* sp. Strain S3.2., *Bacillus* sp. strains 113i, O63 and *Micrococcus* sp. strain S.	Engine oil	100%	Ghazali et al. (2004)
5	*Pseudomonas* sp.	PAH (polycyclic aromatic hydrocarbon)	....	Mrozik et al. (2003)
6	*Pseudomonas stutzeri* strain AN10	Naphthalene	—	Bosch et al. (1999)
	Acinetobacter, Pseudomonas, Bacillus, Flavobacterium, Corynebacterium, and *Aeromonas*	Hydrocarbon	76%	Vasudevan and Rajaram (2001)
7	*Pseudomonas* sp.	Crude oil	72%	Mirdamadian et al. (2010)
8	*Aeromonas hydrophila, Alcaligenes xylosoxidans, Gordonia* sp., *Pseudomonas fluorescens, P. putida, Rhodococcus equi, S. maltophilia* and *Xanthmonas* sp.	Diesel-contaminated soil	89%	Szulc et al. (2014)
9	*Ochrobactrum* sp., *Stenotrophomonas maltophilia,* and *P. aeruginosa*	Crude oil	83.49%	Varjani et al. (2015)
10	*Ochrobactrum* sp., *Enterobacter cloacae*, and *Stenotrophomonas maltophilia*	PAH	74%	Arulazhagan et al. (2010)

levels. Biodegradation in the environment is a natural process, and the microorganisms present in soil and water are capable of carrying out biodegradation efficiently (Table 10.2). However, hydrocarbons inhibit the growth of natural microflora of soil and water. Under hydrocarbon-rich conditions, only those microorganisms survive that are capable of utilizing hydrocarbons as a source of carbon and energy. The majority of microorganisms do not utilize hydrocarbons completely and degrade the hydrocarbon contaminants only partially or higher.

Mirdamadian et al. (2010) isolated petroleum-degrading bacteria from a petroleum-contaminated site through an enrichment culture technique. A total of five bacterial species (*Pseudomonas* (L and W strains), *Rhodococcus*, *Bacillus* and *Micrococcus*) were isolated and identified. These strains were screened for petroleum-degradation ability. Of the tested bacterial species, the highest degradation of crude oil was seen in *Pseudomonas* (72%), followed by *Rhodococcus* (49%). Rouviere and Chen (2003) isolated *Brachymonas petroleovorans* CHX species from wastewater from a petroleum refinery. It has the ability to grow aerobically on cyclohexane, light hydrocarbons (C5–C10), as well as on some aromatic compounds such as toluene and *m*-cresol. Sluis et al. (2002) reported that *Pseudomonas butanovora* can grow with butane due to the presence of a soluble butane monooxygenase enzyme that catalyzes the oxidation of butane to 1-butanol. Enhanced biodegradation of crude oil was observed when bioremediation of crude oil polluted site was done by using carrot peel waste (Hamoudi-Belarbi et al., 2018).

Due to the increased demand of vehicles in India as well as worldwide, there is an increased requirement for petroleum fuel. To meet the requirements, there have been installations of petrol pumps in various places. The tanks for petroleum products are generally buried in the earth that may accidentally break or leak, and the oil may seep into the soil and contaminate both soil and groundwater. During the transport of oil from the gulf and other neighboring countries through sea routes, sometimes accidents and oil spills may be encountered (e.g., Deep Water Horizon oil spill accident in the Gulf of Mexico) (Xue et al., 2015). The chemical constituents of petroleum products have a great impact on aquatic flora and fauna. If the soil is contaminated with hydrocarbons, it causes a reduction in the natural microbial population of soil (Labud et al., 2007). These natural microbial communities are involved in the nitrogen cycle, and due to contamination through hydrocarbons, the nitrogen cycle is disturbed. Crude oil contains hydrocarbon compounds that have carcinogenic, mutagenic and cytotoxic effects on human beings (Ramirez et al., 2017). Bioremediation has many advantages, but it has some limitations also. It is affected by environmental factors and requires sufficient nutrients. Hydrocarbons are not preferable substrates for the microorganisms, but microorganisms in the absence of their preferable nutrient utilize hydrocarbons as a source of carbon and energy (Xu et al., 2018). These are some of the issues associated with bioremediation technology that limits its wide range of application.

REFERENCES

Alexander, M. 1994. *Biodegradation and Bioremediation.* Publishers Academic Press Inc., USA, p. XI.

Arulazhagan, P., Vasudevan, N., Yeom, T. 2010. Biodegradation of polycyclic aromatic hydrocarbon by a halotolerant bacterial consortium isolated from marine environment. *International Journal of Environment, Science and Technology.* 7(4):639–652.

Atlas, R. M., Bartha, R. 1998. Fundamentals and applications. In: *Microbial Ecology*, 4th ed. Benjamin/Cummings Publishing Company, Inc., California, USA, pp. 523–530.

Bosch, R., García-Valdés, E., Moore, E. R. 1999. Genetic characterization and evolutionary implications of a chromosomally encoded naphthalene-degradation upper pathway from pseudomonas stutzeri AN10. *Gene.* 236(1):149–57.

Dong, R., Gu, L., Guo, C., Xun, F., Liu, J. 2014. Effect of PGPR *Serratia marcescens BC-3 and Glomus intraradices* on phytoremediation of petroleum contaminated soil. *Ecotoxicology.* 23(4):674–680.

Ghazali, F. M., Zaliha, R. N., Rahman, A., Salleh, A. B., Basri, M. 2004. Biodegradation of hydrocarbons in soil by microbial consortium. *International Biodeterioration & Biodegradation*. 54:61–67.

Hamoudi-Belarbi, L., Hamoudi, S., Belkacemi, K., Nouri, L., Bendifallah, L., Khodja, M., 2018. Bioremediation of polluted soil sites with crude oil hydrocarbons using carrot peel waste. *Environments*. 5:124.

Hazen, T. C., Tien, A. J., Worsztynowicz, A., Altman, D. J., Ulfig, K., Manko, T. 2000. Biopiles for remediation of petroleum contaminated soils: a Polish case study. In: Sasek, V., Glaser, J. A., Baveye, P. (eds.), *The Utilization of Bioremediation to Reduce Soil Contamination: Problems and Solutions*. NATO Science Series IV, pp. 229–246.

ITOPF. 2006. Oil spills in port, London (https://www.itopf.org/knowledge-resources/documents-guides/document/oil-spills-in-ports-2006/).

Joo, Hung-Sooa, Ndegwa, P. M., Shoda, M., Phae, Chae-Gun. 2008. Bioremediation of oil-contaminated soil using *Candida catenulata* and food waste. *Environmental Pollution*. 156:891–896.

Khan, Z., Anjaneyulu, Y. 2006. Bioremediation of contaminated soil and sediment by composting. *Remediation*. 16(4):109–122.

Labud, V., Garcia, C., Hernandez, T. 2007. Effect of hydrocarbon pollution on the microbial properties of a sandy and a clay soil. *Chemosphere*. 66:1863–1871. doi:10.1016/j.chemosphere.2006.08.021.

Maila, M. P., Cloete, T. E. 2004. Bioremediation of petroleum hydrocarbons through landfarming: are simplicity and cost-effectiveness the only advantages? *Reviews in Environmental Science and Bio/Technology*. 3(4):349–360.

Mirdamadian, S. H., Emtiazi, G., Golabi, M. H., Ghanavati, H. 2010. Biodegradation of petroleum and aromatic hydrocarbons by bacteria isolated from petroleum-contaminated soil. *Journal of Petroleum and Environmental Biotechnology*. 1:102. doi:10.4172/2157-7463.1000102.

Mrozik, A., Piotrwska-Seget, Z., Labuzek, S. 2003. Bacterial degradation and bioremediation of polycyclic aromatic hydrocarbon. *Polish Journal of Environmental Studies*. 12(1):15–25.

Mrozika A., Segetb, Z. P. 2009. Bioaugmentation as a strategy for cleaning up of soils contaminated with aromatic compounds. *Microbiological Research*. 165(5):363–375.

Mueller, J. G., Resnick, S. M., Shelton, M. E., Pritchard, P. H. 1992. Effect of inoculation on the biodegradation of weathered Prudhoe Bay crude oil. *Journal of Industrial Microbiology*. 10:95–102.

Nduka, J. K., Umeh, L. N., Okerulu, I. O., Umedum, L. N., Okoye, H. N. 2012. Utilization of different microbes in bioremediation of hydrocarbon contaminated soils stimulated with inorganic and organic fertilizers. *Journal of Petroleum and Environ Biotechnology*. 3:116. doi:10.4172/2157-7463.1000116.

Nzila, A. 2018. Current status of the degradation of aliphatic and aromatic petroleum hydrocarbons by thermophilic microbes and future perspectives. *International Journal of Environmental Research and Public Health*. 15:2782.

Pye, V. I., Patrick, R. 1983. Groundwater pollution in the United States. *Science*. 221:713–718.

Ramirez, M. I., Arevalo, A. P., Sotomayor, S., Bailon-Moscoso, N. 2017. Contamination by oil crude extraction—refinement and their effects on human health. *Environment Pollution*. 231:415–425.

Rouviere, P. E., Chen, M. W. 2003. Isolation of *Brachymonas petroleovorans* CHX, a novel cyclohexane-degrading beta-proteobacterium. *FEMS Microbiology Letters*. 227(1):101–6.

Shukla, A., Singh, C. S. 2012. Hydrocarbon pollution: effects on living organisms, remediation of contaminated environments, and effects of heavy metals co-contamination on bioremediation. In: Romero-Zerón, Dr. L. (ed.), *Introduction to Enhanced Oil*

Recovery (EOR) Processes and Bioremediation of Oil Contaminated Sites. InTech. ISBN: 978-953-51-0629-6. Available from: http://www.intechopen.com/books/introduction-to-enhanced-oil-recovery-eor-processes-and-bioremediationof-oil-contaminated-sites/heavy-metals-interference-in-microbial-degradation-of-crude-oil-petroleumhydrocarbons-the-challenge.

Sluis, M. K., Sayavedra-Soto, L. A., Arp, D. J. 2002. Molecular analysis of the soluble butane monooxygenase from *'Pseudomonas butanovora'*. *Microbiology*. 148(Pt 11):3617–29.

Spence, J. M., Bottrell, S. H., Thornton, S. F., Richnow, H. H., Spence, K. H. 2005. Hydrochemical and isotopic effects associated with petroleum fuel biodegradation pathways in a chalk aquifer. *Journal of Contaminant and Hydrology*. 79:67–88.

Szulc, A., Ambro˙zewicz, D., Sydow, M., Ławniczak, Ł., Piotrowska-Cyplik, A., Marecik, R., et al. 2014. The influence of bioaugmentation and biosurfactant addition on bioremediation efficiency of diesel-oil contaminated soil: feasibility during field studies. *Journal of Environment Management*. 132:121–128. doi:10.1016/j.jenvman. 2013.11.006.

Thouand, G., Bauda, P., Oudot, J., Kirsch, G., Sutton, C., Vidalie, J. F. 1999. Laboratory evaluation of crude oil biodegradation with commercial or natural microbial inocula. *Canadian Journal of Microbiology*. 45:106–115.

Tse, C., Ma, K. 2016. Growth and metabolism of extremophilic microorganisms. In: Rampelotto P. H. (ed.), *Biotechnology of Extremophiles: Advances and Challenges*. Springer International Publishing, Cham, Switzerland, pp. 1–46.

U. S. EPA. 2001. Toxic release inventory (TRI) Public data release, Washington, D.C. (www.epa.gov/tri/tridata/tri01).

van Beilen, J. B., Funhoff, E. G. 2007. Alkane hydroxylases involved in microbial alkane degradation. *Applied Microbiology and Biotechnology*. 74:13–21.

Varjani, S. J., Rana, D. P., Jain, A. K., Bateja, S., Upasani, V. N. 2015. Synergistic ex-situ biodegradation of crude oil by halotolerant bacterial consortium of indigenous strains isolated from on shore sites of Gujarat, India. *International Journal of Biodeterioration and Biodegradation*. 103:116–124. doi:10.1016/j.ibiod.2015.03.030.

Vasudevan, N., Rajaram, P. 2001. Bioremediation of oil sludge-contaminated soil. *Environment International*. 26:409–411.

Xu, X., Liu, W., Tian, S., Wang, W., Qi, Q., Jiang, P., Gao, X., Li, F., Li, H., Yu, H. 2018. Petroleum hydrocarbon degrading bacteria for the remediation of oil polluted under aerobic conditions: a perspective analysis. *Frontiers in Microbiology*. 9(2885).

Xue, J., Yu, Y., Bai, Y., Wang, L., Wu, Y. 2015. Marine oil-degrading microorganisms and biodegradation process of petroleum hydrocarbon in marine environments: a review. *Current Microbiology*. 71:220–228. doi:10.1007/s00284-015-0825-7.

11 Rhizosphere Bioremediation: Green Technology to Clean Up the Environment

M.H. Fulekar and Jyoti Fulekar*

CONTENTS

11.1 INTRODUCTION

The Hazardous Waste Management and Handling Rules, 1989, categorized the different types of hazardous wastes under the Environment Protection Act, 1986 The wastes generated by industries such as petrochemical industries, pesticides, chemicals, dyes and drug industries and pharmaceuticals consist of organic compounds—halogenated and non-halogenated hydrocarbons, toxic metals and inorganic compounds. These wastes are being treated by physicochemical and biological methods. Industries look to the standard guidelines and established methods

* Corresponding author: Email: mhfulekar@yahoo.com

for the treatment of hazardous waste generated by them. Recent developments in technology involve bioremediation (Anderson et al., 1993).

Bioremediation is the process by which microorganisms are stimulated to degrade complex organic compounds into environmentally friendly substances such as inorganic compounds, biomass and CO_2. The stimulation is achieved by supplying nutrients, electron acceptors and a carbon source. The microorganisms use organic waste as a carbon source and degrade the waste (Fulekar, 2005).

Rhizosphere bioremediation is a plant-assisted bioremediation, which has multiple effects in the bioremediation process due to symbiotic association of mycorrhizal fungi and bacteria along the root zone in the rhizosphere (Wei et al., 2003). The rhizosphere is considered an ecological remediation unit to treat the contaminated soils containing complex organic compounds, including toxic metals, in the rhizosphere. The plants release exudates such as short-chain organic acids, phenolics, sugars, alcohols and a small concentration of high-molecular-weight compounds (enzymes and protein). The plant-released enzymes such as laccase, dehalogenase, nitrilase, nitroreductase and peroxidase are beneficial in enhancing bioremediation.

In the present study, the pot culture technique was employed to develop mycorrhiza in a soil:sand (3:1) medium inoculated using sorghum as the host plant. The physicochemical parameters of the rhizospheric soil were monitored and maintained during the process (Table 11.1). The characterization of mycorrhiza was done estimating spore counts by a wet sieving and decanting method and percentage root colonization by the Phillips and Hayman (1970) method. The final spore count in developed mycorrhizopheric soil was found to be 60 spores per 5 grams of soil with 75–78% of root colonization. The microbial assemblages were abundant in the rhizosphere, comprising total viable counts of bacteria to be 6.6×10^5 and actinomycetes to be 14×10^3. The microbial diversity in this ecological unit was observed. Bacteria such as pseudomonas, bacillus, arthrobacter, azotobacter, rhizobium and azospirillum and fungi such asaspergillus, penicillium and actinomycetes were assessed. The

TABLE 11.1
Physicochemical Characterization of Developing Mycorrhizal Soil Inoculum

Time period in days	pH	Moisture Content (%)	Electrical Conductivity	(meq/100gm) Organic Carbon (gm/kg)	Total Nitrogen	(gm/kg) Total Phosphorus (Pt)(gm/kg)	Organic Phosphorus (Po)(gm/kg)	Potassium(mg/Kg)	Sodium (mg/Kg)	C/N	N/P
0	6.4	42.4	0.38	72	5.8	0.72	0.19	21	2.3	12.41	8.06
15	6.8	40.4	0.37	84	6.2	0.71	0.21	23	2.4	13.55	8.73
30	6.8	48.1	0.45	136	6.2	0.72	0.29	22	2.8	21.94	8.61
45	7.0	46.8	0.44	160	7.3	0.77	0.31	23	2.6	22.07	9.48
60	7.2	45.4	0.42	184	8.3	0.81	0.36	24	2.3	22.17	10.25
75	7.3	42.2	0.34	259	8.4	0.81	0.42	22	3.2	30.83	10.37

mechanism of the symbiotic association in the rhizosphere has been studied. The developed mycorrhizosphere serves as an efficient and effective bioremediation technology. This low-cost technology can be adapted by small-scale industries for the treatment of hazardous waste containing organic compounds as the environmentally sound technology.

11.2 DEVELOPMENT OF MYCORRHIZA

The term mycorrhiza is a combination of two words: *mikes*, a Greek word meaning fungus, and *rhiza*, a Latin word meaning roots. It is an association between plant and fungus that colonizes the cortical tissue of roots during periods of active plant growth (Fulekar, 2005). This symbiosis is characterized by bidirectional movement of nutrients where carbon flows to the fungus and inorganic nutrients move to the plant, thereby providing a critical linkage between the plant root and soil. In infertile soils, nutrients taken up by the mycorrhizal fungi can lead to improved mycorrhizal fungi developed by adopting the "pot culture technique." In the present study, the alluvial surface soil was collected from the field and passed through a 2-mm sieve. The soil was characterized for physicochemical and biological parameters (Table 11.2). The soil and sand were placed in the pot in a ratio of 3:1. A proportionate quantity of soil-based inoculum was added in the medium with sorghum seeds as the host plant in the pots. The pots were then watered daily for seed growth. The physicochemical parameters such as pH, electrical conductivity

TABLE 11.2
Physicochemical Characteristics of Soil and Mycorrhizal Soil

Physicochemical Characteristics of Soil		
Soil Parameters	Values	
	SOIL	MYCORRHIZAL SOIL
pH	6.4	7.3
Moisture Content	42.4	42.2
Electrical Conductivity (meq/100 gm)	0.38	0.34
Organic Carbon (gm/kg)	72	259
Total Nitrogen (gm/kg)	5.8	8.4
Total Phosphorous(Pt) (gm/kg)	0.72	0.81
Organic phosphorus (gm/kg)	0.19	0.42
Potassium (mg/Kg)	21	22
Sodium (mg/Kg)	23	3.2
C/N	12.41	30.83
N/P	8.06	10.37

and total dissolved solids, as well as moisture content and percentage organic carbon, were monitored (Jackson, 1973). The nutrient (Hoagland's solution) was supplied as per the requirement. The composition of the Hoagland's solution was KH_2PO_4 (potassium dihydrogen phosphate): 0.02 M; KNO_3 (potassium nitrate): 1 M; Ca $(NO_3)_2$ (calcium nitrate): 1M; MgSO4 (magnesium sulfate): 1M; trace element solution with Fe-EDTA.

The mycorrhizal fungi and microbial parameters were assessed. The development of mycorrhiza was observed and counted by Philips and Hayman (Philip et al., 1970) using the wet sieving and decanting method (Gerdemann and Nicolson, 1963). There was good development of mycorrhizal soil prepared by this technique for rhizosphere bioremediation (Figure 11.1).

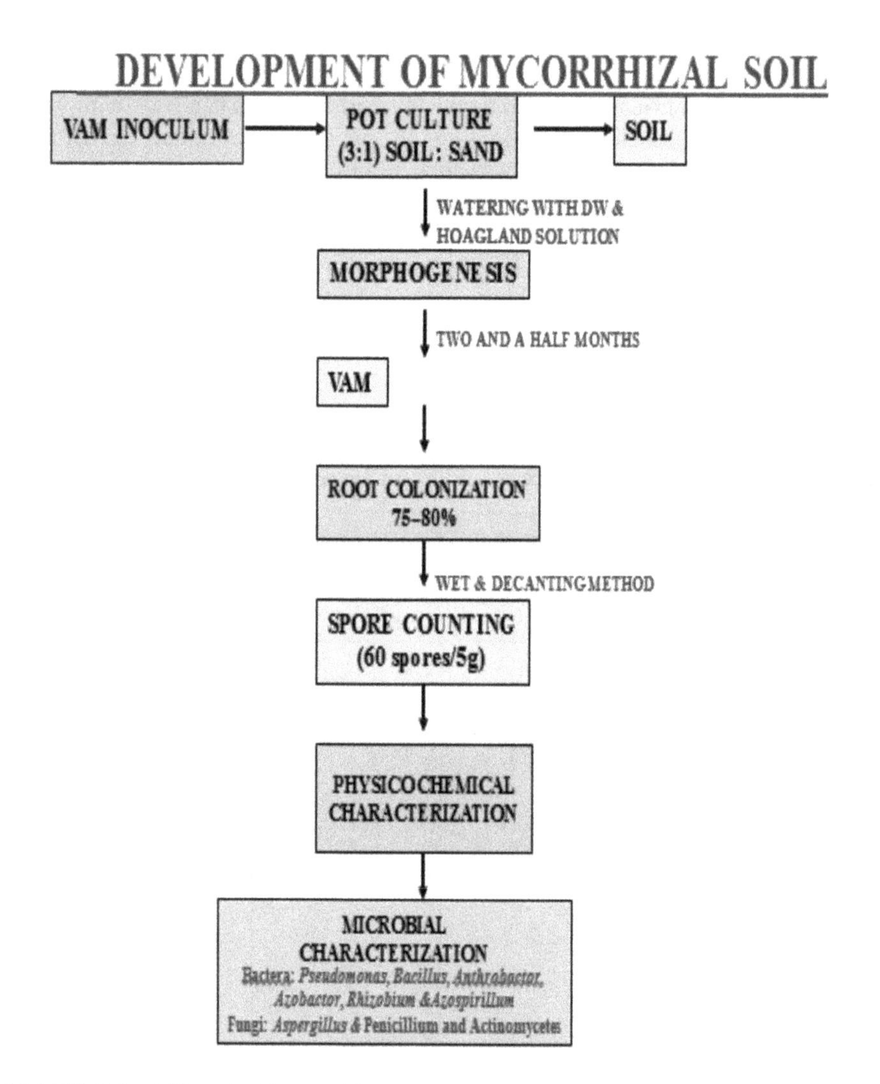

FIGURE 11.1 Life cycle of mycorrhiza.

11.3 LIFE CYCLE OF MYCORRHIZA

The vesicular-arbuscular mycorrhiza (VAM) fungus belongs to the order of Glomales and includes a limited number of genera, i.e., Glomus, Acaulospora, Gigaspora and Scutellospora. This obligate symbiont grows successfully in the presence of host plants. In a dual-culture system, obtained by inoculating a host plant with fungal spores, the fungus completes its life cycle and undergoes a complex morphogenesis. During this process, spores germinate and produce a vegetative mycelium that contacts the host plant root surface. The appressoria then originate hyphae that initiate the infection of the root and subsequently form intracellular hyphae, coils vesicles and, most importantly, a highly branched intracellular structure called an arbuscule (Verma, 1992).

11.4 ECOLOGICAL REMEDIATION UNIT

In the present study, the rhizosphere of the host plant grown in mycorrhizal soil formed a mycorrhizosphere encompassing the plant roots, the root symbiotic mycorrhizal fungi, bacteria actinomycetes and soil in the immediate vicinity of the mycorrhizal roots (Anderson et al., 1993). In the plant-assisted bioremediation, the rhizosphere helps increase soil organic carbon bacteria and mycorrhizal fungi, all factors that encourage degradation of organic compounds and pesticides in soil (Schnoor, 1997). The study observed that plants release exudates in soil that help to stimulate the degradation of organic chemicals by inducing enzyme systems of existing bacterial populations, stimulating growth of the new species that are able to degrade the waste and/or increasing substrate concentrations for all microorganisms. Exudates include short-chain organic acids, sugars, alcohols, phenolics and small concentrations of high-molecular-weight compounds (enzymes and proteins, Schnoor, 1997).

The Environmental Protection Agency (EPA) Laboratory in Athens, Georgia, examined five plant enzyme systems, including dehalogenase, nitroreductase, Nitrilase, laccase and peroxidase, in sediment and soils in proximity to the root (1mm). Dehalogenase enzymes are important in dechlorination reactions of chlorinated hydrocarbons; nitroreductase is needed in the first step for degradation of nitroaromatics, while laccase serves to break aromatic ring structures in organic contaminants. Peroxidase and nitrilase are important in oxidation reactions in rhizosphere soils.

The study shows that the rhizobacteria genera consists of various species belonging to *Acetobacter, Arthrobacter, Azospirillum, Bacillus, Flavobacter, Pseudomonas, Proteus, Rhizobium, Serretia* and others. Also, plant growth–promoting rhizobacteria include a range of organisms that live in close association with the roots (Ghosh et al., 2005).

In the rhizosphere, the importance of biodegradation, where plants help microbial transformations, can be shown in many ways:

- Mycorrhizal fungi associated with plant roots metabolize the organic pollutants.
- Plant exudates stimulate bacterial transformations (enzyme induction).
- Buildup of organic carbon increases microbial metabolization rates (substrate enhancement).
- Plants provide habitat for increased microbial population and activity.
- Oxygen is pumped to roots ensuring aerobic transformations.

11.5 RHIZOSPHERE BIOREMEDIATION

The ecological remediation unit developed by this technique has a symbiotic association of bacteria, fungi and actinomycetes along the root zones. The plant exudate as well as enzymes secreted by the plant and microbial population make it possible to degrade organic compounds to the environmentally friendly compounds, biomass, inorganic compounds and CO_2 (Ghosh et al., 2005). The presence of inorganic compounds to the extent of tolerable limits under that environmental condition undergo bioremediation (Ghosh et al., 2005). Environmental variations take place depending upon the factors prevailing in the field. However, acceleration and enhancement in rhizosphere bioremediation can be done by developing an ecological remediation unit for the degradation of complex organic compounds. The development of such ecological units in the field will prove to be beneficial for the bioremediation of hazardous waste.

Most of the conventional remediation methods for environmental cleanup are physicochemical methods or a combination of physicochemical and biological. The physicochemical treatment methods such as incineration and thermal desorption used for the treatment of the organic contaminants are expensive, labor intensive and noneffective to degrade xenobiotics that result in the generation of secondary pollutants that remain persistent in the environment and are difficult to remove. Increasing costs and limited efficiency of these traditional physicochemical treatments have spurred the development of new remediation technologies that consume less time, are cost-effective, are efficient and are environmentally friendly. Bioremediation is emerging as an effective, innovative and attractive alternative to conventional cleanup technologies due to its relatively low capital costs and inherently aesthetic nature. It is an invaluable tool for wider application in the realm of environmental protection. This technology includes phytoremediation (plants) and rhizoremediation (plant-microbe interaction). Rhizoremediation, which is the most evolved process of bioremediation, involves the removal of specific contaminants from contaminated sites by mutual attraction of plant roots and suitable microbial flora. Thus, rhizosphere bioremediation is an emerging technology that involves interaction between plant roots and associated microbes for environmental cleanup and has become a promising approach for the remediation of contaminated sites.

11.6 PLANT-MICROBE INTERACTIONS IN ROOT ZONE

In the natural environment, many of the limitations for remediation can be overcome by the dynamic synergy that exists between plant root and soil microorganisms. In fact, the activities of rhizospheric microbes and a plant's ability to cope with soil contaminations may be more closely related than previously realized. The microbial activity in the immediate vicinity of the root (rhizosphere) seems to offer a favorable environment for co-metabolism of soil-bound and recalcitrant chemicals (Walton and Anderson, 1990; Shann, 1995). The microbial transformations of organic compounds are usually not driven by energy needs, but a necessity to reduce toxicity for which microbes may have to suffer an energy deficit. Thus, the processes may be helped and driven by abundant energy that is provided by root exudates. Such stimulation of soil microbial communities by root exudates also benefits plants through increased availability of soil-bound nutrients and degradation of phytotoxic soil contaminants (El Shatnawi and Makhadmeh, 2001). Soil microorganisms are also known to produce certain bio-surfactant compounds that may further facilitate the removal/degradation of organic pollutants by increasing their availability to plants (Lafrance and Lapointe, 1998) and may also contribute to reduce the toxicity of heavy metals in cases of mixed pollution (Sandrin et al., 2000).

11.7 MICROBIAL COMPOSITION OF THE RHIZOSPHERE

There is a body of evidence to suggest that exogenous metabolic activities of rhizosphere microorganisms may be controlled by plants for provision of root exudates (Walton et al., 1994). This has led to the speculation that complementary interaction may be part of plant responses to chemical stress in soil (Walton et al., 1994). A strong indication for this has emerged from consistent findings that microbial numbers in a rhizosphere are generally several orders of magnitude than those in nonvegetated soil (Reynolds et al., 1999), a phenomenon that has been observed upto a few millimeters away from the soil-root interface. This interaction, however, may be restricted to a much narrower zone for hydrophobic organic compounds, due to their limited aqueous diffusion, different soil textures, organic matter content, etc. Thus, while Corgie et al. (2003) noted a root effect on phenanthrene degradation that extended <0.6 mm from the surface of ryegrass roots in pure fine sand, Joner and Leyval (2003) found that enhanced phenanthrene degradation extended only <0.6 mm from the root surface of the clover and ryegrass in two industrially polluted soils. The microbial population in a plant rhizosphere is usually composed of diverse and synergistic communities rather than a single strain (Anderson and Coats, 1995). Such diversity further promotes remediation of contaminated soils, as degradation organic compounds may require several organisms with distinctive enzyme systems (Meharg and Cairney, 2000). The composition of a microbial community in a rhizosphere is also known to differ both qualitatively and quantitatively from that in non-rhizospheric soil. It is worth mentioning, however, that the true extent of microbial diversity in the soil or rhizosphere remains to be established. This is because only 10% of soil microbial species can be cultured in the laboratory (Lynch, 2002). Although new methods, such as those based on encapsulation of a single cell with flow cytometry (Zengler et al., 2002), could increase our ability to increase the yet uncultured soil microflora, a number of indirect methods are currently used to establish the biodegradation potential of soil microorganisms (Sorensen et al., 2002; Widada et al., 2002). Many methods rely on PCR application of polymorphic DNA (such as ribosomal 16S, 18S and 23S), or gene-specific sequences (Jackobsen, 1995; Wikstrom et al., 1996). Thus, the occurrence of genetic coding for desulfuration of hydrocarbon in polluted field soil has been shown by PCR combined with denaturing gradient gel electrophoresis (DGGE), while these genes were not detected in unpolluted soils (Duarte et al., 2001). Quantitative PCR has also been used to monitor *Pseudomonas* sp. strain B4 through amplification of the dihydroxybiphenyl dioxygenase gene (*bphC*), and the strain CBS3 by amplification of 4-chlorobenzoate coenzyme A dehalogenase gene (Ducrocq et al., 1999). Other DNA studies on oil-contaminated soils showed that microbial population contained the gene *xylE* for toluene degradation, the gene *nah* for PAH degradation and the gene *ndoB* for naphthalene degradation (Milcic et al., 2001). Diesel-degrading microbial strains have also been isolated from rhizosphere of black poplar and certain herbal plants using the gene *alkB* (alkane hydroxylase) as a marker (Tesar et al., 2002).

The use of mycorrhiza may give plants several advantages; for example, protection against harmful conditions in the soil due to low nutrient level, drought and the presence of toxic pollutants. The mycorrhizal hyphae network is more extensive than the root systems that can increase the uptake of pollutants due to the ability to penetrate relatively small soil pores. As with phytoremediation, the main thrust of research into the use of mycorrhizas has focused on phytoextraction of metal ions from contaminated soil (Leyval et al., 2002). Attention is now diverting to mycorrhizal involvement in the degradation of organic pollutants. Many persistent organic pollutants are known to be degraded by mycorrhizal fungi, including 2, 4-D, atrazine and PCBs (Donnelly et al., 1994; Anderson and Coats, 1995). Some ectomycorrhizal species are particularly efficient in degrading recalcitrant organic compounds, such as BTEX, TNT and PAHs, and their capacity is retained even under symbiotic conditions in the soil (Heinonsalo et al., 2000). Even the less aggressive fungi involved in arbuscular mycorrhiza have been shown to enhance degradation of PAHs (Joner et al., 2001; Joner and Leyval, 2003), though in these cases the fungal effects appear to be indirect through modifications of the bacterial community that enhance degradation rates in rhizosphere. The arbuscular mycorrhiza has also been reported to reduce toxicity of PAHs and/or metabolism in spiked soil due to enhance degradation (Joner et al., 2001).

11.8 RHIZOSPHERE CONTROLS BIODEGRADATION OF ORGANIC COMPOUNDS

The susceptibility of organic pollutants to bacterial metabolism and enzymatic attacks differs widely and is related to molecular structure (Sabljic and Piver, 1992). As a consequence of the coevolution of natural organic compounds and soil microorganisms, xenobiotics that structurally resemble naturally occurring compounds are more likely to be susceptible to biodegradation (Semple et al., 2003). Among other activities, plants can release degradative enzymes into the rhizosphere. Reports are available on the degradation of nitroaromatic compounds (e.g., trinitrotoluene) by plant-derived nitroreductases and laccases at the laboratory scale (Boyajian and Carreira, 1997) and in field tests. Other plant-derived enzymes with the potential to contribute to the degradation of organic pollutants in the rhizosphere include dehalogenase involved in dehalogenizing chlorinated solvents such as hexachloroethane and trichloroethylene, peroxidases degrading phenols and phosphatases cleaving phosphate groups from large organophosphate pesticides (Susarla et al., 2002).

Our knowledge of the relative importance and efficiency of plant extracellular enzymes in the presences of degrading microorganisms is still very limited, but the measured half-life of these enzymes suggest that they may actively degrade organic pollutants for days following their release from plant tissues (Schnoor et al., 1995). Apart from the direct release of degrading enzymes, plants are able to stimulate the activities of microbial degrader organisms/communities. Plant-degrading interactions are thought to be most relevant for the success of rhizodegradation. The mechanism of nonspecific stimulation potentially involved in rhizodegradation includes exudates that serve as analogs or co-metabolites of organic pollutants (Siciliano and Germida, 1998). A pollutant analog includes various known root exudates and allelopathic chemicals excreted by plants in response to a pathogen attack in rhizosphere. Such compounds may be the evolutionary cause for the development of detoxifying microbial communities in the rhizosphere (Siciliano and Germida, 1998). Similarly, one might speculate that phytotoxins produce could serve as pollutant analogs.

11.9 BIOAUGMENTATION

Bioaugmentation is defined as the inoculation of contaminated soil, sediments or sludge with isolated strains or consortia with specific organic compound degrading capabilities to enhance in situ or ex situ bioremediation applications. A bioaugmentation attempt is considered to be successful if the rate and percentage of pollutant removal are increased and if the augmented species is not outgrown by the indigenous populations. Gentry et al. (2004) reviewed several new approaches that could potentially increase the persistence and activity of exogenous microorganisms and/or genes following their introduction into the contaminated site.

These techniques include (Moslemy et al., 2002):

1. Bioaugmentation with cells encapsulated in a carrier such as alginate;
2. Gene bioaugmentation where the goal is for the added inoculant to transfer remediation gene into indigenous microorganism;
3. Rhizosphere bioaugmentation where the microbial inoculant is added to the site along with a plant that serves as a niche for the inoculants growth;
4. Phytoaugmentation where the remediation genes are engineered directly into a plant for use in remediation without microbial inoculant.

11.10 RHIZOSPHERE BIOAUGMENTATION

The mechanism responsible for phytoremediation of contaminated soil is thought to cause an increase in the microbial activity that is credited to the rhizosphere effect. Supporting this hypothesis, the population levels of contaminant-degrading bacteria and the potential of soil to degrade contaminants typically increase during phytoremediation. This can be the reason for the use of microbial inoculants that stimulate phytoremediation and also alter the diversity of the root-associated community to enhance degradation. Therefore, the developing approach for bioaugmentation is to add the microbial inoculant to the soil along with a plant that supports the inoculant's growth; this has become a promising method in the plant-soil system. Microorganisms can degrade numerous organic pollutants due to their metabolic machinery and to their capacity to adapt to inhospitable environments. Thus, microorganisms are major players in site remediation. However, their efficiency depends on many factors, including the chemical nature and concentration of pollutants and their availability to microorganisms. The capacity of microbial population to degrade pollutants within an environmental matrix (e.g., soil, sediments, sludge or wastewater) can be enhanced by the introduction of specific microorganisms to the local population (Gentry et al., 2004). Recently, the use of rhizosphere bacteria and their in situ stimulation by plant root for degrading organic contaminants has also received attention from scientists. A few studies demonstrate that the use of plants enhances the rate loss of organic contaminants (April and Sims, 1990; Walton and Anderson, 1990). The feasibility of using rhizobacteria (*Pseudomonas putida*) was demonstrated for the rapid removal of chlorinated pesticides from contaminated soil and to promote germination of radish seeds in the presence of otherwise phytotoxic levels of herbicide 2,4-dichlorophenoxyacetic acid (2,4-D) (Short et al., 1992) and phenoxyacetic acid (PAA, Short et al., 1992). The investigation was undertaken to determine if strains like Pseudomonas putida PPO301/pRO101 and PPO301/pRO301 could be used to bioremediate 2,4-D amended soils via plant-microbe bioaugmentation. For any bioaugmentation strategy, the relationship of the inoculated microorganism with its new biotic or abiotic environments in terms of survival, activity and mitigation can be decisive in the outcome. This is especially true in a complex and dynamic biotope such as soil (Alvey and Crowley, 1996). The factors such as predation by protists, competition with autochthonous microorganisms for nutrients or electron acceptors and the presence of roots that release organic compounds are biotic elements that adversely affect bioaugmentation. There is increasing evidence from the literature that the best way to overcome these ecological barriers is to look for microorganisms from the same ecological niche as the polluted area (Juhanson et al., 2007). Ecological considerations influence the outcome of bioaugmentation. The ecological background (i.e., the diverse natural life forms that live in communities within the biotope inoculated with an exogenous inoculum) constitutes a major barrier in the successful bioremediation performance of such an inoculum. In bulk soil, carbon is often a limiting resource, and its absence can lead to failure of bioremediation (Bolton et al., 1992).

Rhizosphere microorganisms are known to be more metabolically active than those in the bulk soil, because they obtain carbon and energy from root exudates and decaying organic matter (Bolton et al., 1992). In addition to being more active, the rhizosphere population is more abundant, often containing 108 or more culturable bacteria per gram of soil, whereas the bacterial population on the rhizosphere can exceed 109/g root contributing to the enhanced remediation process (Fredrickson and Elliot, 1985).

11.11 RESEARCH CASE STUDY

11.11.1 Rhizosphere Bioremediation

Rhizosphere bioremediation has been carried out in the greenhouse using the pot culture technique for the bioremediation of pesticides by rye grass at varying concentration of anthracene and chlorpyrifos in a separate setup.

11.11.1.1 Research Case Study I: Rhizosphere Bioremediation of Anthracene

Anthracene was amended in developed mycorrhizal soil in the pot at varying concentrations of 10, 25, 50, 75and 100 mg/kg. The rye grass was grown in each pot by sowing the seeds and watering each pot for grass growth for rhizosphere bioremediation.

The pot culture experiment was continued up to a period of 60 days, and samples were analyzed after 15, 30, 45 and 60 days for physicochemical and microbial degradation of anthracene. The rhizosphere microflora identified includes *Pseudomonas* spp., *Flavobacterium*, *Arthrobacter*, *Bacillus* spp., *Streptomyces* spp., *Achromobacter* sp., *Rhizobium* and *Clostridium*.

11.11.1.2 Research Case Study II: Rhizosphere Bioremediation of Chlorpyrifos

Similarly rhizosphere bioremediation of chlorpyrifos was carried at varying concentrations of 10, 25, 50, 75 and 100 mg/kg in mycorrhizal soil by growing rye grass using the pot culture technique.

Bioaugmentation rhizosphere bioremediation was carried out using identified microflora at higher concentrations, i.e., 100 ppm pesticide amended in soil after the 60-day pot culture experiment. The identified microflora by the 16s RNA technique shows the presence of bacteria: *Bacillus subtilis*, *Bacillus mycoides*, *Bacillus sphaericus*, *Clostridium* sp., *Nocardia* sp., *Pseudomonas* sp., *Acinetobacter* sp. and *fungi*: *Fusarium* sp., *Penicillium* sp., *Aspergillus niger*, *Aspergillus fumigatus*, *Streptomyces* sp. The bacteria culture was used in the bioaugmentation rhizosphere bioremediation at a concentration of 25, 50 and 100 ppm up to a period of 25 days and samples were observed at 0, 7, 14, 21 and 25 days.

Rhizosphere bioremediation was found to be effective and efficient in the removal of chlorpyrifos, converting it into intermediates, and thus achieving complete remediation of pesticides. The rhizosphere bioremediation technology can be transferred from the lab to the land for remediation of pesticides in a soil-water agriculture environment.

REFERENCES

Alvey S, Crowley DE 1996. Survival and activity of an atrazine-mineralizing bacterial consortium in rhizosphere soil. Environ. Sci. Technol., 30:1596–1603.

Anderson TA, Coats JR (eds.) 1995. An overview of microbial degradation in the rhizosphere and its implications for bioremediation. In: Bioremediation: Science and Applications. American Society of Agronomy, Madison, WI, pp. 135–143.

Anderson TA, Guthrie EA, Walton BT 1993. Bioremediation in the rhizosphere. Environ. Sci. Technol., 27:2630–2636.

April W, Sims RC 1990. Evaluation of the use of prairie grasses for stimulating polycyclic aromatic hydrocarbon treatment in soil. Chemosphere, 20(1–2):253–265.

Bolton H, Fredrickson JK, Elliot LF 1992. Microbial ecology of the rhizosphere. In: Metting FB (ed.), Soil Microbiology Ecology, Application in Agriculture, Forestry and Environmental Management. Marcel Dekker, Inc., New York, NY, pp. 27–63.

Boyajian GE, Carreira LH 1997. Phytoremediation: A clean transition from laboratory to marketplace? Nature Biotechnology 15:127–128.

Corgie SC, Joner EJ, Leyval C 2003. Rhizospheric degradation of phenanthrene is a function of proximity of roots. Plant Soil, 275:143–150.

Donnelly PK, Hegde RS, Fletcher JS 1994. Growth of PCB degrading bacteria on compounds from photosynthetic plants. Chemosphere, 28:981–988.

Duarte GF, Rosado AS, Seldin L, de Araujo W, van Elsas JD 2001. Analysis of bacterial community structure in sulfurous oil-containing soils and detection of species carrying dibenzothiophene desulfurization (dsz) genes. Appl. Environ. Microbiol., 67(3):1052–1062.

Ducrocq V, Pandard P, Hallier-Soulier S, Thybaud E, Truffaut N 1999. The use of quantitative PCR (polymerase chain reaction), plant and earthworm bioassays, plating and chemical analysis to monitor 4-chlorobiphenyl biodegradation in soil microcosms. Appl. Soil Ecol., 12(1):15–27.

El Shatnawi MKJ, Makhadmeh IM 2001. Ecophysiology of the plant-rhizosphere system. J. Agro. Crop Sci., 187:1–9.

Environment Protection Act 1986. Ministry of Environment and Forests Department of Environment, Forests and Wildlife. Government of India, New Delhi.

Fredrickson JK, Elliot LF 1985. Colonizing of winter wheat roots by inhibitory rhizobacteria. Soil Sci. Soc. Am. J., 49:1172–1177.

Fulekar MH 2005. Bioremediation technologies for environment. IJEP, 25:358–364.

Gentry TJ, Rensing C, Pepper IL 2004. New approaches for bioaugmentation as a remediation technology. Crit. Rev. Environ. Sci. Technol., 34:447–494.

Gerdemann JW, Nicolson TH 1963. Spores of mycorrhizal Endogone extracted from soil by wet sieving and decanting. Trans. Brit. Mycol. Soc., 46:235–244.

Ghosh B, Korade D, Fulekar MH 2005. Rhizosphere bioremediation—Technology for removal of soil contaminated xenobiotics. Ekologia, 3(2):131–138.

Hazardous Wastes (Management and Handling) Rules 1989. Ministry of Environment and Forests Notification. Government of India, New Delhi.

Heinonsalo J, Jorgensen KS, Haahtela K, Sen R 2000. Effects of Pinus sylvestris root growth and mycorrhizosphere development on bacterial carbon source utilization and hydrocarbon oxidation in forest and petroleum-contaminated soils. Canadian Journal of Microbiology 46:451–464.

Jackobsen CS 1995. Microscale detection of specific bacterial DNA in soil with a magnetic capture-hybridization and PCR amplification assay. Appl. Environ. Microbiol., 61:3347–3352.

Jackson ML 1973. Methods of Soil Chemical Analysis. Prentice Hall of India, New Delhi.

Joner EJ, Johansen A, dela Cruz MAT, Szolar OJH, Loibner A, Portal JM, Leyval C 2001. Rhizosphere effects on microbial community structure and dissipation and toxicity of polycyclic aromatic hydrocarbons (PAHs) in spiked soil. Environ. Sci. Technol., 35(13):2773–2777.

Joner EJ, Leyval C 2003. Rhizosphere gradients of polycyclic aromatic hydrocarbon (PAH) dissipation in two industrial soils, and the impact of arbascular mycorrhiza. Environ. Sci. Technol. 37:2371–2375.

Juhanson J, Truu J, Heinarun E, Heinarun A 2007. Temporal dynamics of microbial community in soil during phytoremediation field experiment. J. Environ. Eng. Landsc. Manag., 15(4):213–220.

Lafrance P, Lapointe M 1998. Mobilization and co-transport of pyrene in the presence of *Pseudomonas aeruginosa* UG2 biosurfactants in sandy soil columns. Ground Water Monit. Remediat., 18:139–147.

Leyval C, Joner E J, del Val C, Haselwandter K 2002. Potential of arbuscular mycorrhizal fungi for bioremediation. In: Gianinazzi S, Schüepp H, Barea JM, Haselwandter K (eds.), Mycorrhizal Technology in Agriculture: From Genes to Bioproducts. Birkhaüser Verlag, Basel, pp. 175–186.

Lynch JM 2002. Resilience of the rhizosphere to anthropogenic disturbance. Biodegradation, 13:21–27.

Meharg AA, Cairney JWG 2000. Ectomycorrhizas—Extending the capacities of rhizosphere remediation? Soil Biol. Biochem., 32:1475–1484.

Moslemy P, Guiot SR, Neufeld RJ 2002. Production of size-controlled gellan gum microbeads encapsulating gasoline-degrading bacteria. Enzyme Microb Technol 30:10–18.

Phillip JM, Hayman DS (1970). Improved procedure for clearing roots and staining parasitic and vesicular arbuscular mycorrhizal fungus for rapid assessment of infection. Trans. Br. Mycol. Soc., 55:158–161.

Reynolds CM, Wolf DC, Gentry TJ, Perry LB, Pidgeon CS, Koenen BA, Rogers HB, Beyrouty CA 1999. Plant enhancement of indigenous soil microorganisms: A low cost treatment of contaminated soils. Polar Rec., 35:33–40.

Sabljic A, Piver WT 1992. Quantitative modelling of environmental fate and impact of commercial chemicals. Environ. Toxicol. Chem., 11:961–972.

Sandrin TR, Chech AM, Maier RM 2000. A rhamnolipid biosurfactant reduces cadmium toxicity during naphthalene biodegradation. Applied and Environmental Microbiology 66:4585–4588.

Schnoor JL 1997. Phytoremediation. Technology Evaluation Report. Ground-Water Remediation Technologies Analysis Center, Pittsburgh, PA.

Schnoor JL, Light LA, McCutcheon SC, Wolfe NL, Carreira LH 1995. Phytoremediation of organic and nutrient contaminants. Environmental Science and Technology, 29(7): 318A–323A.

Semple KT, Morriss AWJ, Paton GI, 2003. Bioavailability of hydrophobic contaminants in soils: Fundamental concepts and techniques for analysis. Eur. J. Soil Sci., 54:809–818.

Shann JR 1995. The role of plants and plant/microbial systems in the reduction of exposure. Environ. Health Perspect., 103:13–15.

Short KA, King RJ, Seidler RJ, Olsen RH 1992. Biodegradation of phenoxyacetic acid in soil by Psuedomonas putida PPO301(pRO 103), a constitutive degrader of 2,4-dichlorophenoxyacetate. Mol. Ecol., 1:89–94.

Siciliano SD, Germida JJ, 1998. Mechanism of phytoremediation: Biochemical and ecological interactions between plant and bacteria. Environ. Rev., 6:65–79.

Susarla S, Medina VF, McCutcheon SC 2002. Phytoremediation: An ecological solution to organic chemical contamination. Eco. Eng., 18:647–658.

Tesar M, Reichenauer TG, Sessitsch A 2002. Bacterial rhizosphere populations of black poplar and herbal plants to be used for phytoremediation of diesel fuel. Soil Biol. Biochem., 34:1883–1892.

Verma SVS 1992. Prospects of improving nutritive value of poultry feeds through biotechnology. Poult. Guide, 29:37–39.

Walton BT, Anderson TA 1990. Microbial degradation of trichloroethylene in the rhizosphere: Potential application to biological remediation of waste sites. Appl. Environ. Microbiol., 56:1012–1016.

Walton BT, Gunthrie EA, Holyman AM 1994. Toxicant degradation in the rhizosphere. In: Anderson TA, Coats JR (eds.), Bioremediation Through Rhizosphere Technology. ACS Symposium Series 563. American Chemical Society, Washington, DC, pp. 11–26.

Wei, WS, Qixing, Z, Kaisong Z, Jidong, L 2003. Roles of rhizosphere in remediation of contaminated soils and its mechanisms. Yingyong Shengtai Xuebao, 14:143–147.

Widada J, Nojiri H, Omori T 2002. Recent developments in molecular techniques for identification and monitoring of xenobiotic-degrading bacteria and their catabolic genes in bioremediation. Appl. Microbiol. Biotechnol., 60:45–59.

Wikstrom P, Wiklund A, Andersson AC, Forsman M 1996. DNA recovery and PCR quantification of catechol 2,3-dioxygenase genes from different soil types. J. Biotechnol., 52:107–120.

Zengler K, Toledo G, Rappe M, Elkins J, Mathur EJ, Short JM, Keller M 2002. Cultivating the uncultured. Proc. Natl. Acad. Sci. USA, 99:15681–15686.

12 Macrophytes as Bioremediators of Toxic Inorganic Pollutants of Contaminated Water Bodies

*Abdul Barey Shah and Rana Pratap Singh**

CONTENTS

12.1 INTRODUCTION

For efficient management and remediation of water in aquatic ecosystems, knowing the quality of water and any changes produced therein due to anthropogenic activities is the preliminary step to establish an effective cleaning system (Shah et al., 2015; Shah and Singh, 2016). Aquatic biodiversity is a significant component of aquatic ecosystems and is affected by the quality of the water. A variety of aquatic macrophytes (rooted, free-floating, submerged and emergent) dwelling in riverine systems have demonstrated the potential to accumulate different toxic pollutants inside their tissues and are being used to clean the pollution levels (Rai and Tripathi, 2009; Rawat et al., 2012; Souza et al., 2013). The plants are differentially tolerant to various inorganic pollutants as they have the capacity to assimilate or scavenge them. This promising potential of plants has been developed as a vast field of phytotechnological studies for their evaluation in phytoremediation for the removal of toxic pollutants from contaminated soil and water (Bauddh and Singh, 2012; Rai et al., 2015).

* Corresponding author: Email: cceseditor@gmail.com

Phytoremediation that involves the use of plants to remove viz-a-viz detoxify pollutants is gaining global interest by using selective, improved and value-added aquatic plants. The principal component of this technique is the selection of appropriate plant species. To be a better phytoremediator, plants should possess (i) tolerance to higher levels of metals; (ii) higher uptake rate of pollutants; (iii) high transportation of absorbed metals through shoot system; and (iv) effective mechanism to tolerate increased metal levels in different plant parts (Singh et al., 2012). Phytoremediation of polluted waters using aquatic macrophytes is recognized as eco-friendly, cost-effective and energy-efficient technology employing both the biological processes of plants and engineering tools for effective remediation (Abou-Elela et al., 2013; Rai et al., 2013). Accumulation of metals by aquatic macrophytes has been reported to be proportional to the greater metal content in water, which reflects that these species can serve as agents for ecological surveys as in situ biomonitors of water quality due to their capability to accumulate metals in tissue parts so as to predict the level and kind of environmental pollution (Shah et al., 2015). The effect of circulation on the remediation of polluted waters has been shown to enhance the treatment capacity, increase the hydraulic retention time (HRT), reduce the area required for treatment and facilitate operation at higher loads.

The application of biotechniques coupled with engineering methods has recently been gaining attention and is being emphasized for remediation and protection of environments (Uysal and Taner, 2009; Arini et al., 2012). The potential of mono and mixed cultures of free-floating and submerged plants for the confiscation of multimetals and other inorganic contaminants under the effect of different factors such as water velocity, HRT and changing physicochemical characteristics of water prevailing in natural aquatic ecosystems is not well documented (Shah et al., 2015). Furthermore, developing the knowledge of metal-binding sites within the plants, metal localization and class of compounds in which the metals are transported and distributed inside the plant parts needs to be extensively discussed (Shuvaeva et al., 2013).

12.2 ANALYSIS OF FACTORS AFFECTING THE REMOVAL OF METALS AND OTHER INORGANIC POLLUTANTS FROM WATER

The bioavailability of metals in water for accumulation by aquatic plants is determined and influenced by several factors of the water column such as temperature, pH, chemical speciation, redox potential, salinity, sediment type, organic matter and seasonal changes (Shuping et al., 2011). Both root and shoot tissues of aquatic macrophytes have been demonstrated for the metal accumulation and translocation under natural ecosystems. Generally, metal concentrations have been reported to be much higher in the roots as compared to the shoot of the plants (Rieumont et al., 2007; Rai, 2012). The reason behind this may be ascribed to low movement of metals from root to shoot due to the development of complex biochemical compounds with the—COOH group, which inhibits metal translocation to higher plant parts (Cardwell et al., 2002). However, at certain sites, it was observed that plants accumulated metals in higher proportions in shoots than in roots. Kim et al. (2003) stated differential metal accumulation by different plant species, which may be due to compartmentalization and translocation in the inner system of plants.

While comparing the metal-accumulating capabilities of different aquatic plants under natural conditions, it has been observed that free-floating aquatic plants generally accumulated a greater amount of metals than rooted and emergent plants, with the exception of *Hydrilla verticillate*, a submerged plant (Ebrahimpour and Mushrifah, 2008; Shah et al., 2015). The reason for greater metal accumulation by submerged aquatic plants like *H. verticillata* is that they readily absorb available elements directly from the water through foliar surfaces (Baldantoni et al., 2004) (Table 12.1). The efficacy of phytoremediation differs significantly between species as diverse mechanisms of ion uptake are active in each species, based on their morphological, physiological, genetic and anatomical characteristics (Rahman and Hasegawa, 2011). Under controlled conditions combined with the effect of water circulation, the metal-accumulating potential by aquatic macrophytes in mono and mixed cultures has been reported to be ameliorated to a large extent. Circulation of multimetal contaminated water on different days of treatment had a significant effect on the enhanced metals removal and concurrent accumulation by the plants than under noncirculating water conditions. Intermittent circulation increases the number of passages of water through the treatment beds in comparison to other conventional techniques, where the movement of water occurs only once. Therefore, the HRT in the treatment systems is enhanced and the elimination of pollutants is likely to be increased, as it is documented that the

TABLE 12.1
Reports on the Removal of Different Pollutants by *Pistia stratiotes* and *Hydrilla verticillata* under Different Treatment Conditions

Plant Species	Type of Treatment System	Type of Pollutants Removed	References
Pistia stratiotes	Aquaculture wastewater treatment	Turbidity, COD, TKN, NO_3, NH_4, and PO_4	Akinbile and Yusoff (2012)
Pistia stratiotes	Wastewater treatment	NO_3 and PO_4	Aina et al. (2012)
Pistia stratiotes	Municipal sewage	Mn, Zn, Pb, Cr, and Cu	Tewari et al. (2008)
Pistia stratiotes	Constructed wetland	Pb, Cd, Fe, Ni, Cr, and Cu	Khan et al. (2009)
Pistia stratiotes	Constructed water detention system	Suspended solids, N, and P	Lu et al. (2010)
Pistia stratiotes	Hydroponic culture	Nitrate, nitrite	Rawat et al. (2012)
Pistia stratiotes	Hydroponic culture	Cd and Pb	Veselý et al. (2011a and 2011b)
Pistia stratiotes	Hydroponic culture	Cd	Das et al. (2013)
Hydrilla verticillata	Simulated field conditions	As	Srivastava et al. (2011)
Hydrilla verticillata	Hydroponic culture	As, Zn and Al	Bakar et al. (2013)
***Pistia stratiotes* L.**	Natural conditions	Cd, Cr and Cu	Shah et al. 2014

efficacy of remediation of pollution in constructed wetlands is largely determined by HRT (Maltais-Landry et al., 2009). Furthermore, intermittent recirculation of contaminated water in the bed has the following benefits: (1) a prolonged contact between pollutants and biomass in the treatment bed and (2) an augmentation of the natural reaeration during each circulation stage. Differences in the bioaccumulation rates and translocation factors among the two plant species reflect the abundance of the metals studied and the intrinsic abilities of the plants to sequester the metals (Anning et al., 2013).

The inorganic nutrients in water were nitrate, nitrite, ammonium and phosphate that degrade the quality of water and deplete the dissolved oxygen present in water. Nitrite, a natural component of the nitrogen cycle in ecosystems, has been recognized as a potential problem for its presence in the environment due to its well-established toxicity to animals (Sinha and Nag, 2011). Therefore, it is necessary to eliminate it from water so as to reduce its toxic effect to the humans and animal consumers of the water, since they cannot assimilate nitrite like plants and bacteria (Alonso and Camargo, 2009). Our study pertaining to the changes in the inorganic nutrients reveals that there is a significant decrease in the concentration of inorganic pollutants from the treatment system. Similar results pertaining to our study were reported by Rawat et al. (2012). The potential rate of uptake of nutrients by plant is limited by the growth rate (net productivity) and the concentration of nutrients in the plant tissues (Vymazal, 2007). Furthermore, it has been observed that a decrease in the concentration of nitrate in the water column could be due to the enhanced plants uptake by their roots rather than microbial denitrification (Bindu et al., 2008; Kadlec and Wallace, 2009). For the removal of ammonia and its species in the constructed wetland, denitrification has been reported to be the major pathway (Rai et al., 2013).

The mechanisms behind the remediation of phosphorus from water have been reported to occur by complexation, sorption, precipitation and assimilation into plant and microbial biomass. Wetland plants play a pivot role in the removal of phosphate either by direct uptake and/or establishment of favorable conditions for microorganisms that use phosphorus as a nutrient (Mbuligwe, 2004). Plants efficiently remove different nutrient concentrations from water, which has been accounted to be 15–80% for N and 24–80% for P (Rawat et al., 2012). Differential uptake and variable remediation of nutrients from water bodies and constructed treatment systems have been ascribed to certain factors such as variable potential of different species, nutrient loading rates in water streams and columns and climatic differences in various studies (Vymazal, 2007).

12.3 CONSTRUCTED WETLAND SYSTEMS AS EFFICIENT TOOLS IN THE PHYTOREMEDIATION OF POLLUTED WATER

Wetland ecosystems act as natural filters and have been effectively used for the treatment of wastewaters, aimed at the removal of toxic pollutants, including heavy metals, through absorption by plants. Currently various wetlands are being engineered, constructed and designed in such a way so as to exploit natural processes involving in situ vegetation, soil and associated microbial interactions to assist in the treatment of wastewaters (Khan et al., 2012). These constructed wetland systems are designed to take advantage of many natural processes that occur in natural wetlands, but in a more controlled environment. Application of constructed wetlands is a practical solution for treating contaminated water by simulating natural wetlands, due to their cost-effectiveness, lesser operation and maintenance requirements, with very less input of energy (Varnell et al., 2009). For the treatment of polluted waters generating from different sources such as domestic, highways, mining and industrial sectors, constructed wetlands offer a highly efficient and an effective alternative system for traditional wastewater treatments (Khan et al., 2009). The usage of constructed wetlands has been recognized in different sectors, ranging from rehabilitating those areas where wetlands were previously located to serving very definite purposes such as treatment of wastewater (Rai et al., 2015). Constructed wetlands have been efficiently used to remove various pollutants and nutrients from wastewater (Vymazal, 2007; Bindu et al., 2008) and also successfully used to treat wastewater with high nutrient concentrations (Gottschall et al., 2007). In recent years, the focus on the use of constructed wetlands has been for removing toxic metals from wastewater and drinking water sources (Maine et al., 2004; Hadad et al., 2006; Jayaweera et al., 2008).

The constructed wetland system possesses different types of aquatic macrophytes that not only accumulate pollutants directly into their tissues but also act as catalysts for various purification reactions that usually occur in the rhizosphere of the plants (Jenssen et al., 1993). Substrate interactions in constructed wetlands are able to remove most of the metals from contaminated water (Liu et al., 2007). The development of a temporary or permanent oxygen-deficient environment in wetland soils helps in heavy metal immobilization in the highly reduced sulfite or metallic forms. Furthermore, enhanced remediation of metals is carried by plants through adsorption, cation exchange, filtration, and root-induced chemical changes in the rhizosphere (Liu et al., 2007). However, the treatment processes in constructed wetlands and subsequent metal remediation of the contaminated sites are subject to various conditions and factors such as water and sediment pH, mobilization and uptake of metals from the soil, confiscation and compartmentalization within the root system of plants, efficacy of xylem loading and transport (transfer factors), distribution between metal sinks in the aerial parts of plants, absorption and storage in leaf cells and plant growth and rate of transpiration (Hadad et al., 2006). Although emergent macrophytes are the most common plants used in constructed wetlands for wastewater treatment, the design of the systems in terms of media as well as the flow regime varies. The most common constructed wetland systems are designed with a horizontal subsurface flow (HF constructed wetlands), but vertical flow (VF constructed wetlands) systems are becoming more popular (Vymazal, 2005). Among various types of constructed wetlands, horizontal subsurface flow constructed wetlands (HSSFCWs) are most widely used and are an efficient alternative to more conventional wastewater treatment processes. In a typical HSSFCW, wastewater is held at a constant depth and water flows horizontally below the surface of the bed, thereby efficiently removing pollutants, organic matter and pathogens. Recently, angular HSSFCWs have been employed for sewage treatment using *Colocasia esculenta* and a reduction in various pollution parameters such as dissolved solids, biochemical and chemical oxygen demand and inorganic nutrients (nitrate, nitrite, ammonium and phosphate) has been successfully achieved against treatment of sewage in the control bed (Chavan and Dhulap, 2012). Thus, constructed wetlands can also be used for removing organic as well as inorganic pollution load of wastewater.

12.4 ROLE OF FLOATING WETLAND BEDS IN WASTEWATER REMEDIATION

Floating wetlands systems for water treatment are a novel and innovative alternative of wetlands and pond technology that offer great potential for treatment of polluted waters (Ladislas et al., 2015). Floating treatment wetlands (FTWs) are an evolving variant of constructed wetland technology, which comprises emergent wetland plants developing hydroponically on structures. The floating islands develop an efficient plant of hungry microbes, waiting for nutrients to consume, whose by-product is food for other organisms in the complex mesh of a wetland (Chaudhuri et al., 2014). This technique is accurate sustainability, which appears a far cry from the many plotted and dysfunctional water cleansing structures that are common today. For stormwater holding ponds, the floating islands have impressive potential to purify nutrient-rich water and create multidimensional habitat at the same time (Maltais-Landry et al., 2009).

The National Institute of Water and Atmospheric Research (NIWA) in New Zealand evaluated the application of floating wetlands for enhanced stormwater treatment and stated, “In existing systems, the FTW may become a low cost option to upgrade existing stormwater ponds for removal of fine particles and associated metals.” It depicts that further research is mandatory to identify key treatment processes in floating wetland systems. Primary studies are revealing that cattail (*Typha latifolia*) is not the only potential wetland hero for water purification, but also there are the carbon-sequestering abilities of the microbes within these systems (Tanner and Headley, 2011). The floating islands are a “stacked function tool” for water management. Municipalities with a floating wetland in the form of retention pond near the commercial sector of town can afford the following functions at very low costs: cleanup of water by the removal of soluble nutrients like nitrogen and phosphorous, providing habitat for flora and fauna, in a way that is visible and aesthetically pleasing; creating a new riparian edge that can be planted with indigenous wetland plants; mobilizing the local community to get involved in planting the islands; creating a completely new form of environmental stewardship; sequestering atmospheric carbon dioxide and other greenhouse gases; offering municipal water managers another tool for water treatment which must be cost-effective, eco-friendly, unremarkable and attractive (Hubbard, 2010). Artificially created floating wetlands have been used with varying success for a number of applications to date, such as water quality improvement, habitat enhancement and aesthetic purposes in ornamental ponds. In terms of water quality improvement, the main applications of FTWs reported to date have been for the treatment of stormwater, combined stormwater sewer and overflow sewage (Tanner and Headley, 2011). Previous studies have shown that floating wetlands can be effective systems for removing the dissolved metals present in runoff. However, such processing methods are designed for implementation in situ, directly on the surface of existing retention basins (Sukias et al., 2010). Other studies have reported on systems floating on the surface of runoff retention basins (airport, residential areas) or lakes or rivers. Although floating wetlands are increasingly used to treat a variety of types of wastewater, the construction of the system remains mainly empirical, and a research effort is needed to define design parameters more precisely. Most studies of floating treatment systems have analyzed chemical oxygen demand (COD), biochemical oxygen demand (BOD), nutrients (N and P) and suspended solids (TSS). To our knowledge, the role played in the decontamination process by the microbial biofilm that develops on the surface of roots suspended in the water column has been little studied. Previous research has shown that *Juncus* and *Carex* are able to grow and accumulate Ni, Cd and Zn under hydroponic conditions in the laboratory (Ladislas et al., 2013). The plant roots flourish through the floating mat and into the water below. As well as storing nutrients directly from the water column (rather than the bottom sediments), the roots develop a large surface area for adsorption and biofilm attachment (Tanner and Headley, 2011). Because FTWs can stand deep and changing water levels, they can be used in conditions where use of conventional surface-flow wetlands with bottom-rooted emergent aquatic macrophytes would be unacceptable. As such, FTWs assimilate the nutrient attenuation proficiencies of wetlands with the elasticity of deeper pond systems, and so increase the variety of conditions where wetland ecotechnologies can be applied for water quality improvement (Headley and Tanner, 2008).

12.5 CONCLUSION

The presence of toxic chemicals in biospheric media such as water has necessitated the need for the development of cost-effective, eco-friendly and sustainable water treatment technologies. Macrophytes growing in the water column of aquatic ecosystems have a tremendous potential to remediate water in a sustainable way. The use of macrophytes as bioremediators of contaminated water is a growing field of research. Different biotechniques employing the application of aquatic macrophytes in engineered constructed wetland systems, floating wetlands, constructed ponds and natural wetlands are currently in use for the management and treatment of water polluted with toxic inorganic substances.

REFERENCES

Abou-Elela, S. I., Golinielli, G., Abou-Taleb, E. M., Hellal, M. S. (2013). Municipal wastewater treatment in horizontal and vertical flows constructed wetlands. Ecological Engineering. 61:460–468.

Aina, M. P., Kpondjo, N. M., Adounkpe, J., Chougourou, D., Moudachirou, M. (2012). Study of the purification efficiencies of three floating macrophytes in wastewater treatment. International Research Journal of Environmental Sciences. 1(3):37–43.

Akinbile, C. O., Yusoff, M. S. (2012). Assessing water hyacinth (*Eichhornia crassopes*) and lettuce (*Pistia stratiotes*) effectiveness in aquaculture wastewater treatment. International Journal of Phytoremediation. 14(3):201–211. DOI: 10.1080/15226514.2011.587482.

Alonso, A., Camargo, J. A. (2009). Effects of pulse duration and post-exposure period on the nitrite toxicity to a freshwater amphipod. Ecotoxicology and Environmental Safety. 72(7):2005–2008.

Anning, A. K., Korsah, P. E., Addo-Fordjour, P. (2013). Phytoremediation of wastewater with *Limnocharis flava, Thalia geniculata* and *Typha latifolia* in constructed wetlands. International Journal of Phytoremediation. 15(5):452–464.

Arini, A., Feurtet-Mazel, A., Morin, S., Maury-Brachet, R., Coste, M., Delmas, F. (2012). Remediation of a watershed contaminated by heavy metals: A 2-year field biomonitoring of periphytic biofilms. Science of the Total Environment. 425:242–253. DOI: 10.1016/j.scitotenv.2012.02.067.

Bakar, N. I. A., Mansor, M., Harun, N. Z. (2013). Approaching vertical greenery as public art: A review on potentials in urban Malaysia. Journal of Architecture and Environment. 12(1):1–26.

Baldantoni, D., Alfani, A., Di Tommasi, P., Bartoli, G., De Santo, A. V. (2004). Assessment of macro and microelement accumulation capability of two aquatic plants. Environmental Pollution. 130(2):149–156. DOI: 10.1016/j.envpol.2003.12.015.

Bauddh, K., Singh, R. P. (2012). Cadmium tolerance and its phytoremediation by two oil yielding plants *Ricinus communis* (L.) and *Brassica juncea* (L.) from the contaminated soil. International Journal of Phytoremediation. 14(8):772–785.

Bindu, T., Sylas, V. P., Mahesh, M., Rakesh, P. S., Ramasamy, E. V. (2008). Pollutant removal from domestic wastewater with taro (*Colocasia esculenta*) planted in a subsurface flow system. Ecological Engineering. 33(1):68–82.

Cardwell, A., Hawker, D., Greenway, M. (2002). Metal accumulation in aquatic macrophytes from southeast Queensland, Australia. Chemosphere. 48:653–663.

Chaudhuri, D., Majumder, A., Misra, A. K., Bandyopadhyay, K. (2014). Cadmium removal by *Lemna minor* and *Spirodela polyrhiza*. International Journal of Phytoremediation. 16(11):1119–1132.

Chavan, B. L., Dhulap, V. P. (2012). Sewage treatment with constructed wetland using *Panicum maximum* forage grass. Journal of Environmental Science and Water Resources. 1(9):223–230.

Das, S., Goswami, S., Talukdar, A. D. (2014). A study on cadmium phytoremediation potential of water lettuce, Pistia stratiotes L. Bulletin of Environmental Contamination and Toxicology. 92(2):169–174. DOI: 10.1007/s00128-013-1152-y

Ebrahimpour, M., Mushrifah, I. (2008). Heavy metal concentrations (Cd, Cu and Pb) in five aquatic plants species in Tasik Chini, Malaysia. Environmental Geology. 54:689–689.

Gottschall, N., Boutin, C., Crolla, A., Kinsley, C., Champagne, P. (2007). The role of plants in the removal of nutrients at a constructed wetland treating agricultural (dairy) wastewater, Ontario, Canada. Ecological Engineering. 29(2):154–163. DOI: 10.1016/j.ecoleng.2006.06.004

Hadad, H. R., Maine, M. A., Bonetto, C. A. (2006). Macrophytes growth in a pilot-scale constructed wetland for industrial wastewater treatment. Chemosphere. 63:1744–1753.

Headley, T. R., Tanner, C. C. (2008). Floating wetlands for stormwater treatment: Removal of copper, zinc and fine particulates. Technical Report 2008-030, Auckland Regional Council Auckland, NZ, p. 38. (ARC07231, November 2008).

Hubbard, R. K. (2010). Floating vegetated mats for improving surface water quality. In: Shah V. (ed.), Emerging Environmental Technologies. Springer, New York, pp. 211–244.

Jayaweera, M. W., Kasturiarachchi, J. C., Kularatne, R. K., Wijeyekoon, S. L. (2008). Removal of aluminium by constructed wetlands with water hyacinth (*Eichhornia crassipes* (Mart.) Solms) grown under different nutritional conditions. Journal of Environmental Science and Health Part A. 42(2):185–193.

Jenssen, P. D., Mæhlum, T., Krogstad, T. (1993). Potential use of constructed wetlands for wastewater treatment in northern environments. Water Science and Technology. 28(10):149–157. DOI: 10.2166/wst.1993.0223

Kadlec, R. H., Wallace, S. D. (2009). Treatment Wetlands, 2nd ed. Taylor & Francis, Boca Raton, FL. ISBN: 978-1-56670-526-4.

Khan, M. Z., Mondal, P. K., Sabir, S. (2012). Aerobic granulation for wastewater bioremediation: A review. The Canadian Journal of Chemical Engineering. 91(6):1045–1058. DOI: 10.1002/cjce.21729

Khan, S., Ahmad, I., Shah, M. T., Rehman, S., Khaliq, A. (2009). Use of constructed wetland for the removal of heavy metals from industrial wastewater. Journal of Environmental Management. 90(11):3451–3457. DOI: 10.1016/j.jenvman.2009.05.026

Kim, C.-G., Bell, J. N. B., Power, S. A. (2003). Effects of soil cadmium on *Pinus sylvestris* L. seedlings. Plant and Soil. 257(2):443–449. DOI: 10.1023/a:1027380507087

Ladislas, S., Gérente, C., Chazarenc, F., Brisson, J., Andrès, Y. (2015). Floating treatment wetlands for heavy metal removal in highway stormwater ponds. Ecological Engineering. 80:85–91. DOI: 10.1016/j.ecoleng.2014.09.115

Ladislas, S., Gérente, C., Chazarenc, F., Brisson, J., Andrès, Y. (2013). Performances of two macrophytes species in floating treatment wetlands for cadmium, nickel, and zinc removal from urban stormwater runoff. Water, Air, & Soil Pollution. 224(2):1408.

Liu, J., Dong, Y., Xu, H., Wang, D., Xu, J. (2007). Accumulation of Cd, Pb and Zn by 19 wetland plant species in constructed wetland. Journal of Hazardous Materials. 147(3):947–953.

Lu, Q., He, Z. L., Graetz, D. A., Stoffella, P. J., Yang, X. (2010). Phytoremediation to remove nutrients and improve eutrophic stormwaters using water lettuce (*Pistia stratiotes* L.). Environmental Science and Pollution Research. 17:84–96. DOI: 10.1007/s11356-008-0094-0

Maine, M. A., Suñé, N. L., Lagger, S. C. (2004). Chromium bioaccumulation: Comparison of the capacity of two floating aquatic macrophytes. Water Research. 38(6):1494–1501.

Maltais-Landry, G., Maranger, R., Brisson, J., Chazarenc, F. (2009). Nitrogen transformations and retention in planted and artificially aerated constructed wetlands. Water Research. 43(2):535–545.

Mbuligwe, S. E. (2004). Comparative effectiveness of engineered wetland systems in the treatment of anaerobically pre-treated domestic wastewater. Ecological Engineering. 23(4):269–284.

Rahman, M. A., Hasegawa, H. (2011). Aquatic arsenic: Phytoremediation using floating macrophytes. Chemosphere. 83:633–646.

Rai, P. K. (2012). An eco-sustainable green approach for heavy metals management: Two case studies of developing industrial region. Environmental Monitoring and Assessment. 184(1):421–448.

Rai, P. K., Tripathi, B. D. (2009). Comparative assessment of *Azolla pinnata* and *Vallisneria spiralis* in Hg removal from GB Pant Sagar of Singrauli Industrial region, India. Environmental Monitoring and Assessment. 148(1–4):75–84.

Rai, U. N., Tripathi, R. D., Singh, N. K., Upadhyay, A. K., Dwivedi, S., Shukla, M. K., Nautiyal, C. S. (2013). Constructed wetland as an ecotechnological tool for pollution treatment for conservation of Ganga river. Bioresource Technology. 148:535–541.

Rai, U. N., Upadhyay, A. K., Singh, N. K., Dwivedi, S., Tripathi, R. D. (2015). Seasonal applicability of horizontal sub-surface flow constructed wetland for trace elements and nutrient removal from urban wastes to conserve Ganga river water quality at Haridwar, India. Ecological Engineering. 81:115–122.

Rawat, S. K., Singh, R. K., Singh, R. P. (2012). Remediation of nitrite contamination in ground and surface waters using aquatic macrophytes. Journal of Environmental Biology. 33:51–56.

Rieumont, S. O., Lima, L., De la Rosa, D., Graham, D. W., Columbie, I., Santana, J. L., Sanchez, M. J. (2007). Water hyacinths (*Eichhornia crassipes*) as indicators of heavy metal impact of a large landfill on the Almendares River near Havana, Cuba. Bulletin of Environment Contamination and Toxicology. 79: 583–587.

Shah, A. B., Singh, R. P. (2016). Monitoring of hazardous inorganic pollutants and heavy metals in potable water at the source of supply and consumers end of a tropical urban municipality. International Journal of Environmental Research. 10(1):149–158.

Shah, A. B., Rai, U. N., Singh, R. P. (2015). Correlations between some hazardous inorganic pollutants in the Gomti River and their accumulation in selected macrophytes under aquatic ecosystem. Bulletin of Environmental Contamination and Toxicology. 94:783–790.

Shah, M., Hashmi, H. N., Ali, A., Ghumman, A. R. (2014). Performance assessment of aquatic macrophytes for treatment of municipal wastewater. Journal of Environmental Health Science and Engineering. 12:106.

Shuping, L. S., Snyman, R. G., Odendaal, J. P., Ndakidemi, P. A. (2011). Accumulation and distribution of metals in *Bolboschoenus maritimus* (Cyperaceae), from a South African river. Water, Air, & Soil Pollution. 216:319–328.

Shuvaeva, O. V., Belchenko, L. A., Romanova, T. E. (2013). Studies on cadmium accumulation by some selected floating macrophytes. International Journal of Phytoremediation. 15(10):979–990. DOI: 10.1080/15226514.2012.751353

Singh, D., Gupta, R., Tiwari, A. (2012). Potential of duckweed (*Lemna minor*) for removal of lead from wastewater by phytoremediation. Journal of Pharmacy Research. 5(3):1578–1582.

Sinha, S. N., Nag, P. K. (2011). Air pollution from solid fuels. In: Nriagu J. O. (ed.), Encyclopedia of Environmental Health (Vol. 1). Elsevier, Amsterdam, The Netherlands, pp. 46–52.

Souza, F. A., Dziedzic, M., Cubas, S. A., Maranho, L. T. (2013). Restoration of polluted waters by phytoremediation using *Myriophyllum aquaticum* (Vell.) Verdc., Haloragaceae. Journal of Environmental Management. 120:5–9.

Srivastava, S., Shrivastava, M., Suprasanna, P., D'Souza, S. F. (2011). Phytofiltration of arsenic from simulated contaminated water using *Hydrilla verticillata* in field conditions. Ecological Engineering. 37(11):1937–1941. DOI: 10.1016/j.ecoleng.2011.06.012

Sukias, J. P. S., Yates, C. R., Tanner, C. C. (2010). Assessment of floating treatment wetlands for remediation of eutrophic lake waters—Maero Stream (Lake Rotoehu). NIWA Client Report for Environment Bay of Plenty, HAM2010-104, NIWA, Hamilton Dec 2010.

Tanner, C. C., Headley, T. R. (2011). Components of floating emergent macrophyte treatment wetlands influencing removal of stormwater pollutants. Ecological Engineering. 37:474–486. DOI: 10.1016/j.ecoleng.2010.12.012

Tewari, A., Singh, R., Singh, N. K., Rai, U. N. (2008). Amelioration of municipal sludge by Pistia stratiotes L.: Role of antioxidant enzymes in detoxification of metals. Bioresource Technology. 99(18):8715–8721. DOI: 10.1016/j.biortech.2008.04.018

Uysal, Y., Taner, F. (2009). Effect of pH, temperature, and lead concentration on the bioremoval of lead from water using *Lemna minor*. International Journal of Phytoremediation. 11:591–608.

Varnell, C. J., Thawaba, S., Brahana, J. V. (2009). Constructed wetlands for the pre-treatment of drinking water obtained from coal mines. The Open Environmental Engineering Journal. 2:1–8.

Veselý, T., Tlustoš, P., Száková, J. (2011a). Organic salts enhanced soil risk elements leaching and bioaccumulation in *Pistia stratiotes*. Plant Soil and Environment. 57:166–172.

Veselý, T., Tlustoš, P., Száková, J. (2011b). The use of lettuce (*Pistia stratiotes* L.) for rhizofiltration of a highly polluted solution by cadmium and lead. International Journal of Phytoremediation. 13:859–872.

Vymazal, J. (2005). Horizontal sub-surface flow and hybrid constructed wetlands systems for wastewater treatment. Ecological Engineering. 25(5):478–490. DOI: 10.1016/j.ecoleng.2005.07.010

Vymazal, J. (2007). Removal of nutrients in various types of constructed wetlands. Science of the Total Environment. 380(1):48–65.

13 Restoration and Conservation Strategies of Historical Monuments

*Chandrahas N. Khobragade**,
Madhushree M. Routh, and Suchita C. Warangkar

CONTENTS

* Corresponding author: Email: profcnkbt@rediffmail.com and cnkhobragade@gmail.com

13.1 INTRODUCTION

Culture and cultural heritage has long been an important component in the unification of human civilization as well as retaining their significant diversity. At the same time, it is the key factor that distinctly defines our identity because it reflects our glorious past (Greieken et al., 1998). The preservation of our cultural heritage involves the prevention of deterioration and proper conservation of important historical monuments. There has been awareness of the importance of preserving our cultural heritage for a long time, but this idea was more crystallized in the nineteenth century worldwide. It included a series of resolutions, recommendation, conventions and organizations like UNESCO, ICOMOS (International Council of Monuments and Sites), ICOM (International Council of Museums), ICCROM (International Centre for the Study of the Preservation and Restoration of Cultural Property), etc. (Caneva et al., 1991). Increasing environmental pollution has endangered the survival of historical monuments for many decades. Our cultural heritage is made up of almost all type of materials present in nature and used by people to produce several types of artifacts from very simple monocomponents to complex structures integrating inorganic and organic matters.

These cultural heritage objects, even if made with more resistant stones and other materials, are gradually influenced by several environmental parameters (Khobragade et al., 2006). In particular, historical monuments located in tropical wet and dry climates (10–20°C of the equator) undergo the process of biodeterioration due to environmental factors, such as high temperature, high relative humidity levels and heavy rainfall followed by cold, that favor the growth and sustenance of a wide variety of living organisms on stone surfaces. A wide variety of deteriogens act on stone monuments in tropical environments due to particularly favorable environmental conditions in those regions. Most of the organisms can cause direct and indirect effects or damage to different types of stones (Griffin et al., 1991; Kumar and Kumar, 1999).

13.2 TROPICAL REGIONS

The parts of world that reside between the Tropic of Cancer (23.5° N) and the Tropic of Capricorn (23.5° S) are called humid tropics (Kumar and Kumar, 1999). More than one-third of the earth's total surface is covered within this zone that is the most extensive climatic area of earth. Latitude is the major deciding factor that affects the climatic condition of a particular area by controlling the amount of solar radiation. Humid tropics receive a large amount of solar radiation throughout the year. This wet or humid tropical climate is widespread in numerous islands and coastal Southeast Asia (Mink, 1983).

13.3 ECOLOGICAL ASPECTS OF BIODETERIORATION

The natural properties of the stones determine the development of specific biological species on its surface. These properties are mineral constituents, pH, relative percentage of various molecules, salinity, moisture content and its structure (Camuffo et al., 1986). In addition, there are certain environmental factors, such as temperature, relative humidity, light conditions, atmospheric pollution levels, wind, rainfall, etc., that influence the development of specific biological species on the surface of the stones of the monument. All living organisms are broadly classified as autotrophs and heterotrophs on the basis of their nutrition (Crispim and Gaylarde, 2005). Mainly autotrophs get all the required nutrients from the inorganic surfaces as compared to heterotrophs. The environmental factors of tropical regions such as high temperature and high relative humidity levels are favorable for the growth and development of most of the microorganisms. Atmospheric pollutants like SO_2 and NO_2 are found to be helpful for the growth of certain heterotrophic organisms (Otto and Morales, 2006).

13.4 METHOD OF IDENTIFYING BIODETERIOGENS

A good understanding of the morphological and physiological characteristics of biological organisms involved in biodeterioration is necessary before identifying them. Only then is it possible to understand the role of these biological members in the stone-decaying process of monuments (Mansch and Bock, 1998; Sterflinger, 2010).

13.4.1 Characterizing Biodeteriogens

There are two types of biodeteriogens. The macro-biodeteriogens like lichens, mosses, liverworts and higher plants are easily identified by visual observation and microscopic interventions, but, on the other hand, the micro-biodeteriogens such as bacteria, actinomycetes, fungi and algae are not easily identifiable. These microorganisms need to be isolated and characterized by certain enriched or pure culture techniques (May, 2003; McNamara and Mitchell, 2005; Saiz-Jimenez and Laiz, 2000; Videla et al., 2000). The use of those traditional phenotypic methods for the identification of such agents has many disadvantages. Only less than 10% of the microorganisms, particularly bacteria, present on stones can be cultured on standard media. Culturing is a time-consuming task, and phenotypic identification methods are more extensive than genotypic identification methods. Phenotypic identification methods are well established and are based on modern technology and are easy to implement (McNamara and Mitchell, 2005; Sutton and Cundell, 2004). In recent years, molecular-based techniques, or genotypic identification methods, have been successfully used to examine the biological diversity in deteriorated artifacts.

13.4.1.1 Observation of Biodeteriogens in *In Situ* Conditions

Many modern scientific techniques such as optical, electronic and stereomicroscopic examinations of monumental samples sometimes give insight into *in situ* observation of such micro-biodeteriogens. Many times fluorescence microscopy (FM) used with fluorescent dyes and antibodies or scanning electron microscopy (SEM) gives useful information for identifying microorganisms with their activity *in situ*. Different microorganisms can be isolated and identified from field samples by a number of microbial methods such as enrichment culture techniques, staining, enzymatic techniques, etc. (Heyrman and Swings, 2003). Quantitative estimation of these organisms can be gathered through count methods like plate count and most probable number.

13.5 TECHNIQUES USED TO STUDY BIODETERIORATION

A number of scientific techniques have been used to study the phenomenon of biodeterioration of monumental stones. The methods include several types of microscopy such as SEM, polarized light microscopy (PLM), FM, etc. Other techniques such as energy dispersive spectroscopy (EDS), f-ray diffraction (XRD), thin layer chromatography (TLC), gas chromatography-mass spectroscopy (GC-MS) and infrared spectroscopy (IR) help in the qualitative identification of organic and inorganic chemical species (Webster and May, 2006). Techniques like videomicroscopy, remission spectroscopy and respiration bell-jar methods have been used for *in situ* biodeterioration observation (Becker et al., 1994; Gürtler and Stanisich, 1996; Scherer et al., 2001).

There are also certain techniques available that help in determining microbial activity that occurred in the geological past. This technique is known as the isotopic fractionation technique. Enzymes released by dead cells on the rocks can be tested by modified assay of dehydrogenase activity (DHA) (Warscheid et al., 1990).

Deterioration of stone monuments in tropical regions like India due to the biodeteriogens has long been recognized, but this subject received attention of conservation scientists only just a few years ago. This process is carried out by higher plants as well as lower microorganisms in collaboration with number of biodeterioration of stone monuments in tropics, which may be broadly classified into three categories:

i. Biophysical (abrasion, mechanical stress)
ii. Biochemical (solubilization, new reaction products)
iii. Aesthetic (colored patches or patinas or crusts) (Kumar and Kumar, 1999)

All these processes may act independently or simultaneously depending upon the biodeteriogens, nature of stone and environmental conditions.

13.5.1 Biophysical Deterioration

Biophysical deterioration is initiated by the pressure exerted on the surrounding surface due to the growth and penetration of an organism or its parts. Hyphae and root-like attachment devices deeply penetrate into the stone through preexisting cracks or crevices, causing pressure that leads to physical damage of the stone. This process of damage slowly gets aggravated by fragmentation due to periodic loosening of attachment devices during repeated wet and dry cycles. After the stone is damaged as a result of biophysical factors, it becomes more prone to other deterioration factors such as biochemical factors (Scheerer et al., 2009).

13.5.2 Biochemical Deterioration

In biochemical deterioration, the organism uses the stone surface as a source of nutrition; hence, it is also called an assimilatory process. On the other hand, the organism produces a number of metabolites that react with the stone surface chemically, and here the process is called a dissimilatory process. Most of the plants and autotrophic microorganisms produce acids that can dissolve some stones. Heterotrophic organisms produce organic acids that are capable of dissolving stones by the leaching of cations. These inorganic and organic acids form salts through microbial action that causes biochemical deterioration of monumental stones by decomposing its minerals, and the chelates formed cause stress in the pores, resulting in the formation of cracks. The insoluble salts or chelates may concentrate or precipitate on the stone surface as crusts. The carbon dioxide produced by the process of respiration changes into carbonic acid in an aqueous environment. This carbonic acid can dissolve carbonates such as limestones and marble (Kumar and Kumar, 1999).

13.5.3 Aesthetic Deterioration

The biodeterioration of stone, which is readily visible, is very crucial. The growth of these biodeteriogens on the stone surface alters their appearance. The growth pattern of the microorganisms and plants on the surface of monumental stones creates a poor aesthetic picture that gradually leads to biophysical and biochemical deterioration (McNamara and Mitchell, 2005).

13.5.4 Bacterial Biodeterioration

Bacteria are considered a group of motile or immotile prokaryotic unicellular or colonial organisms of various shapes. They may be autotrophic as well as heterotrophic. Due to their simple growth patterns, they develop easily on outdoor stone objects that have high water content. Three bacterial groups are identified as potential biodeteriorators of stone structures in a tropical climate. They include chemoautotrophic sulfur-oxidizing and nitrifying bacteria, photoautotrophic cyanobacteria and heterotrophic bacteria including actinomycetes. They are very active in stones like limestones and calcareous sandstones. Stones are attacked by autotrophic sulfur oxidizing bacteria under aerobic conditions (Khobragade et al., 2006). They oxidize sulfur-containing nutrients to sulfuric acid. Sulfuric acid further reacts with constituents of stone to form sulfates, which forms crusts on the stones' surfaces. The sulfates, when dissolved in rainwater, get precipitated within the pores of the stone. There they get recrystallized, and this puts tremendous stress on the pore walls, which results in stone damage. Sulfates in soil are reduced to sulfides by sulfur-reducing bacteria, and these sulfides are then passed into the stones by capillary action (Scheerer et al., 2009).

Autotrophic nitrifying bacteria oxidize ammonia to nitrite and nitrate ions; these may result in nitric acid formation. Due to these phenomena of stone dissolution, powdering appears as efflorescence on the stone surface. Little attention has been paid to heterotrophic bacteria, although they are highly present in tropical regions. They secrete biogenic acids that have chelating abilities and may cause stone dissolution through the mobilization of cations as Fe^{+3}, Ca^{+2}, Mn^{+2}, Al^{+3} and Si^{+4}.

Cyanobacteria are often found growing on the surface stones of historical monuments. They generally cause aesthetic damage to the stone monuments by producing various colored microbial films on their surfaces (Crispim and Gaylarde, 2005). These films are rich in adsorbed inorganic materials, such as quartz, calcium carbonate and clay, and detritus, such as dead cells and microbial by-products. The slimy surface created by these bacteria facilitate the adherence of pollen, oil, airborne particles, dust, coal ash, etc., which results in a hard crust that is very difficult to eradicate. As cyanobacteria have photosynthetic pigments, they do create a microenvironment where respiration and photosynthesis produce acids as by-products. These acid by-products cause biochemical deterioration on a large scale. Cyanobacteria also participate in biophysical deterioration of stone by a repeated cycle of drying and moistening. This loosens the mineral grains present in the surface of the stones (Mansch and Bock, 1998).

13.5.5 Bacterial Biofilms

Damage of stone by bacterial biofilms is very crucial because biofilms are the best media for further bacterial colonization as they retain humidity and provide a number of nutrients. Most of the biofilms developed on the stones are of heterotrophic bacteria, although chemolithotrophic bacteria have also been described in some cases (Crispim et al., 2003). Biofilms also play a role in carbonate precipitation because they adsorb irons and provide nucleation sites. *Myxococcus xanthus* is able to produce coherent carbonate cementing; thus, it can be used as effective biomineralization processes to restrict monumental stone decay (Khobragade et al., 2006).

13.5.6 Fungal Biodeterioration

Fungi lack chlorophyll—they are unicellular or multicellular filamentous and chemoautotrophic in nature. Basically stones are inorganic in nature so they generally do not favor the growth of fungi on their surface, but when the stone contains organic residues, it favors the growth of fungi on their surface. Several species of fungi have been isolated from monumental stones of tropical regions (Hirsch et al., 1995). Fungal hyphae extensively penetrate limestones and start intervening adjacent stones also, thus causing biophysical deterioration. The biochemical action of fungi is more important because they are believed to be the most potential contributors of decay of limestones by producing oxalic acid and citric acids. These chelating acids can leach calcium, iron and magnesium from the stone surface (Sterflinger, 2010).

13.5.7 Algal Biodeterioration

These are eukaryotic, unicellular or multicellular organisms of various shapes and sizes. They contain different types of pigments such as chlorophyll, carotenoids, xanthophylls, etc. Dampness, warmth, light and inorganic nutrients present on the stone surface provide the best conditions for algae to grow (Tomaselli et al., 2000). According to the affinity of growth, these algae are divided into two types: epithelic algae, which grow on the exposed surfaces, and endolithic algae, which grow in the interior of the substrate. Algae are never solely responsible for stone decay, but the most prominent type of damage caused by algae on stone monuments is loss of aesthetic value. They form a recognizable patina of sheets varying in extent, thickness, consistency and color. Due to varied coloration, they create a remarkable deterioration. Algae may also cause biochemical deterioration by producing a variety of metabolites such as organic acids. These acids actively dissolve stone constituents (Kovacik, 2000).

13.5.8 Lichen Biodeterioration

Lichen is formed by the symbiotic association of chlorophyta and a fungus. They are highly resistant to desiccation and high temperatures. Lichens penetrate the attachment device of thallus into the pores, preexisting cracks and openings. These cracks and openings may gradually widen the gap due to an increase in the mass of the thallus during growth (Gehrmann et al., 1989). Foliose and Crustose lichens are considered to be the most potent biodeteriogens. Carbonic acid produced by lichens plays a great role in the biochemical deterioration of the concerned rocks where they are attached. Many chelating agents like lecanoric acid and polyphenolic compounds are found in the thallus of lichens, which affect the stone quality in various ways. Apart from corrosion, many of these acids are also capable of forming organo-metal complexes, e.g., by chelating cations (Ca^{2+}, Fe^{3+}, Mn^{+}, Mg^{2+}), and when they are removed from the solution, allow continuous dissolution, resulting in considerably greater corrosive damage to the material than other deteriogens (Seaward et al., 1989; Warscheid and Braams, 2000).

13.5.9 Mosses and Liverworts Biodeterioration

Bryophytes are considered to be the amphibians of the plant kingdom, as they grow favorably in moisture-rich environments. They are phototrophic organisms and don't have a well-defined plant body and they are thallus like in structure. They cause aesthetic biodeterioration of stones generally. Mosses grow well where there is humus deposit, and when they die they gather more humus that supports the growth of higher plants that are more harmful (Kumar and Kumar, 2000).

13.5.10 Higher Plants

They are autotrophic with specialized organogenesis like roots, stems and leaves. The mechanism of biodeterioration by higher plants in the stone surface is quite complex. Biophysical deterioration is mainly caused due to the rise in thickening of

TABLE 13.1
Effect of Organisms on Limestones

Organism	Condition	Production	Result	Damage
Chemoautotrophic sulfur oxidizing bacteria	Aerobic	Sulfuric acid	Sulfate crust	Stone surface deterioration
Autotrophic nitrifying bacteria	Aerobic	Nitrite and nitrate ions	Nitric acid	Stone dissolution
Heterotrophic bacteria	Aerobic	Biogenic acids	Chelating agents	Stone dissolution
Cyanobacteria	Aerobic	Microbial films	Etching of mineral components	Dissolves stone binding minerals
Fungi	Aerobic	Oxalic acid citric acid	Leaching of metabolic cations	Corrosion of stone minerals and dissolution
Algae	Aerobic	Colored pigments, organic acids	Stone deterioration	Increases solubility and alter physicochemical properties
Lichens	Aerobic	Carbonic acid, lecanoric acid, polyphenolic compounds	Chelating agents	Stone dissolution

the roots inside the stone, which results in increasing pressure and further damage of the stones (Griffin et al., 1991). The acidity of root tips is responsible for the leaching of minerals and chelating of root exudates that cause biochemical deterioration (Miller et al., 2006).

13.5.11 Animals

Animals also play a major role in biodeterioration of stones. The members of the phylum Arthropod such as the red mite (*Balaustiu murorum*), Acari (*Phauloppia lucorum*) and Psocid (*Cerobasis lucorum*) are observed on the stone surfaces supporting algal and lichenic growth (Lewin and Charola, 1981). The microfauna feeds on detached lichen particles and helps in the diffusion and propagation of new colonies and as such hastens the process of stone deterioration. The avifauna—mostly pigeons, crows and other birds—drop excrements containing high amounts of nitrate and phosphate compounds that provide organic substances to the stone surface as nutritive substrata for heterotrophic microflora like bacteria and fungi. Marble stones are likely to get dark black spots due to dense mosquito excreta deposits (De Silva, 1975) (Table 13.1).

13.6 PREVENTIVE METHODS

Several preventive and remedial methods are suggested in the literature, and generally the combination of these two methods often gives better results. Remedial methods directly aim at eliminating all biodeteriogens, whereas preventive methods include all activities that are aimed at inhibiting biological attack at stone monuments (Miller et al., 2008; Saiz-Jimenez, 1993). Environmental parameters such as humidity, temperature, light and rainfall can be maintained in an indoor atmosphere, whereas outdoors, it is rarely feasible. Nutritive depositions that are not related to the composition of stone such as organic deposits, dirt and bird droppings can be reduced. The physicochemical parameters of the stone surface can be altered by applying preservative treatments (Ascaso et al., 2002; Orial et al., 1993).

13.6.1 Routine Maintenance

The general design of the monumental stones cannot be changed because it is restricted in the ethics of monument conservation. So, routine preventive measures that can control humidity and eliminate the causes of dampness in buildings can be taken into account. Measures such as repairing of roofs, gutters and other water shedding systems and the introduction of damp proofing systems can be started. Installation of marrow flashing strips of thin-gauge copper has been shown to have a long-term biological inhibiting effect on the stone walls (Griffin et al., 1991).

13.6.2 Periodic Cleaning of Dust, Dirt, Spores and Seeds

Dust deposits of various organic substances, bird dropping, and many unwanted particles become the source of nutrition for microorganisms on the surface of stones. If measures could be taken to effectively clean the initial establishments of lichens, mosses, fungi and higher plants, then these situations can be avoided to a large extent (Fitzner and Heinrichs, 2001).

13.6.3 Water Repellent Treatments

In tropical environments, synthetic polymers and resins have been used as protective coatings against water aggregation in stone monuments. Many field experiences have suggested that the growth of microorganisms is usually associated with moisture retention, and certain water repellents may be effectively used to increase the effective life of biocidal treatment. Application of 2% of polymethyl methacrylate solution in toluene and sandstone after a wet cleaning and biocidal treatment is effective in inhibiting biological growth for at least five years on several Indian monuments.

13.6.4 Stone Surface Cleaning

It is necessary to remove biological growth before applying biocidal treatment to the stones that may be encrusted by a heavy growth of lichen, fungi, algae, bacteria or other higher plants. Traditionally, conservators apply physical methods for removing them with the help of tools such as stiff bristles, axes, spatulas, scrapers and brushes, etc. Microorganisms that are associated with the aesthetic biodeterioration can be cleaned by dry or wet scrubbing or brushing and washing with water. Mosses and lichens are generally cleaned by low-pressure washing and application of certain chemicals to destroy them. Algae generally die from lack of moisture.

13.6.5 Application of Biocide

A biocide is any chemical that is able to kill or inhibit the growth of living organisms. These chemicals should not be potentially harmful to wildlife and humans, however. The biocide should always be compatible with the surface on which it is applied. It should not alter the nature, composition and appearance of stone (Denyer, 1990). Some of these chemical biocides are reported to create damage in the stone, so biocides should be carefully chosen keeping in mind that it should not alter the stone quality in any way. Periodic qualitative and quantitative evaluation of the efficacy of biocides on substrata should be carried out. Biocide biodeteriogens are products that are commercially available both as active principle or formulates and cover a wide range of chemical classes, from very simple inorganic compounds, such as Na and Ca hypochlorite, to very complex organic ones, such as the quaternary ammonium compounds (Preventol R50, Neo-Desogen). They can possess a strictly specific mode of action, such as the urea derivatives (diuron, Karmex), that block the photosynthetic process, or a broad toxic spectrum, like the organotin compounds (TBTO) (Caneva et al., 1991; May et al., 1993). Biocides intended for use on historical monuments and rock sites must be not only effective against biological growths, but they should also not cause any damage to the stone material either by direct action or by leaving deposits on it which may result in successive damage (Gehrmann et al., 1989) (Table 13.2).

13.6.6 Stone Status Checking

Stones are basically classified into silicious, calcareous and argillaceous according to their chemical composition. Silicious stones contain primarily silica, whereas carbonate and alumina are the main constituents of calcareous and argillaceous stones, respectively. The disintegration or decay of stone is commonly known as weathering and is caused by three kinds of agents such as physical, mechanical and chemical or organic. Heat and cold, air in the form of wind and, water in the form of rain or ice are the mechanical agents; various acids present in the atmosphere as well as produced by the growth of microorganisms are the chemical agents; and the organic agents are the vegetative growth that persists in damp and shady places. The most susceptible stones that are prone to dissolving actions of various weathering agents are limestone, sandstone and granite containing feldspar. Carbonic acid, which is present in the atmosphere in an amount of 400 parts of acid to 1,000,000 parts of air, when combined with water creates a corroding action on the carbonates that are the principal constituents of the stones. Many microorganisms also produce carbonic acid that causes discoloration and patching of the stones. Nitric acid, which is frequently present as a constituent of the atmosphere, exerts the most destructive actions on the limestones. Sulfuric acid, which is present in the atmosphere of cities to an extent of 250 parts in 1,000,000 parts of air, especially affects stones containing feldspar, potash, soda or lime and creates small holes that further deteriorate the stone's quality.

Biological infections and the intensity of biodeterioration processes are strongly influenced by water availability. Salts contaminate the monumental stones that are

TABLE 13.2
Commonly Found Biodeteriogens in Historical Monuments of Tropical Areas

Organism	Species	Types of Surface
Chemoautotrophs	*Thiobacillus* sp.	Sandstone, andesite
	Nitrosomonas sp.	Quartzite, soapstone
Chemoheterotrophs	*Micromonospora* sp.	Sandstone and limestone
	Bacillus sp.	
	Desulfovibrio sp.	
	Psuedomonas sp.	
	Micrococcus sp.	
Photoautotrophs	*Chlorogloea* sp.	Marble, sandstone, limestone
	Chrococoecus sp.	
	Oscillatoria sps	
	Gleocapsa lividia	
Fungi	*Cladosporium* sp.	Quartzite, soapstone
	Aspergillus flavus	Andesite, limestone
	Fusarium sp.	
	Penicillium notatum	
Algae (chlorophytes)	*Dermococcus* sp.	Marble, sandstone, limestone, andesite
	Pleurococcus sp.	
	Oocystis	
Algae (bacillariophytes)	*Navicula* sp.	Andesite, sandstone, limestone
	Pinnularia sp.	
Lichen (crustose)	*Arthopyrenia* sp. *Caloplaca* sp.	Limestone, sedimentary rocks
Lichen (foliose)	*Candellaria* sp. *Dinnaria* sp.	Sandstone
	Parmelia sp.	
Mosses	*Barbula indica*	Limestone
Higher plants	*Euphorbia* sp.	Andesite, sandstone
	Acacia arabica, *Ficus religiosa*	

situated near the shorelines and tropical regions. Wind transports salt from soil and water, which largely damages stones in three mechanisms, like crystallization of salts from solution, hydration of salts and thermal expansion of salts. Salts contaminating the stones lead to its crumbling and powdering (Grieken et al., 1998). This is determined by material-specific parameters, like porosity and permeability; environmental conditions of the site; and exposure of the object (Warscheid and Braams, 2000). The mechanical test of a stone provides data from which a fair estimation of the durability of the stone may be made. The American Society for Testing and Materials has recommended many methods for determining the apparent conditions of stones used in various monuments.

13.6.6.1 *In Situ* Diagnosis of Monumental Stones

The *in situ* diagnosis of monumental stones should be rapid in order to analyze the condition of the stone structure and the cause of deterioration and to find out the treatment and conservation method according to the suitability of climatic conditions. These types of maintenance possibilities are known as preventive conservation. A wide range of diagnostic techniques are available nowadays such as portable spectroscopy units, portable XRF (Thornbush and Viles, 2006), combined XRD/XRF, x-ray tomography and light detection and ranging scanners (LiDAR). All these methods are broadly of two types: physical and chemical. The physical properties of a stone can be tested by studying different physical parameters such as color, reflectance, temperature, water or moisture content, water absorption, surface roughness, surface morphology, strength and apparent density, etc.

13.6.6.2 Aggregate Crushing Value

This is significant because it provides a relative measure of the resistance of a stone to crushing under a gradually applied compressive load. This value is generally calculated to know the tension-bearing capacity of a particular stone or aggregate. It is calculated by the following formula (Duggal and Puri, 1991):

Aggregate crushing value (percent) = $W_2/W_1 \times 100$,
where total weight of dry stone = W_1 gm;
weight of portion passing 2.36 mm sieve = W_2 gm.

13.6.6.3 Aggregate Impact Value

The property of a material to resist impact is known as toughness. The stone aggregates should have sufficient toughness to resist their disintegration due to impact of any external factors. This characteristic is measured by the impact value test. It may also be called a test of measure of resistance to sudden impact or shock, which may differ from its resistance to a gradually applied compressive load (Duggal and Puri, 1991).

Aggregate impact value (percent) = $W_2 \times 100/W_1$ percent,
where total weight of dry sample taken = W_1 gm;
weight of portion passing 2.36 mm sieve = W_2 gm.

According to the aggregate impact value calculations, the stones are classified into four types as shown in Table 13.3.

TABLE 13.3
Classification of Stones According to Aggregate Impact Value

Aggregate Impact Value	Classification
<10%	Exceptionally strong
10–20%	Strong
10–30%	Satisfactory
>35%	Weak

13.6.6.4 Specific Gravity of Stones

Specific gravity is the ratio of the density of a substance to the density (mass of the same unit volume) of a reference substance at defined conditions of temperature and pressure. This is significant because the specific gravity of deteriorated stones is always less than a virgin stone used to study the extent of biochemical deterioration. The American Society for Testing Materials has recommended the following method for the determination of the specific gravity of stone. The formula for its calculation is as follows (Duggal and Puri, 1991):

$$S = A/B - C,$$

where

S = specific gravity.
A = weight of dried sample.
B = weight of the wet sample.
C = weight of the basket holding the sample to immerse it in water.
B – C = weight of the water displaced by the stone sample when immersed in water.

13.6.6.5 Absorption Test

The absorption of water per cubic foot of rock is determined by the absorption test. It is calculated as per the formula provided below:

$$x = W_2 - W_1/W - W_1 \times 62.24,$$

x = pounds of water absorbed per cubic foot of stone.
W = weight in grams.
W_1 = weight in grams of sample in water just after immersion.
W_2 = weight in grams of sample in water after being immersed for 48 hours.
62.24 = weight in pounds of a cubic foot of distilled water at 25°C.

13.7 DISCUSSION

In the last 50 years, our understanding of the interaction between microorganisms and stone materials has advanced greatly, and this is only because of the rapid improvements in scientific methodologies and awareness among multidisciplinary scientists to work for our cultural heritage conservation. Biodeterioration is a broad phenomenon that occurs along with other causes of destruction. This decay phenomenon is also similar to other physical and chemical causes. Therefore, it is difficult to measure the amount of destruction caused by biological agents—in particular, microorganisms. But it is accepted that microorganisms cause stone deterioration, and in the presence of microorganisms, this process is more rapid than physical and chemical deterioration of stone. Most of the research states that lichens, algae, fungi, bacteria and higher plants destroy the stone quality of our historical monuments. Microbial activity is always not correlated with the number of microorganisms on the stones, but it is yet to be calculated the limit above which the organisms become pathogenic to stones. Intact stones show lesser amount of microorganisms in comparison to weathered stones because they have suitable locations like cracks and crevices. More studies needed where the interaction between various organisms in a single environment is reported. Microbial biofilms contribute to the deterioration and discoloration of various monumental stones in tropical regions. There are many different modes of investigation because organisms can act in a synergistic way in the deterioration of stones. A multidisciplinary approach should be inculcated when studying microbial impact on historical monuments. It is necessary to understand these concepts in order to determine the mechanisms of deterioration as well as the methods of control. Microbiologists, who are specialists in materials technology; chemists; and geologists have to work together in order to avoid the use of protective treatments that can cause more damage to the building material of monuments than the microorganisms.

13.8 CONCLUSION

The need for a sustained campaign through media publicity, exhibition, seminars and education down to village level; provision of adequate funds from government agencies; training of personnel for restoration work of heritage monuments; use of new technology for restoration work; exchange of best practices for restoration and further maintenance of monuments; strict enforcement of heritage restoration rules formulated by legislative bodies; and effective coordination among the agencies and the public are necessary steps to create the common awareness in society to preserve these excellent architectures that reflect our glorious past.

ACKNOWLEDGEMENTS

We are thankful to Prof. Pandit Bhalchandra Vidyasagar, Honorable Vice Chancellor, SRTM University, Nanded, Maharashtra, India, for his kind support.

REFERENCES

Ascaso, C., Wierzchos, J., Souza-Egipsy, V., De los Rıos, A., & Rodrigues, J. D. (2002). In situ evaluation of the biodeteriorating action of microorganisms and the effects of biocides on carbonate rock of the Jeronimos Monastery (Lisbon). *International Biodeterioration & Biodegradation*, *49*(1), 1–12.

Becker, T. W., Krumbein, W. E., Warscheid, T., & Resende, M. A. (1994). Investigations into microbiology. In *IDEAS—Investigations into devices against environmental attack on stones: A German-Brazilian project, 147-85*. Geesthacht, Germany: GKSS-Forschungszentrum Geesthacht GmbH.

Camuffo, D., Bernardi, A. & Ongaro, A. 1986. The challenges of the microclimate and the conservation of works of art. Remes Rencontres Internationals pour la Protection du Patrimoine Cultural, Avignon, 215–228.

Crispim, C. A., & Gaylarde, C. C. (2005). Cyanobacteria and biodeterioration of cultural heritage: A review. *Microbial Ecology*, *49*(1), 1–9.

Crispim, C. A., Gaylarde, P. M., & Gaylarde, C. C. (2003). Algal and cyanobacterial biofilms on calcareous historic buildings. *Current Microbiology*, *46*(2), 0079–0082.

De Silva, R. H. (1975). Rock painting in Sri Lanka. In *Preprint of IIC congress on conservation in archeological and the applied arts* (pp. 69–74). Stockholm: International Institute for Conservation of Historic and Artistic Works.

Denyer, S. P. (1990). Mechanisms of actions of biocides. *International Biodeterioration, 26*, 397–417.

Duggal, A. K., & Puri, V. P. (1991). *Laboratory manual in highway engineering* (pp. 1–09). New Delhi: Wiley Eastern Limited.

Fitzner, B., & Heinrichs, K. (2001). Damage diagnosis at stone monuments—Weathering forms, damage categories and damage indices. *ACTA-Universitatis Carolinae Geologica, 1*, 12–13.

Gehrmann, C. K., Petersen, K., & Krumbein, W. E. (1989). *Silicole and calcicole lichens on Jewish tombstones: Interactions with the environment and biocorrosion.* VIth International Congress on Deterioration and Conservation of Stone. Supplement.= VIe Congrès International sur l'altération et la conservation de la pierre. Torun, 12-14.09. 1988, pp. 33–38. Toruń, Poland: Uniwersytet Mikolaja Kopernika w Toruniu.

Greieken, R. V., Delalieux, F., & Gysels, K. (1998). Cultural heritage and the environment. *Pure and Applied Chemistry, 70*(12), 2327–2331.

Griffin, P. S., Indictor, N., & Koestler, R. J. (1991). The biodeterioration of stone: A review of deterioration mechanisms, conservation case histories, and treatment. *International Biodeterioration, 28*(1), 187–207.

Gürtler, V., Stanisich, V. A. (1996). New approaches to typing and identification of bacteria using the 16S-23S rDNA spacer region. *Microbiology, 142*, 3–16.

Heyrman, J., & Swings, J. (2003). Modern diagnostic techniques on isolates. *Coalition, 6*, 9–13.

Hirsch, P. F. E. W., Eckhardt, F. E. W., & Palmer, R. J., Jr. (1995). Fungi active in weathering of rock and stone monuments. *Canadian Journal of Botany, 73*(S1), 1384–1390.

Kamh, G. M. E. (2011). Salt weathering, bio-deterioration and rate of weathering of dimensional sandstone in ancient buildings of Aachen City, Germany. *International Journal of Water Resources and Environmental Engineering, 3*(5), 87–101.

Khobragade, C. N., Srinivasarao, R., Borkar, P. S., & Yangade, R. S. (2006). Microbially induced impact on physico-chemical properties of porous limestones: A case study from Kandhar fort. *Current Science, 91*(10), 1318–1320.

Kovacik, L. (2000). Cyanobacteria and algae as agents of biodeterioration of stone substrata of historical buildings and other cultural monuments. *New Millennium International Forum on Conservation of Cultural Property*, Institute of Conservation Science for Cultural Heritage, December 5–8, pp. 44–58.

Kumar, R., & Kumar, A. V. (1999). *Biodeterioration of stone in tropical environments: An overview.* Los Angeles, CA: Getty Publications.

Lewin, S. Z., & Charola, A. E. (1981). Plant life of stone surfaces and its relation to stone conservation. *Scanning Electron Microscopy, 1*, 563–568.

Mansch, R., & Bock, E. (1998). Biodeterioration of natural stone with special reference to nitrifying bacteria. *Biodegradation, 9*(1), 47–64.

May, E. (2003). Microbes on building stone—For good or ill? *Culture, 24*, 5–8.

May, E., Lewis, F. J., Pereira, S., Tayler, S., Seward, M. R. D., & Allsopp, D. (1993). Microbial deterioration of building stone—A review. *Biodeterioration Abstracts, Cab International*, 7(2), 109–123.

McNamara, C. J., & Mitchell, R. (2005). Microbial deterioration of historic stone. *Frontiers in Ecology and the Environment, 3*, 445–451.

Miller, A., Dionísio, A., & Macedo, M. F. (2006). Primary bioreceptivity: A comparative study of different Portuguese lithotypes. *International Biodeterioration & Biodegradation, 57*(2), 136–142.

Miller, A. Z., Laiz, L., Gonzalez, J. M., Dionísio, A., Macedo, M. F., & Saiz-Jimenez, C. (2008). Reproducing stone monument photosynthetic-based colonization under laboratory conditions. *Science of the Total Environment, 405*(1), 278–285.

Mink, J. F. (1983). Groundwater hydrology in agriculture in the humid tropics. In *Hydrology of the Humid Tropical Regions with Particular Reference to the Hydrological Effects of Agriculture and Forestry Practice*. London: Wiley.

Orial, G., Castanier, S., Le Metayer, G., & Loubière, J. F. (1993). *The biomineralization: A new process to protect calcareous stone; applied to historic monuments*. Biodeterioration of Cultural Property 2: Proceedings of the 2nd International Conference on Biodeterioration of Cultural Property, October 5–8, 1992, held at Pacifico Yokohama (Pacific Convention Plaza Yokohama), pp. 98–116. International Communications Specialists.

Otto, B., & Morales, O. (2006). Cyanobacterial diversity and ecology on historic monuments in Latin America. *Micobiologia, 48*, 188–195.

Saiz-Jimenez, C. (1993). Deposition of airborne organic pollutants on historic buildings. *Atmospheric Environment: Part B. Urban Atmosphere, 27*(1), 77–85.

Saiz-Jimenez, C, & Laiz, L. (2000) Occurrence of halotolerant/halophilicbacterial communities in deteriorated monuments. *International Biodeterioration and Biodegradation, 46*, 319–326.

Scheerer, S., Ortega-Morales, O., & Gaylarde, C. (2009). Microbial deterioration of stone monuments—An updated overview. *Advances in Applied Microbiology, 66*, 97–139.

Scherer, G. W., Flatt, R., & Wheeler, G. (2001). Materials science research for the conservation of sculpture and monuments. *MRS Bulletin, 26*(01), 44–50.

Seaward, M. R. D., Giacobini, C., Giuliani, M. R., & Roccardi, A. (1989). The role of lichens in the biodeterioration of ancient monuments with particular reference to Central Italy. *International Biodeterioration, 25*(1), 49–55.

Sterflinger, K. (2010). Fungi: Their role in deterioration of cultural heritage. *Fungal Biology Reviews, 24*(1), 47–55.

Sutton, S. V. W., & Cundell, A. M. (2004). Microbial identification in the pharmaceutical industry. *Pharmacopeia Forum, 30*, 1884–1894.

Thornbush, M. J. & Viles, H. A. (2006). Use of portable X-ray fluorescence for monitoring elemental concentrations in surface units on roadside stone at Worcester College, Oxford. In: Fort, R. et al (eds.) *Heritage, Weathering and Conservation*. Taylor and Francis, London: 613–620.

Tomaselli, L., Lamenti, G., Bosco, M., & Tiano, P. (2000). Biodiversity of photosynthetic micro-organisms dwelling on stone monuments. *International Biodeterioration & Biodegradation, 46*(3), 251–258.

Videla, H. A., Guiamet, P. S., & de Saravia, S. G. (2000). Biodeterioration of Mayan archaeological sites in the Yucatan Peninsula, Mexico. *International Biodeterioration and Biodegradation, 46*, 335–341.

Warscheid, T., & Braams, J. (2000). Biodeterioration of stone: A review. *International Biodeterioration & Biodegradation, 46*(4), 343–368.

Warscheid, T., Petersen, K., & Krumbein, W.E. (1990). A Rapid Method to Demonstrate and Evaluate Microbial Activity on Decaying Sandstone. *Studies in Conservation, 35*(3), 137–147.

Webster, A, & May, E. (2006). Bioremediation of weathered-building stone surfaces. *Trends in Biotechnology, 24*(6), 255–260.

14 Wastewater Treatment: Common Effluent Treatment Plant—Case Study

*Ashita Rai and M. H. Fulekar**

CONTENTS

* Corresponding author: E-mail: mhfulekar@yahoo.com

14.1 INTRODUCTION

Water is an essential component for sustainability and is crucial for social and economic development, energy and food production for the survival of human beings (United Nations Department of Economic and Social Affairs [UNDESA], 2015). Water links society with the environment and is at the core of adaptation to climate change. Water covers 70% of the earth's surface, which makes its continuous and uninterrupted supply in the future necessary (World Water Development Report [WWDR], 2015). Of the 70% of total water on earth, oceans occupy 97% and freshwater is extremely scarce, holding only 3%. Two-thirds of freshwater is entrapped in ice sheets and ice caps or are inaccessible for use (Table 14.1). About 1.1 billion people in the world do not have an accessible water supply, and 2.7 billion people face water scarcity for a month every year (World Wildlife Fund [WWF], 2019). Inadequate sanitation exposes approximately 2.4 billion people to water-borne diseases such as cholera and typhoid fever. At least 2 million people die every year from diarrhea (World Health Organization [WHO], 2017). It is predicted that by 2050, half of the world will be exposed to water scarcity (Boretti, A. & Rosa, L., 2019). The sixth Sustainable Development Goal lays emphasis on ensuring available and sustainable water management and sanitation to everyone (United Nations Development Programme [UNDP], 2015).

14.1.1 STATUS OF WATER ON EARTH (TABLE 14.1)

TABLE 14.1
Distribution Status of Water

Source of Water	Volume of Water (Cubic kilometer)	Percent Freshwater	Percentage of Total Water
Oceans, seas and bays	1,338,000,000	–	96.5
Ice caps, glaciers and permanent snow	24,064,000	68.7	1.74
Groundwater	23,400,000	–	1.7
Freshwater	10,530,000	30.1	0.76
Saline water	12,870,000	–	0.94

(Continued)

TABLE 14.1 *(Continued)*
Distribution Status of Water

Source of Water	Volume of Water (Cubic kilometer)	Percent Freshwater	Percentage of Total Water
Soil moisture	16,500	0.05	0.001
Ground ice and permafrost	300,000	0.86	0.022
Lakes	176,400	–	0.013
Fresh	91,000	0.26	0.007
Saline	85,400	–	0.006
Atmosphere	12,900	0.04	0.001
Swamp water	11,470	0.03	0.0008
Rivers	2,120	0.006	0.0002
Biological water	1,120	0.003	0.0001
Total	1,386,000,000	–	100

Source: Gleick, P. H. (1996).

14.1.2 Water Pollution

The deterioration of water quality leads to pollution, which arises from several sources, such as runoff of pesticides and fertilizers from agricultural fields and human and industrial activities (Natural Resource Defense Council [NRDC], 2018). Leaching of these pollutants into the soil horizon affects groundwater aquifers. These impacts are instant, when the water quality degrades and makes it unsuitable for regular use and sometimes contaminants from industries move and biomagnify along the food chain. Rapid growth in population, urbanization, industrial development and modern agriculture requires a huge quantity of freshwater and generates a large volume of effluents due to industrial processing and other activities (Cosgrove, W.J. & Loucks, D.P., 2015). With the exponential population growth and industrial development, pressure on the water supply has increased significantly. Human-related activities like urban and industrial development, modern agriculture and population growth have caused the degradation of water quality. Unavailability of water has significantly affected water quality improvement and water pollution control measures. Researchers and scientists are focusing on water pollution control measures. Thus, water quality improvement is an utter need for the abatement of water pollution and eradication of water scarcity (Food and Agriculture Organization [FAO], 2000).

14.2 CHARACTERISTICS OF WASTEWATER

Industrial waste may be toxic, hazardous and contain varying concentrations of organic and inorganic compounds. The contaminants may be biodegradable or non-biodegradable, or it may contain both types of contaminants. It is important to characterize the wastewater before employing treatment techniques. The characteristics of wastewater include important parameters discussed in Section 14.2.1.

14.2.1 Oxygen Demand

At least 3 mg/L of dissolved oxygen is essential for maintaining the biological activity in wastewater. Aerobic bacteria require dissolved oxygen to decompose dissolved contaminants (National Research Council [NRC], 1993). A high pollutant load will require a higher dissolved oxygen demand. Biological and chemical oxygen demand determines the pollutant load in the industrial effluents. BOD determines the amount of oxygen required for the oxidation of contaminants by the microorganisms (Show, K.-Y., 2008). Its main limitation is that it takes 5 days at 20°C and 3 days at 27°C for the oxidation of the contaminants. COD estimates chemical oxidation of contaminants by strong acids. The BOD/COD ratio indicates the portion of contaminants that are degradable in the effluent (Choi, Y.-Y et al., 2017). A BOD/COD ratio below 2–2.5 shows readily biodegradable contaminants in the effluent (Płuciennik-Koropczuk, E., & Myszograj, S., 2019). Normal domestic effluent may contain about 200mg/L of BOD, 345 mg/L of biodegradable COD and 450 mg/L of COD.

14.2.2 Suspended Solids

Suspended and colloidal particles in effluent increase turbidity and block pipes and filters used during the treatment process. Suspended solids in human-generated waste effluent may range between 50 and 300 mg/L, and can be as high as 1500 mg/L in some instances. Normal domestic wastewaters contain 200 mg/L of total suspended solids (TSS) and 80% volatile suspended solids (VSS) (Fulekar, M. H., 2010). The availability data summarizes that 65% of VSS in wastewater are biodegradable and exert a five-day BOD of 100 mg/L (Fulekar, M. H., 2010).

14.2.3 Nutrients

Nutrients form building blocks for the growth and development of microorganisms and ensure the synthesis of enzymes and other cofactors for their metabolism. Hence, nutrients play an essential role in water quality management. Depending upon the quantity required by the microorganisms, nutrients fall into two categories, viz. macronutrients (nitrogen, phosphorus, sulfur, potassium, calcium and magnesium) and micronutrients (iron, boron, copper, manganese, zinc, chromium and cobalt) (Fulekar, M. H., 2010). Higher concentrations of these essential nutrients cause toxicity. Nitrogen and phosphorus are essential from the pollution point of view. A high level of phosphorous facilitates algal growth and results in eutrophication and reduces sunlight and the dissolved oxygen level in water bodies, leading to water quality degradation. For effective biodegradation of contaminants, a balanced carbon, nitrogen and phosphorus ratio is essential, preferably 100:10:1 or 100:5:1(Treatment Plant Operator [TPO], 2018).

14.2.4 Alkalinity and pH

pH denotes the acidic or basic characteristic of water. pH ranging between 6.5 and 8.5 is beneficial for the proper treatment and quality control of industrial effluents and causes no significant impact on the environment. Alkalinity provides the buffering capacity to maintain the normal pH values. The combined action of alkalinity and pH has a detrimental impact on irrigation and other human-related activities. Alkalinity is generated by the carriage water and protein and urea in the wastewater. The alkalinity in the carriage water is related to sodium and calcium carbonates. Metabolization of protein and urea generate ammonical nitrogen. The ammonia reacts with the carbon dioxide to produce ammonium bicarbonate alkalinity. The alkalinity keeps the pH from shifting very quickly when acids are produced by metabolism (Fulekar, M. H., 2010). Alkalinity and pH provide the most reliable information about the characteristic of wastewater.

14.2.5 Temperature

Temperature regulates microbial activity and plays a significant role in their growth. The rate of metabolism changes by a factor of 2 with every 10°C temperature change between 5°C and 40°C. The temperature favored by psychrophilic organisms has a range from 15°C to 20°C. A mesophilic organism favors a temperature range from 30°C to 37°C and thermophilic ranges between 55°C and 60°C (Fulekar, M. H., 2010). The increase in temperature above 40°C kills the mesophilic microbes. Thermophilic reactors work more efficiently in the degradation of the contaminants and are preferred over mesophilic reactors. Thermophilic microbes double their rate of metabolism until the temperature exceeds 65°C. The thermophiles may survive to even higher temperatures, but the rate of endogenous respiration as well as synthesis creates a real problem for survival. Anaerobic degradation of wastewater by thermophilic microorganisms is a better alternative for the treatment and water quality.

14.2.6 Microorganisms

The population of microorganisms in the biological treatment process can either be natural or developed to act on specific compounds in the waste. Both prokaryotic and eukaryotic organisms have potential for biological treatment of toxic wastes. Eukaryotes, which include protozoa, fungi and most groups of algae, have a highly organized cell structure. On the other hand, prokaryotes, which include bacteria and blue green algae, have a much simpler cell structure with a classical nucleus. Among these, bacteria are the most important for biological treatment. Other microorganisms, however, feed on dispersed bacteria and fine particulates that do not settle well and therefore help in improving the effluent characteristics. They are secondary in importance to bacteria.

14.2.6.1 Bacteria

The bacteria are single-celled organisms which metabolize soluble organics of the wastewater. Surface enzymes convert suspended organics to soluble organics. A very large population of bacteria is necessary to stabilize the organics in wastewater, because a bacterium weighs approximately 10^{-12} g each. Most of the soluble

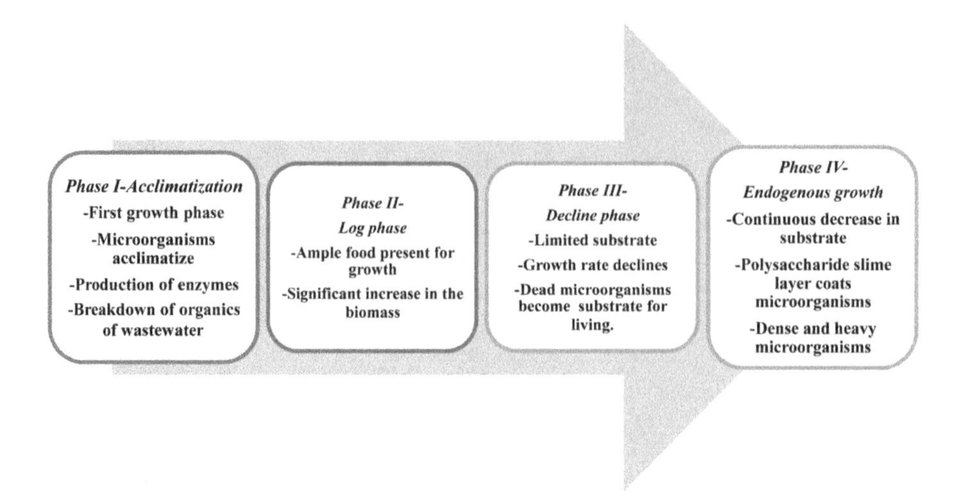

FIGURE 14.1 Growth phases of microorganisms.

organics are metabolized by organic-solubilizing bacteria, and the inorganic content is metabolized by inorganic-solubilizing bacteria. Hence, their niches do not overlap and they coexist in the environment. Normal wastewater contains 10^5 to 10^7 bacteria per mL (Fulekar, M. H., 2010).

The microbial population in a biological system passes through four distinct growth phases (Figure 14.1). After the waste and microorganisms are combined, the reproductive rate of microorganisms is quite low, and the number of cells in the system increases gradually.

14.2.7 Energy Reactions

14.2.7.1 Aerobic Condition

Dissolved oxygen is a common electron acceptor in aerobic processes. The solubility of oxygen, however, is limited in water, i.e. only 9.1 mg/L at 20°C and 1 atm air pressure. Aerobic metabolism requires oxygen to be used as the ultimate electron acceptor and yields the maximum amount of energy to the microbes for metabolism. Aerobic metabolism produces carbon dioxide and water as end products in addition to the cell mass. These two end products are the most oxidized forms of carbon and hydrogen and can be discharged to the environment without creating a further oxygen demand.

Bacteria can reduce the nutrients in the solution to a very low level under aerobic conditions, although they cannot remove 100% of the nutrients. The rate of aerobic metabolism is normally controlled by the rate of oxygen transfer. Mixing is the most important factor affecting the rate of aerobic metabolism because it determines the rate of dispersion of nutrients, microbes and dissolved oxygen.

The end product of protein metabolism is ammonical nitrogen, but it is not the most stable form of nitrogen in the process:

- Nitroso bacteria metabolize ammonical nitrogen to nitrite, and nitrobacteria convert it further to nitrate. Nitrate nitrogen is the most stable form of nitrogen.

Thiobacillus metabolize hydrogen sulfideto sulphate, similar to the nitrogen-solubilizing bacteria. Both nitrogen-solubilizing bacteria and sulfur-oxidizing bacteria utilize CO_2 and H_2O for nutrients to form cellular components.

14.2.7.2 Anaerobic Conditions

In oxygen-deficient conditions, the bacteria metabolize nutrients anaerobically. Under anaerobic conditions, nitrate, sulphate, iron and carbon dioxide may serve as terminal electron acceptors. Thus, the ultimate electron acceptor is shifted from dissolved oxygen to other materials with a lower energy yield and a lower cell mass production per unit of nutrients metabolized.

14.2.8 Denitrification

Nitrates are electron acceptors under anaerobic conditions because nitrates produce almost as much energy for the bacteria as dissolved oxygen. Nitrogen gas is relatively insoluble in water and thus it is lost to the atmosphere as tiny bubbles. The process of nitrate reduction in nitrogen gas is known as denitrification. It should be noted that if the bacteria needs nitrogen for cell protoplasm and no ammonia nitrogen is available, the bacteria will reduce nitrates to ammonia for cell synthesis. Nitrate will not be reduced to ammonia for energy reaction. This is an important difference of utility of nitrogen by bacteria in denitrification.

14.2.9 Sulphate Reduction

Desulfo bacteria can use sulphates as their electron acceptors for cell synthesis reactions. The sulphate-reducing bacteria convert the sulphate to thio sulphates, sulphates, sulfur dioxide, free sulfur, or hydrogen sulfide. The available sulphates and substrates determine which bacteria will predominate and how it will proceed to metabolic reactions. In the presence of excess nutrients and less sulphate, reduction favors the formation of hydrogen sulfide under normal conditions of wastewater. Hydrogen sulfide has greater solubility in water than oxygen, but it forms gas that diffuses out from the water.

14.2.10 Organic Acid

Bacteria often use the nutrients as electron acceptors for dissolved oxygen deficiency. They oxidize a portion of the nutrients to obtain energy while reducing a portion of the nutrients. The bacteria lose some of their energy in the ultimate electron transfer and have very little energy for cell synthesis. This results in limited cell mass production and a considerable amount of reduced nutrients unmetabolized. The oxygen demand potential of the end products is almost equal to that of the initial nutrients. But the metabolism may change the form of nutrients present. The primary metabolic end products are the organic acids. The organic acids can also be reduced to aldehydes and alcohols upto a limited extent (Fulekar, M. H., 2010).

14.2.11 Methane and Carbon Dioxide Production

The organic acid in an anaerobic condition will accumulate, and methanogenic bacteria will act on it to metabolize acetate to methane and carbon dioxide. A second group of methanogenic bacteria can reduce carbon dioxide to form methane and water. Hydrogen will be transferred from the bacteria metabolizing the organic acid to acetic acid. Methane is insoluble in water and is released in the atmosphere. Similarly, carbon dioxide will also exceed the solubility and will be lost in the atmosphere. The methanogenic bacteria can reduce the organics to a relatively lower level, producing a low organic effluent.

14.3 COMMON EFFLUENT TREATMENT PLANT

Small scale industries (SSI) generally find it difficult to establish and operate individual effluent treatment plants due to their limited size and scale of operations. However, as SSI are generally located at industrial areas/estates, the concept of a common effluent treatment plant (CETP) for the characteristic improvement of industrial waste has been developed to achieve satisfactory treatment through collective efforts (MOEF). The idea of CETP is principally based on the cooperative approach among SSI. Its main purpose is to reduce treatment cost and promote environment protection. CETP has following objectives:

- Treatment cost reduction for the SSI.
- Efficient and trained manpower.
- Minimal land requirement for CETP.
- Monitoring of waste discharge.
- Treatment of large volume of waste.
- Reuse and recycle of waste.

14.3.1 Status of CETP in India

SSI produce effluents of undesirable characteristics, and their treatment in a separate unit is economically unfeasible as the operation and capital cost is high. The treatment and disposal capacity of these SSI is also inefficient, and as a result, they discharge their effluent in freshwater systems and createsoil-water pollution. The discharged effluents do not comply with the prescribed standards of pollution monitoring boards/agencies. Hence, the government of India launched the concept of CETP to combat the water treatment problems. State Pollution Boards were directed by the Ministry of Environment and Forest to check the possibility of establishing CETP in their respective states. According to the reports of the Central Pollution Control Board in 2011–2012, 193 CETPs have been installed across the country serving 212 industrial areas/estates.

14.3.2 Advantages of CETP

- Economical: Less capital and operational cost.
- Effective: Large volume of effluents is treated to comply with the standards.
- Resource: Less land and other resource requirement.
- Manpower: Technically trained manpower is employed.
- Reuse and recycle of waste effluent and sludge.

14.4 CASE STUDY: GUJARAT

The idea of CETP emerged for the characteristic improvement of untreated or partially treated industrial effluent. The CETP receives the effluent from member industries with treatment costs for effluent and employs the treatment measures discussed in Sections 14.4.1–14.4.4.

14.4.1 Preliminary Treatment

The preliminary treatment of effluent comprises of the removal of coarse, floating, heavy biodegradable contaminants and stabilization of the effluent by equalization or chemical addition (Figure 14.2). For better operation and efficient treatment, the large particles are removed typically by coarse screens, grit chambers and comminution of large objects.

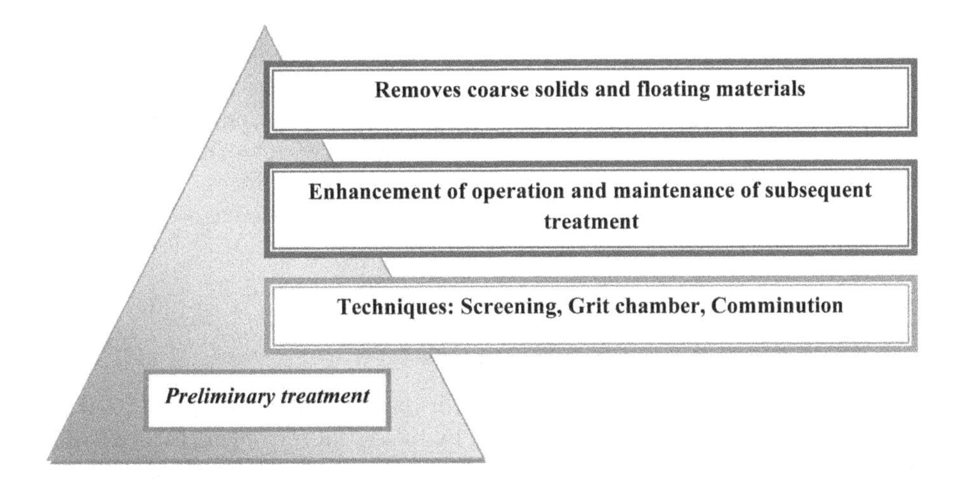

FIGURE 14.2 Preliminary treatment of wastewater.

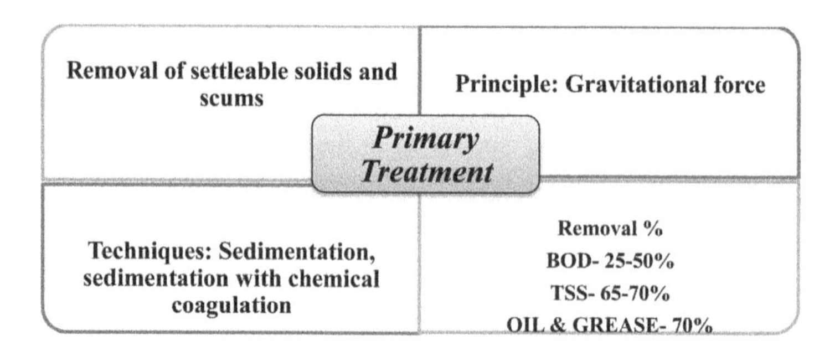

FIGURE 14.3 Primary treatment of wastewater.

14.4.2 Primary Treatment

The primary treatment aims at the purging of settleable organic and inorganic matter under the influence of gravity and removal of scum by skimming (Figure 14.3). This method helps in the removal of 25–50% of BOD, 65–70% suspended and 70% oil and grease from preliminary treated effluents. Nutrients, organic and inorganic contaminants can also be removed. Colloidal and dissolved solids cannot be removed by primary treatment.

14.4.3 Secondary or Biological Wastewater Treatment Processes

The secondary treatment technique utilizes the potential of prokaryotic or eukaryotic microorganisms to control water quality (Figure 14.4). It removes organics and dissolved solids from wastewater. They are more efficient than chemical methods. Secondary treatment can be categorized into two types: aerobic and anaerobic. Aerobic treatment is carried out in an oxygen-rich environment by aerobic microorganisms. Organic contaminants are oxidized and microbial diversity proliferates on the supplement of oxygen and hence increases the degradation efficiency of the process. Aerobic processes differ from each other by the mode and rate of oxygen supply. An aerobic process is performed in oxygen-deficient condition by anaerobes. The activated sludge process is frequently used in CETPs.

14.4.4 Tertiary or Advanced Wastewater Treatment

Tertiary treatment for wastewater is applied where contaminants are of particular characteristics and cannot be treated by biological treatment (Figure 14.5). So, specific treatment is required. Mostly this advanced treatment follows

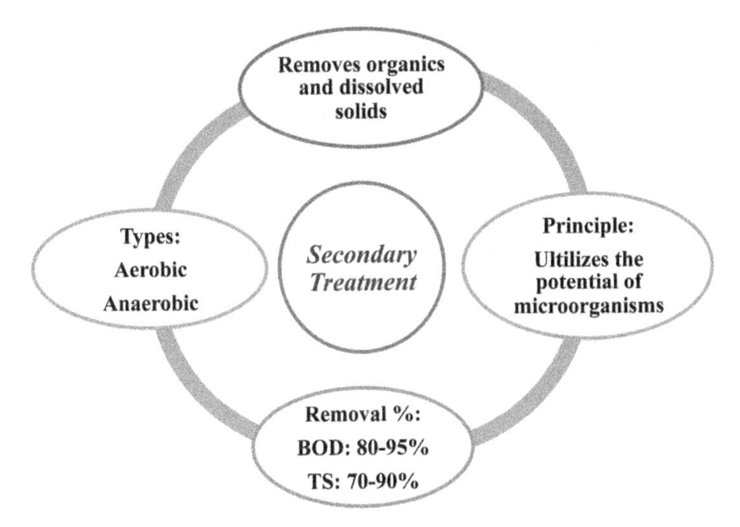

FIGURE 14.4 Secondary treatment of wastewater.

secondary treatment. It is the final step in the treatment process. Occasionally, tertiary treatment is employed in combination with primary and secondary treatment. The treated water can be of drinking water quality. With tertiary treatment, 99% of the contaminants can be treated. It usually consists of disinfection, ion exchange and reverse osmosis.

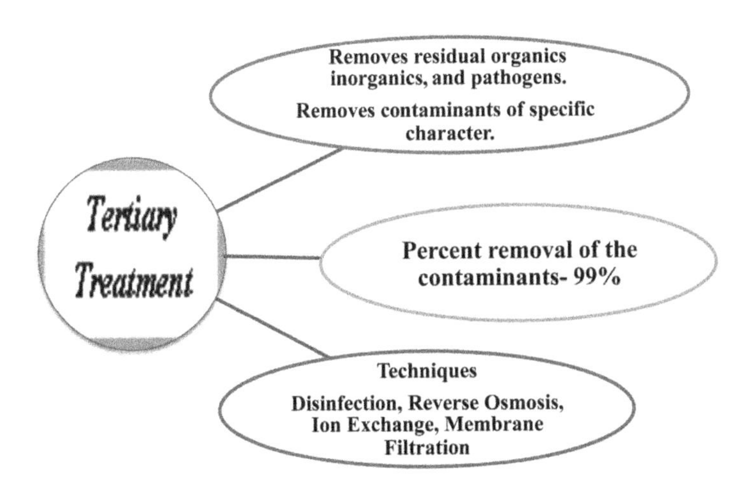

FIGURE 14.5 Tertiary treatment of wastewater.

14.5 STATUS OF CETP IN GUJARAT (GPCB, 2017)

Gujarat is an industrial state having 33 operational CETP plants. CETP plants differ in operation for the treatment of industrial effluent. The CETP receives the industrial effluent and mix with sewage and employs the treatment as categorized as CETP I (Vapi). A CETP operating by receiving treated industrial waste of other CETP plants is categorized as CETP II (Ankleshwar). A CETP plant in which only primary and secondary treatment is employed using sewage microbial biomass is categorized as CETP III (Vatva).

The Hazardous Waste Management and Handling Rule 2016 has categorized different types of hazardous wastes. The wastes generated by these industries like petrochemical, pesticide, chemicals, dyes and drugs and industrial and pharmaceutical are treated using physicochemical and biological methods. Industries are looking forward for standard guidelines and established methods for the treatment of hazardous waste generated by industries.

The standards of physicochemical parameters have been prescribed under the related act. The industries generating the hazardous waste are treating the effluent/waste to comply with the standard before discharging into the soil-water environment. In practice, the large group of industries have the treatment facilities to treat the generated waste and comply with the standard before letting out into the environment. The middle group of industries have the treatment facilities for the treatment of effluent/waste to comply with the prescribed standard; however, some middle group of industries have partial treatment facilities and effluent/waste is let out to CETP. Besides, SSI cannot afford to have treatment facilities for generated effluent/waste. The CETP thus receives effluent/waste from member industries for treatment employing primary, secondary, and tertiary treatment of the complex waste.

In the present study, three CETPs have been selected for characterization of industrial effluent. The status of physicochemical parameters in compliance with the standard has been assessed. The CETPs have been selected under the project to bring out the physicochemical and microbial status to comply with the standard at the point treated wastewater discharge is let out in soil-water environment.

14.6 DESCRIPTION OF SELECTED COMMON EFFLUENT TREATMENT PLANTS

14.6.1 CETP I: Vapi

The Vapi Industrial Estate (VIA) under Gujarat Industrial Development Corporation (GIDC), developed in phases, was started four decades back in 1967–1968. VIA covers an area of 1140 hectares and includes 759 industries. Most of the industries are small scale. Most of the member industries are dye- and dyestuff-based, pesticide, chemical and pharmaceutical. The other includes packaging, engineering, plastic,

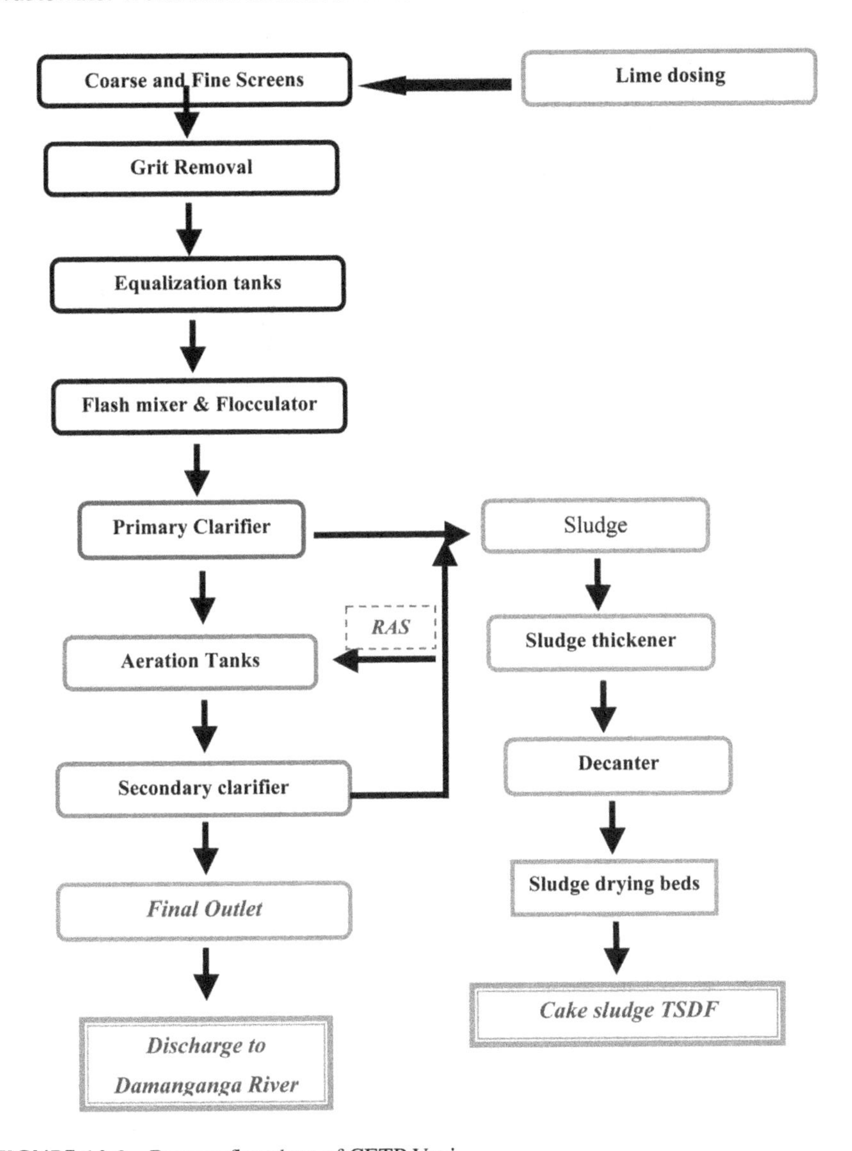

FIGURE 14.6 Process flowchart of CETP Vapi.

textiles, food processing and printing ink. CETP Vapi mixes sewage with the industrial waste, thereby increasing the complexity of the waste (Figure 14.6). CETP Vapi is a simple, conventional plant treating effluents using physicochemical and biological approaches. CETP Vapi treats 55 million liters per day (MLD) of effluent daily using primary and secondary measures and discharge effluent in Damanganga River (Vapi Green Enviro Limited [VGEL], 2011).

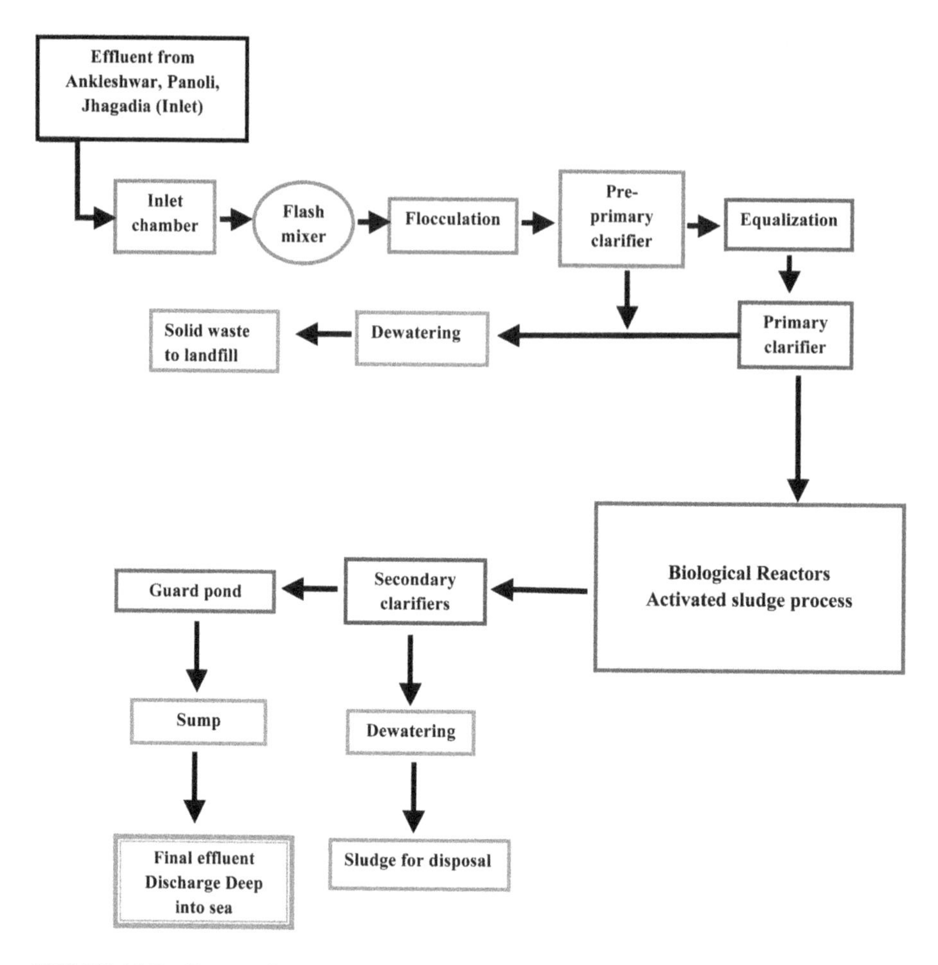

FIGURE 14.7 Process flowchart of CETP Ankleshwar.

14.6.2 CETP II: Ankleshwar

Narmada Clean Tech Limited (NCTL) developed under GIDC is a final effluent treatment plant for the effluents of CETPs established in Ankleshwar, Jhagadia and Panoli. The Final Effluent Treatment Plant (FETP) receives the complex waste from member CETPs and discharges it through a 50-km pipeline into the deep sea (Figure 14.7). The treatment processes include primary and secondary approaches (NCT, 2019).

14.6.3 CETP III: Vatva

GIDC developed Vatva Industrial Estate, which houses more than 1800 industrial units. The complex waste received by the CETP comprises mainly dyes, dye-intermediates, pigments, fine chemicals and other such organics. The treatment capacity of Vatva is 16 MLD. CETP Vatva also treats its complex waste using primary and secondary methods and discharges the effluent in Sabarmati River (Figure 14.8) (Green Environment Co-Operative Services Society Ltd [GESCSL], 2018).

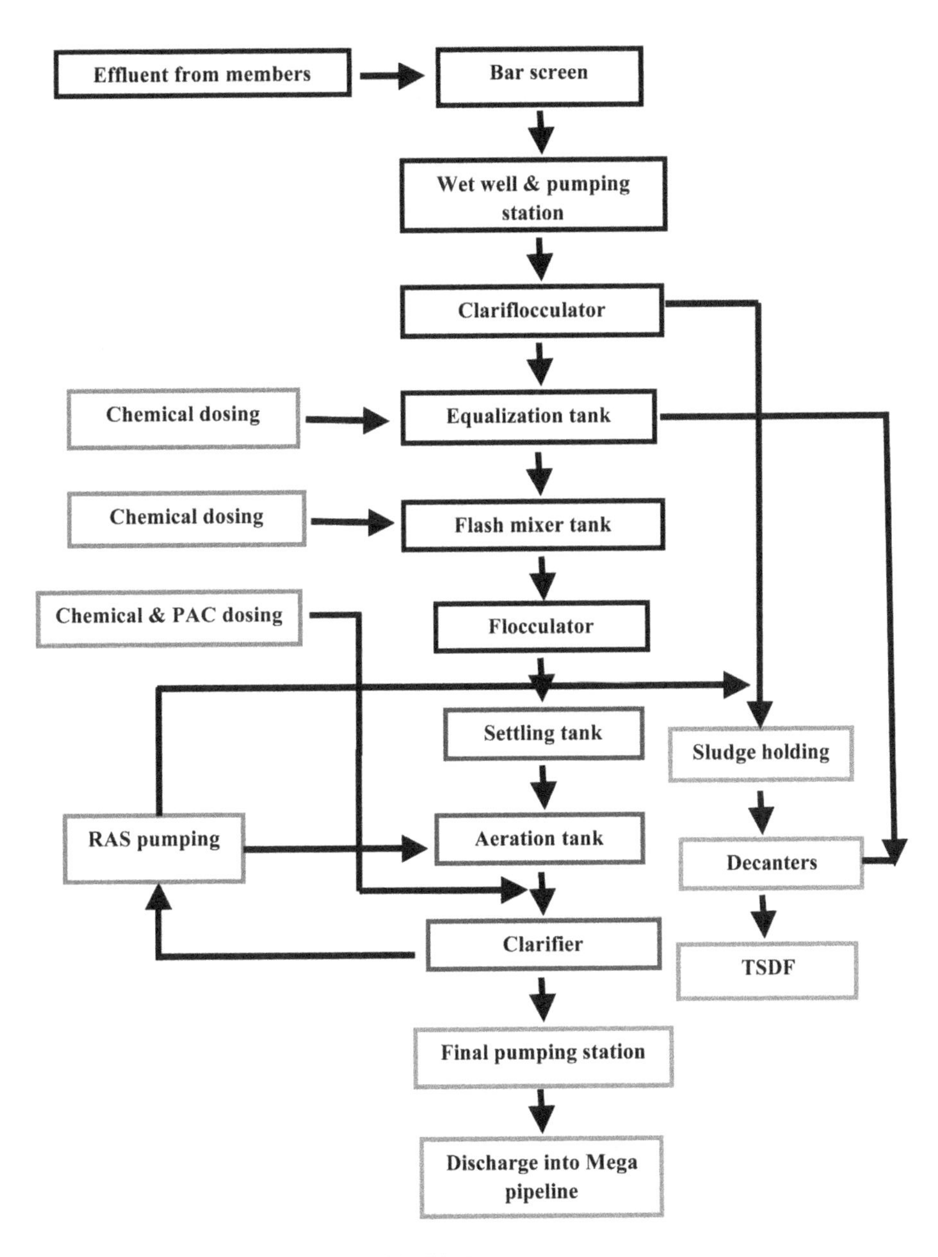

FIGURE 14.8 Process flowchart of CETP Vatva.

14.7 STANDARDS FOR DISCHARGE OF ENVIRONMENTAL EFFLUENTS (TABLE 14.2)

TABLE 14.2
Wastewater Discharge Standard

		Standards			
S. No.	**Parameter**	**Inland Surface Water**	**Public Sewers**	**Land for Irrigation**	**Marine Coastal Areas**
1	Suspended solids mg/L, Max.	100	600	200	(a) For process wastewater 100 (b) For cooling water effluent 10% above total suspended matter of influent
2	pH	5.5 to 9.0	5.5 to 9.0	5.5 to 9.0	5.5 to 9.0
3	Temperature	shall not exceed 5°C above the receiving water temperature	–	–	shall not exceed 5°C above the receiving water temperature
4	Biochemical Oxygen demand 1 [3 days at 27°C] mg/L max.	30	350	100	100
5	Chemical oxygen demand, mg/L, max.	250	–	–	250
6	Ammonical nitrogen (as N), mg/L Max.	50	50	–	50
7	Total Kjeldahl nitrogen (as NH_3) mg/L, Max.	100	–	–	100
8	Nitrate nitrogen	10 mg/L	–	–	20 mg/L
9	Dissolved phosphates (as P), mg/L Max.	5.0	–	–	–
10	Sulphide (as S) mg/L Max.	2.0	–	–	5.0
11	Fluoride (as F) mg/L Max.	2.0	15	–	15

Source: General Standards, Central Pollution Control Board, New Delhi.

TABLE 14.3
Selected CETP Discharge Status

	CETP Vapi		CETP Ankleshwar		CETP Vatva	
Parameters	**Discharge Standard**	**Discharge**	**Discharge Standard**	**Discharge**	**Discharge Standard**	**Discharge**
pH	6.5–8.5	7.76	6–9	6.97	6.5–8.5	7.42
COD	250 mg/L	268 mg/L	500 mg/L	466 mg/L	250 mg/L	618 mg/L
BOD	30 mg/L	33 mg/L	100 mg/L	22 mg/L	30 mg/L	109 mg/L
SS	100 mg/L	95 mg/L	100 mg/L	107 mg/L	100 mg/L	142 mg/L
NH_3-N	50 mg/L	46 mg/L	50 mg/L	48 mg/L	50 mg/L	13 mg/L
Phenolic compounds	1 mg/L	0.53 mg/L	5 mg/L	0.61 mg/L	1 mg/L	0.5 mg/L

Source: December 2018 Discharge Status, Gujarat Pollution Control Board.

14.8 DISCHARGE STATUS OF CETP (TABLE 14.3)

14.9 CONCLUSION

The wastewater discharge characterization and treatment comprising primary, secondary and tertiary treatment makes water suitable for its intended use. The industries have their own treatment processes and comply with prescribed standards and discharge in the soil-water environment. Small industries cannot afford to have their own treatment facilities and become members of a CETP. The CETP receives complex waste from member industries and collectively treats a large volume of effluent using primary and secondary processes, whereas tertiary treatment is not employed often. The compliance with prescribed standards has become major problem for CETPs and is contaminating the environment. The case study highlights treatment processes of three different CETPs, viz. CETP I mixes industrial waste with sewage and further treats the effluent employing primary and secondary treatment methods. CETP II receives treated CETP effluent and further treats it on a large scale. CETP III receives dyestuff waste and treats it using primary and secondary methods. The Gujarat Pollution Control Board (GPCB) data highlights that selected CETPs do not comply with the prescribed standards and discharge effluent in the water body, thereby polluting the environment. To comply with the prescribed standards, tertiary or advanced treatment methods should be installed at the CETPs to improve water quality and resolve the water scarcity problem.

REFERENCES

Boretti, A., Rosa, L. (2019). Reassessing the Projections of the World Water Development Report. *npj Clean Water*. 2, 15.

Central Pollution Control Board. (1993). General Standards for Discharge of Environmental Pollutants Part-A: Effluents. Retrieved from https://www.cpcb.nic.in/GeneralStandards.pdf

Choi, Y.-Y., Baek, S.-R., Kim, J.-I., Choi, J.-W. (2017). Characteristics and Biodegradability of Wastewater Organic Matter in Municipal Wastewater Treatment Plants Collecting Domestic Wastewater and Industrial Discharge. *Water*. 9(6), 409.

Cosgrove, W. J., Loucks, D. P. (2015). Water Management: Current and Future Challenges and Research Directions. *Water Resources Research*. 51, 4823–4839.

Food and Agriculture Organization (FAO). (2000). Retrieved from http://www.fao.org/3/a-bl035e.pdf

Fulekar, M. H. (ed.) (2010). Biotechnology—Pollution Abatement. In: *Environmental Biotechnology*. CRC Press, Boca Raton, FL.

Gleick, P. H. 1996. Water Resources. In: Schneider, S. H. (ed.), *Encyclopedia of Climate and Weather* (Vol. 2). Oxford University Press, New York, pp. 817–823.

Gujarat Pollution Control Board (GPCB). (2018). Status of CETPs in Gujarat. Retrieved from https://www.gpcb.gov.in/status-of-cepts-in-gujarat.htm

Ministry of Environment and Forest of India. (2010). Common Effluent Treatment Plants (CETPs). Retrieved from http://www.moef.nic.in/divisions/cpoll/cept.pdf

Narmada Clean Tech (NCT). (2019, April). Retrieved from http://www.nctc.co.in/Layout1.aspx?mm=2&lid=9&sm1=1

National Research Council (NRC). (1993). In Situ Bioremediation: When Does it WORK? In: *Principles of Bioremediation*. The National Academics Press, Washington DC.

Natural Resource Defense Council (NRDC). Water Pollution: Everything You Need to Know. Retrieved from https://www.nrdc.org/stories/water-pollution-everything-you-need-know

Płuciennik-Koropczuk, E., Myszograj, S. (2019). New Approach in COD Fractionation Methods. *Water*. 11(7), 1484.

Show, K.-Y. (2008). Seafood Wastewater Treatment. In: Klemes, J., Smith, R., Kim, J.-K. (eds.), *Handbook of Water and Energy Management in Food Processing*. Woodhead Publishing, Sawston, United Kingdom, pp. 776–801.

The Green Environment Co-Operative Services Society Ltd (GESCSL). (2018). Vatva Industries Association. Retrieved from http://www.vatvaassociation.org/?page_id=1179

Treatment Plant Operator (TPO). (2018). Optimal Nutrient Ratios for Wastewater Treatment. Retrieved from https://www.tpomag.com/whitepapers/details/optimal_nutrient_ratios_for_wastewater_treatment_sc_001fa

United Nations Department of Economic and Social Affairs (UNDESA). (2015). Water and Sustainable Development. Water for Life Decade. Retrieved from https://www.un.org/waterforlifedecade/water_and_sustainable_development.shtml

United Nations Development Programme (UNDP). (2015). Sustainable Development Goals. Retrieved from https://www.undp.org/content/undp/en/home/sustainable-development-goals.html##targetText=The%20Sustainable%20Development%20Goals%20(SDGs,peace%20and%20prosperity%20by%202030

Vapi Green Enviro Limited (VGEL). (2011). Retrieved from http://www.vgelvapi.com/cetp+process+flow-pg-24.html

World Health Organization (WHO). (2017). Diarrhoeal Disease. Retrieved from https://www.who.int/news-room/fact-sheets/detail/diarrhoeal-disease

World Water Development Report (WWDR). (2015). Water for a Sustainable World. Retrieved from https://sustainabledevelopment.un.org/content/documents/1711Water%20for%20a%20Sustainable%20World.pdf

World Wildlife Fund (WWF). (2019). Water Scarcity. Retrieved from https://www.worldwildlife.org/threats/water-scarcity

15 Nanomaterials-Based Wastewater Treatment Technology: An Overview

*Dimple P. Dutta**

CONTENTS

15.1 INTRODUCTION

Water has always been an invaluable resource for sustaining life on earth. However, with the burgeoning population and the exponential increase in material aspirations, sustainable efforts need to be made to make this basic necessity available to all. Currently, almost one-seventh of the world population is deprived of clean potable water, and the number will increase exponentially in coming years if the problem is not rectified with correct measures. The depletion and pollution of water resources are due to the erosion of soil, the introduction of saltwater to potable water, and unhygienic sanitation procedures, which is also accompanied by surface water contamination by the disposal of hazardous chemical contaminants from multiple industries. Wastewater remediation can be an effective method to conserve and optimize this precious natural resource. Over the years, wastewater remediation has been done using different physical processes where contaminants have been coagulated, flocculated, and removed; they have also been trapped in membranes used for filtrations and adsorbed on sorbent materials [1–5]. Chemical processes leading to oxidation of the contaminants are also a tried-and-tested method. However, each technique has its own benefits and problems. Coagulation and flocculation techniques cannot eliminate various microbes and pharmaceutical complexes [6, 7]. Membrane fouling is a hindrance in the case of membrane separation techniques. Ion exchange, as well

* Corresponding author: Email: dimpled@barc.gov.in

as reverse osmosis, is expensive, but it is an effective method to treat heavy metal ion contamination. Biological treatments are mostly unable to remove a gamut of toxic components, and their separation from the effluent is a tedious process [8]. Hence, in order to arrest this ever-increasing threat to safe water, it is essential to develop inexpensive and novel materials and methods to address these problems. In this regard, environmental nanotechnology is being increasingly considered as an effective technique for wastewater remediation.

Nanomaterials are typically defined as materials having dimensions less than 100 nm. At this scale, materials change their properties, which depend on their size and which are generally different from the bulk material. Some of these wastewater remediation applications benefit from the high specific surface area exhibited by these nanomaterials, which shortens the dissolution time and leads to greater reactivity and robust sorption. Some nanomaterials also exhibit properties of quantum confinement, localized surface plasmon resonance, and superparamagnetism, which can be used to improve the effectiveness of the remediation technique. In recent years, various nanocatalysts, nanostructured catalytic membranes, dendrimers, zeolites, and nanosorbents have shown potential for wastewater remediation. In this chapter, an effort has been made to review and summarize the different nanotechnology-based wastewater treatment processes reported in recent times.

15.2 NANOCATALYSTS FOR WASTEWATER REMEDIATION

Nanocatalysts have increased surface catalytic activity, which is attributed to its higher surface area, and hence they prove to be more effective in wastewater treatment. Semiconductor materials, few metals, viz. Au, Ag, Pt, Pd, etc., and certain bimetallic materials in their nanoform show remarkable catalytic efficiency. TiO_2, which can be classified as a low-cost, nontoxic semiconductor material, shows excellent photocatalytic properties, which help in wastewater treatment [9]. On ultraviolet (UV) light irradiation, electron/hole (e^-/h^+) pairs are formed in semiconductor TiO_2, which migrate to the surface more efficiently in the case of nanostructures. Various reactive oxygen species (ROS), such as OH radicals, superoxide radicals, etc., are formed which can deactivate bacteria, fungi, algae, and viruses and also lead to degradation of various organic pollutants. Photodegradation results in total oxidative and reductive degradation of both inorganic and organic pollutants and yields CO_2, water, and mineral acids as the final products. The photocatalytic properties of TiO_2 nanostructures can be tailored by altering their morphology and dimensions, restricting the recombination of e^-/h^+ pairs by noble metal doping, and increasing the extent of pollutant adsorption and surface reactivity by surface modification. The changes in the morphology of TiO_2 from spherical nanoparticles to nanotubes have been reported to enhance the photodegradation of organic contaminants [10]. This is due to the shorter migration path of electron-hole pairs to the walls of the tube and quicker transfer of more reactants on the surface of nanotubes. The limitation of TiO_2 is its minimal absorption of visible light, and this problem has been taken care of by tuning its band gap via transition metal doping and also doping of nitrogen, carbon, sulfur, or fluorine atoms [11–14].

Tungsten trioxide (WO_3) is another material which has been proven effective in photocatalytic wastewater treatment. WO_3 absorbs visible light and has a lower band gap (<450 nm) compared to TiO_2 which helps in the generation of more electron-hole pairs [15]. The photocatalytic activity can be further enhanced via doping of Pt in WO_3, which helps in extracting the electrons from the conduction band of WO_3 and hence results in more effective e^-/h^+ separation by preventing their recombination [16].

ZnO is also a viable photocatalyst for water remediation with an added advantage of being water soluble, and it also serves as an antibacterial agent [17]. The band gap of ZnO is 3.37 eV, which makes it sensitive to UV light. Band gap tuning of ZnO has been done via transition metal ion doping, as well as doping of nitrogen and sulfur atoms, to facilitate visible light activity.

Doping of silver (Ag) in TiO_2 enhances its photocatalytic activity by acting as electron scavengers which lead to a decrease in electron-hole recombination in TiO_2. On a similar note Ag doped/dispersed $AgVO_3$, $BiPO_4$, $ZnTiO_3$, and $Bi_4Ti_3O_{12}$ nanocatalysts exhibit enhanced visible light photocatalytic degradation of organic pollutants as well as antibacterial activity [18–21].

However, it should be noted that certain problems are associated with the use of nanocatalysts. The small size makes its removal from the medium in which it is dispersed a very tedious process. For practical applications in industry, it is generally agglomerated to form larger micron-sized lumps or has to be loaded on porous substrates so that during water treatment via a through-flow technique, the material does not get dispersed in the output flow which can jeopardize the process. Hence, research on developing reusable photocatalytic support material is also an integral part of this field.

15.3 NANOSTRUCTURED CATALYTIC MEMBRANES

The membranes function as a filter which allows only particles smaller than its mesh size to pass through. The effectiveness of the membrane depends on the material it is made of, and its performance can be enhanced by adding functional nanomaterials, which can improve its strength, penetrability, heat resistance, pollutant removal capability, and self-cleaning properties. Nanostructured catalytic membranes have immense application in wastewater treatment. They provide an even distribution of catalytic centers and faster kinetics of the reaction which is profitable enough for industrial applications. Inorganic semiconductor compounds like TiO_2 have been deposited on membrane surfaces, which can increase its efficiency by coupling in the photocatalytic effect. Polymer membranes have been coated with catalysts, which enhance the pollutant degradation reaction without the following complicated step of catalyst recovery from the purified product. Fe^0 as well as noble metal supported Fe^0 has been anchored on polymeric membranes, which proves to be extremely effective for the degradation of chlorinated contaminants [22]. Fe^0 donates an electron, which acts as a reducing agent and oxidizes various contaminants to a nontoxic form, and the noble metals help in increasing the kinetics of the reaction. Polymeric membranes composed of polyvinylidene fluoride (PVDF), cellulose acetate, polysulfone, and chitosan have been impregnated with metal nanoparticles, which improve the reactivity of the membrane surface. It is a symbiotic process since the membrane surface restricts the tendency of the nanoparticles to agglomerate, which is also beneficial for increasing its degradation efficiency for toxic contaminants.

15.4 NANOZEOLITES FOR WASTEWATER REMEDIATION

Zeolites are mostly aluminosilicate materials with a porous network and have a 3D crystalline structure made up of SiO_4/AlO_4 tetrahedra which are connected through their corners. The Si^{4+}/Al^{3+} exchange is an important phenomenon in these materials, as this leads to a buildup of excess negative charge in the system, which can be counterbalanced by the uptake of positive ions. Hence, zeolites find application as ion exchangers which establish its important role in water purification. Ag is known to have antibacterial properties. The incorporation of Ag in such a zeolite matrix makes it effective for the disinfection of water. When contaminated water is passed through such zeolite columns, the Ag gets eluted out as some other cations take their position, and the released Ag can kill pathogens [23]. The efficiency of any adsorbent is dependent on its surface area, chemical composition, and polarity of the surface. Hence, zeolites with nanopores have large surface area, more accessible reactive sites, and shorter diffusion pathways, which render them more effective compared to zeolites with micropores [24]. The potential of synthetic nanozeolites and zeolitic material modified with quaternary amines to eliminate pharmaceutical wastes and nonionic organic solutes, respectively, from wastewater has also been demonstrated. Natural zeolites like clinoptilolite are being gradually replaced by its lab-synthesized surface-modified version, which exhibits a heavy metal adsorption capacity that is ten times more than that observed in the natural version [25]. Nanoforms of zeolites with composition $Na_nAl_nSi_{96-n}O_{192}{\cdot}16H_2O$ $(0 < n < 27)$ commonly known as ZSM-5 have shown remarkable efficiency in trapping toxic Hg(II) from contaminated water [26].

15.5 NANODENDRIMERS FOR WASTEWATER REMEDIATION

Dendrimers have a branched chain appearance and belong to the group of synthetic macromolecules. They have a 3D structure with a central unit with innumerous regular branches made of identical building blocks and a variety of functional groups attached to the fringes. They can be designed for the adsorption of organic pollutants as well as heavy metal ions via complex formation, surface modifications leading to better electrostatic interactions, and inclusion of functional groups, which promote hydrogen bonding and hydrophobic effect [27]. Their high surface area, uniform molecular structure, presence of various surface functional groups, and internal pores render them invaluable for application in water remediation. Poly(amidoamine) (PAMAM) is a type of dendrimer that has been demonstrated to remove toxic Cu(II) from wastewater. The decreased viscosity in this dendrimer makes it an excellent choice for water remediation, as it needs less pressure and consequently energy to operate [28]. Apart from PAMAM, poly(propyleneimine) (PPI) is also considered to be a good dendrimer system for water remediation. There are reports on the deposition of nanoparticles of Ag, palladium, and platinum on PAMAM and PPI, which finds application in the reduction of toxic 4-nitrophenol [29]. Micelles can be classified as a type of dendrimer with hydrophobic interior shells and having exterior branches embellished with hydroxyl/amine functional groups. They serve a dual function since hydrophobic centers encapsulate the organic contaminants, whereas the outer surface has an affinity for heavy metal ion adsorption.

15.6 NANOSORBENTS

Among the various possible techniques, adsorption of heavy metal ions and organic contaminants from wastewater by a suitable sorbent material is a really simple, reliable, and effective process. Adsorption is a surface phenomenon and follows the same mechanism for the removal of both types of pollutants. When wastewater flows through the porous channel of the sorbent material, there is an intermolecular force of attraction between the solid surface and passing fluid phase, which traps some of the solute particles from the fluid phase to adhere to the surface of the sorbent material. The trapped solute clinging onto the surface of the sorbent material is called adsorbate, and the process is called adsorption [30]. The formation of an adsorbed phase whose composition is very different from the fluid phase, which serves as a carrier for it, leads to the process of separation by adsorption technology. In a bulk material, the bonding requirements predicted by the valency of the constituent atoms are all satisfied by formation of either ionic/covalent/metallic bond with other atoms. However, the surface atoms have unfulfilled bonding requirements, and therefore they exhibit an affinity for the adsorbate. The process of adsorption can be that of physisorption, which is characterized by the formation of weak van der Waals forces. Stronger bonding is observed in case of chemisorption, which results due to formation of covalent bonds. In the intermediate process, where neither physisorption nor chemisorption is the undisputed procedure, the bonding possibility is mostly governed by electrostatic attraction. During the last two decades, there has been a spate of research on adsorption of organic pollutants on inexpensive natural adsorbents like clay, zeolites, and agricultural/biomass wastes. However, it is a challenge to grade their relative competency since the efficiency of the sorption process depends critically on the porous nature of the material, its surface area, functional groups on the surface, and physical strength. Apart from the intrinsic properties of each individual sorbent in wastewater treatment which affects the sorption process, tweaking the experimental conditions can also enhance the efficiency of the process. Hence, there is a persistent need for novel adsorbents fulfilling the criteria of greater sorption of contaminants in shorter time accompanied with good reusability and ease of separation. Nanomaterials with its small size and high surface area are the perfect candidates for developing inexpensive and environmental benign water purification units. They can be tailored to exhibit selective sorption of materials and enhanced capacity of sorption. Though activated carbon is the most studied adsorbent, they are rendered ineffective for the sorption of many antibiotics and pharmaceuticals, as these bulky constituents cannot be accommodated in its micropores. Carbon nanotubes (CNTs) are better sorbents for these bulky organic molecules since they have a similar surface area associated with much larger pores which are easily accessible [31]. CNTs also show better sorption for heavy metal ions like Cu^{2+}, Pb^{2+}, Cd^{2+}, and Zn^{2+} compared to activated carbon, which has been attributed to their conveniently positioned adsorption sites and shorter intraparticle diffusion distance [32]. Various metal oxides, viz. Fe_2O_3, TiO_2, Al_2O_3, etc., have been established as good inexpensive sorbents for heavy metals and radionuclides. The sorption is the outcome of complex formation between dissolved metal ion contaminants and the oxygen present in metal oxides. The first step is quick adsorption of metal ions on the surface of metal oxide particles, which is then subsequently followed by slow intraparticle diffusion along the micropore walls. Obviously, nanostructured materials have much better adsorption capacity and impressive kinetics attributed to its higher specific surface area, shorter intraparticle dispersion distance, and substantially larger number of surface reaction sites. The process can be made more efficient by incorporating superparamagnetic nano-maghemite and nano-magnetite. These iron oxides are attracted to external magnetic fields and can be easily separated from the reaction medium using a simple magnet. These magnetic nanoparticles can also be used in a core-shell nanoparticle structure, where the shell provides the desired function of adsorption of contaminants through surface modifications, while the magnetic core takes care of the ease of separation. Apart from oxides, several metal nanomaterials have been found to be good adsorbents of arsenic, lead, mercury, copper, cadmium, chromium, and nickel ion contaminants from water, and they demonstrate sorption capacities equivalent to that of activated carbon. Regeneration of these metal, metal oxide, and CNT sorbents is possible by altering the pH of the solution, which leads to desorption of the adhering contaminants from the sorbent surface.

Currently, a new class of metal organic frameworks (MOFs) materials with nanopores is being developed for application in the selective adsorption of organic pollutants. However, their synthesis is quite complicated and leads to an escalation in the cost of water treatment. Very recently, various metal tungstates and molybdates nanoparticles (M = Ba, Mn, Cu, etc.) have been synthesized, and they have shown exceptional sorption affinity for cationic dyes and heavy metal ions like Cu^{2+} [33–36].

15.7 CONCLUSION

Water pollution is an area of major concern for environmentalists, and nanotechnology can be a game changer in wastewater treatment in the near future. Nanosorbents, nano-photocatalysts, and nanocatalytic membranes are already making headway in pilot testing and to some extent commercialization stages. The key areas where researchers should focus include use of less expensive materials, good scalability, and efficient regeneration and removal of the nanomaterials from the water body to be treated so as to reduce risk to the environment and humans alike. A synergistic relationship between research institutions, industry, and the government can provide positive solutions to our wastewater treatment challenges.

REFERENCES

1. Parsons, S. A., Jefferson, B. *Introduction to Potable Water Treatment Processes.* Blackwell Publishing: Oxford, United Kingdom, 2006.
2. Shu, H. Y., Huang, C. R., Chang, M. C. *Chemosphere.* 1994, **29**, 2597–2607.
3. Allégre, C., Moulin, P., Maisseu, M., Charbit, F. *J. Membrane Sci.* 2006, **269**, 15–34.

4. Pagga, U., Taeger, K. *Water Res.* 1994, **28**, 1051–1057.
5. Santhy, K., Selvapathy, P. *Bioresour. Technol.* 2006, **97**, 1329–1336.
6. Westerhoff, P., Yoon, Y., Snyder, S., Wert, E. *Environ. Sci. Tech.* 2005, **39**, 6649–6663.
7. Vieno, N., Tuhkanen, T., Kronberg, L. *Environ. Technol.* 2006, **27**, 183–192.
8. Urase, T., Kikuta, T. *Water Research.* 2005, **39**, 1289–1300.
9. Hashimoto, K., Irie, H., Fujishima, A. *Jap. J. Appl. Phys.* Part 1. 2005, **44**, 8269–8285.
10. Macak, J. M., Zlamal, M., Krysa, J., Schmuki, P. *Small.* 2007, **3**(2), 300–304.
11. Subramanian, V., Wolf, E., Kamat, P. V. *J. Phys. Chem. B.* 2001, **105**, 11439–11446.
12. Livraghi, S., Paganini, M. C., Giamello, E., Selloni, A., Di Valentin, C., Pacchioni, G. *J. Am. Chem. Soc.* 2006, **128**, 15666–15671.
13. Lin, L., Lin, W., Zhu, Y. X. et al. *J. Mol. Catal. A-Chem.* 2005, **236**, 46–53.
14. Umebayashi, T., Yamaki, T., Itoh, H., Asai, K. *Appl. Phys. Lett.* 2002, **81**, 454–456.
15. Kominami, H., Yabutani, K., Yamamoto, T., Kara, Y., Ohtani, B. *J. Mater. Chem.* 2001, **11**, 3222–3227.
16. Kim, J., Lee, C. W., Choi, W. *Environ. Sci. Tech.* 2010, **44**, 6849–6854.
17. Dimapilis, E. A. S., Hsu, Ching-S., Mendoza, R. M. O., Lu, Ming-C. *Sus. Environ. Res.* 2018, **28**, 47–56.
18. Singh, A., Dutta, D. P., Ballal, A., Tyagi, A. K., Fulekar, M. H. *Mater. Res. Bull.* 2014, **51**, 447–454.
19. Fulekar, M. H., Singh, A., Dutta, D. P., Roy, M., Ballal, A., Tyagi, A. K. *RSC Adv.* 2014, **4**(20), 10097–10107.
20. Dutta, D. P., Singh, A., Tyagi, A. K. *J. Environ. Chem. Eng.* 2014, **2**, 2177–2187.
21. Dutta, D. P., Tyagi, A. K. *Mater. Res. Bull.* 2016, **74**, 397–407.
22. Wu, L. F., Ritchie, S. M. C. *Environ. Prog.* 2008, **27**, 218–224.
23. Petrik, L., Missengue, R., Fatoba, M., Tuffin, M., Sachs, J. Report to the Water Research Commission. 2012, No KV 297/12.
24. Almeida, L. C., Garcia-Segura, S., Bocchi, N., Brillas, E. *Appl. Catal. B: Environ.* 2011, **103**, 21.
25. Kang, Y., Shan, W., Wu, J., Zhang, Y., Wang, X., Yang, W., Tang, Y. *Chem. Mater.* 2006, **18**, 1861–1866.
26. Alijani, H., Beyki, M. H., Mirzababaei, H. N. *Desalin. Water Treat.* 2015, **55**, 1864–1875.
27. Crooks, R. M., Zhao, M. Q., Sun, L., Chechik, V., Yeung, L. K. *Acc. Chem. Res.* 2001, **34**, 181–190.
28. Savage, N., Diallo, M. S. *J. Nanopart. Res.* 2005, **7**, 331.
29. Esumi, K., Isono, R., Yoshimura, T. *Langmuir.* 2004, **20**, 237.
30. Rashed, M. N. http://dx.doi.org/10.5772/54048.
31. Dutta, D. P., Venugopalan, R., Chopade, S. *Chem. Select.* 2017, **2**, 3878–3888.
32. Lu, C., Chiu, H., Bai, H. *J. Nanosci. Nanotech.* 2007, **7**, 1647–1652.
33. Dutta, D. P., Singh, A., Ramkumar, J., Bhattacharya, K., Tyagi, A. K., Fulekar, M. H. *RSC Adv.* 2013, **3**(44), 22580–22590.
34. Dutta, D. P., Mathur, A., Ramkumar, J., Tyagi, A. K. *RSC Adv.* 2014, **4**, 37027–37035.
35. Dutta, D. P., Singh, A., Ramkumar, J., Bhattacharya, K., Tyagi, A. K., Fulekar, M. H. *Adv. Porous Mater.* 2014, **2**, 237–245.
36. Dutta, D. P., Rathore, A., Ballal, A., Tyagi, A. K. *RSC Adv.* 2015, **5**, 94866–94878.

16 Nanomaterials Development and Their Environmental Applications

*Ashita Rai and M. H. Fulekar**

CONTENTS

* Corresponding author: E-mail: mhfulekar@yahoo.com

16.1 INTRODUCTION

Nanotechnology, coined in 1974 by Professor Norio Taniguchi of Tokyo Science University, describes the development of materials at a nanoscale. Professor Richard P. Feynman in 1959 stated that there is enough space at the bottom (Feynman, 2011). The size range of interest lies between 1 and 100 nm to the atomic level because the material in this range displays different structural behavior and properties. The reasons for this changed behavior are increased surface area, which results in improved chemical reactivity of materials and facilitates them to be used as a functional nanocatalyst. And also the quantum effect affects the electrical, magnetic and optical behaviors of developed materials (The Royal Society, 2004). The electrical property of materials extends between metallic and semiconducting materials and is also determined by the diameter of the nanoscale materials. Vibrations in the nano-based materials are tenfold higher than the metals, which results in greater thermal conductivity (Solairajan et al., 2014). Nanomaterials can withstand extreme strain. The synthesis method of nanomaterials checks the size in the range of 1–100 nm. There are two methods for the development of nanoparticles: bottom up and top down. Nanotechnology encompasses the design, development, production and characterization of nanomaterials between 1 and 100 nm ranges with modified electrical, optical and mechanical attributes than the original materials. Research and development of nanoparticles involves the manufacturing of nanoparticles at atomic, molecular or macromolecular scale ranging between 1 and 100 nm. Developed nanostructures and devices with enhanced novel properties have been utilized in various fields.

Nanotechnology-based advanced materials and manufacturing comprise the development of novel materials, devices and systems. These innovative structures have improved and enhanced structural and functional properties, which aids in sustainable environmental development. Nowadays, the industrial sector has already attributed sophisticated innovations working on smaller devices such as computer chips and carbon nanofibers with the increasing competition in the twenty-first century.

16.2 NANOMATERIALS

Materials with less than 100 nm possessing novel size-oriented characteristics, different from their larger counterparts, are considered nanomaterials. Nano-sized particles have enhanced properties, namely a high surface area. The increased surface area promotes rapid dissolution, faster reactivity and higher absorption of contaminants present in the environment. Recent developments in nanotechnology viz. nanocatalysts, zeolites and dendrimer have proved their potential in removing pathogens and toxicants (organic, inorganic, metal ions) from the industrial effluents. In this chapter, the synthesis of nanoparticles, in particular, titanium dioxide and doped titanium dioxide, as well as silver nanoparticles by plant and bacteria, is highlighted. The development of novel materials with enhanced and improved structure and function is critical for a growing competitive society for achieving overall sustainable development. According to material manufacturing industries, novel materials with enhanced attributes are of more a priority than the steps involved in their processing and manufacturing. Research and development is focused on the development of such materials and requires indelible control of integral properties and manufacturing of nanomaterials in consideration with possible effects on human health and environment. Improved nanomaterials developed using biotechnological and nanotechnological approaches are given importance. For improving the reliability and efficiency of developed materials, emphasis is placed on the in-depth analysis of structural design and their characteristics. Manufacturing sectors must focus on in-depth analysis for the development of materials rather than resource-centric manufacture. In-depth–based materials can be processed depending upon the attributes viz. manufacturing, design and characterization together with its consumption and future impacts on the environment.

16.3 ADVANCES IN NANOTECHNOLOGY

Advancements in nanotechnology indicate that the present water quality issue can be checked and settled using nano-sized materials such as nanoparticles, nanocomposite, bio-nanocomposite, nano-sized membrane filters and photocatalyst, and they reflect a remarkable effect on the quality of water (Diallo and Savage, 2005). Nano-based approaches render economical, productive, efficient and long-lasting methods for water quality improvement as nanomaterials for water treatment and purification can be less contaminating and involve lower capital, labor and other resources (Pandey et al., 2011). Generally, nanomaterials are categorized into four classes for water quality improvement, viz. metal containing nanoparticles, carbonaceous nanomaterials, zeolites and dendrimers. These nanomaterials have an extensive range of physicochemical attributes, which ensures their efficacy in water treatment (Tiwari et al., 2008). Nanostructures offer huge potential as a water purifier and redox-centric medium because of greater surface to volume area and their structural, optical, electrical and catalytic properties (Obare and Meyer, 2004). Titanium dioxide in the presence of sunlight can effectively degrade organic contaminants in water (Goswami et al., 1997). Nano-sized materials can selectively degrade halogenated organic compounds from waste effluents via dehalogenation, even in the presence of other contaminants (Yuan and Keane, 2003).

16.4 ENVIRONMENTAL NANOTECHNOLOGY

Degradation of the environment is of grave concern to human beings. Manufacturing technologies allow operations that pollute the environment and cause harmful effects. Environmental nanotechnology plays a significant role in reshaping present environmental science and technology. Nanoscale technology has rendered new innovative and economically efficient methods for the identification, degradation and catalysis of contaminants. Expectations have been placed over nanostructures for pollution abatement and improvement of environment leading toward sustainability. Developing countries face scarcity of water both qualitatively and quantitatively. Nanotechnology can play a pivotal role in rendering safe, secure and efficient measures for cleaner water. Nanomaterials can be potentially used as nanosorbents when combined with magnetic particles, which facilitate separation of particles. Photocatalysts are useful for the degradation of the pollutants. Nanofilters are being used for the remediation of organic compounds and metals from industrial effluents. Focus is given on research and development of safer, economical and efficient nano-based techniques and technologies for the remediation of pollutants from industrial effluents. Extrusive studies have been carried out on titanium dioxide as a photocatalyst in the degradation of organic pollutants. Many nano-sized structures have been used as detectors of contaminants. Environmental nanotechnology deals with pollution abatement by minimizing raw material utilization and degradation of waste. On the other hand, it focuses heavily on economical and energy-centric development. Green synthesis for the development of nanomaterials and nano-based approaches renders a way for waste minimization, lesser hazard, efficient and safer green technology. Yet there is a wide array of arguments regarding safety and probable environmental effects of nanomaterials.

16.4.1 Physical Methods

Physical methods for nanomaterial development comprise evaporation-condensation and laser ablation. Nanomaterials such as Ag, Au, PbS and CdS have been developed by the evaporation-condensation technique. Physical methods have advantages over other methods as they do not have solvent-based contaminants, and nanoparticles are distributed uniformly (Kruis et al., 2000). According to Jung et al. (2006) silver nanomaterials can be developed using a ceramic heater. The liberated vapor may condense at a faster rate due to the intensive temperature gradient in the surface area of the heater, and nanomaterials are developed in higher concentrations. Physical techniques can be also used for long-term nanomaterial production for inhalation toxicity study and calibrators for nano-based instruments. Kabashin and Meunier (2003) synthesized silver nanomaterials using laser ablation of a bulk metal solution. Nanomaterials developed using this technique depend on the wavelength of laser influencing bulk metal, time period of laser ablation, power of laser impinging the bulk metal and solution with or without surfactants (Tarasenko et al., 2006). This technique is preferred over other methods, as it does not require chemical reagents and it generates pure and contaminant-free nanomaterials.

16.4.2 Chemical Methods

Reducing agents, such as Tollens' reagent, are being used for the synthesis of silver nanoparticles by reducing the silver ions present in aqueous and non-aqueous solutions and facilitate the development of silver metal ions, which agglomerate into oligomeric clusters and ultimately form silver nanoparticles (Merga et al., 2007).

Protective agents are used during metal nanomaterial development to stabilize a dispersive nanomaterial that absorbs and binds upon the other nanomaterial and prevents agglomeration (Oliveira et al., 2005). Surfactants with functional groups such as thiols, amine and alcohol can be helpful in stabilizing particles and prevent their agglomeration. Currently, Tollens' method has been used for the development of silver nanomaterials for size control in which silver ions undergo reduction by carbohydrate in the presence of ammonia. Chemical synthesis involves the use of hazardous chemicals and it is not economical, so emphasis is being laid on environmentally sustainable methods for the synthesis of nanomaterials (Kharissova et al., 2013).

16.4.3 Conventional Methods

Conventional techniques of nanomaterial development comprise crushing, grinding, milling and air micronization. They have many disadvantages, namely the desired size may not be obtained and it may interfere with the physicochemical property of nanomaterials. Many compounds can be chemically reactive, thermally labile, explosive, and hazardous, which cannot be processed using conventional methods. It can be tricky to process dyes, pigments, and polymers using the conventional method as they may degrade under high temperature and pressure.

16.4.4 Sonochemical Method

16.4.4.1 Preparation of Titanium Dioxide Nanoparticles

The sonochemical method can be used for the synthesis of titanium dioxide. Sonochemical approaches rely on acoustic cavitation formed from the uninterrupted generation, magnification and violent disintegration of bubbles formed in solution. In this technique, titanium isopropoxide is mixed with distilled water. The pH of the solution is regulated at 10 by adding ammonium hydroxide. The system is irradiated with 100 W/cm^{2}of ultrasonic radiation at 20 KHz for 60 minutes. A titanium horn is placed at a depth of 6 cm. Precipitate develops after the ultrasonication and then the obtained precipitate is washed twice with distilled water and finally with ethanol. The precipitate is centrifuged at 10,000 rpm for 5 minutes. Then the precipitate is annealed at 500°C for 5 hours in a muffle furnace to obtain titanium dioxide nanoparticles (see Figure 16.1).

16.4.4.2 Preparation of Titanium Dioxide (TiO_2) Nanoparticles Doped with Silver (Ag)/Iron (Fe)

Doping of TiO_2 with Ag is performed to increase the photocatalytic efficiency of the nanomaterials. Titanium isopropoxide and silver nitrate or ferric nitrate can be used as precursors. In this method, the precursor is a mixture of 2 mL of titanium isopropoxide and silver nitrate (1% molar mass) and 20 mL of distill water.

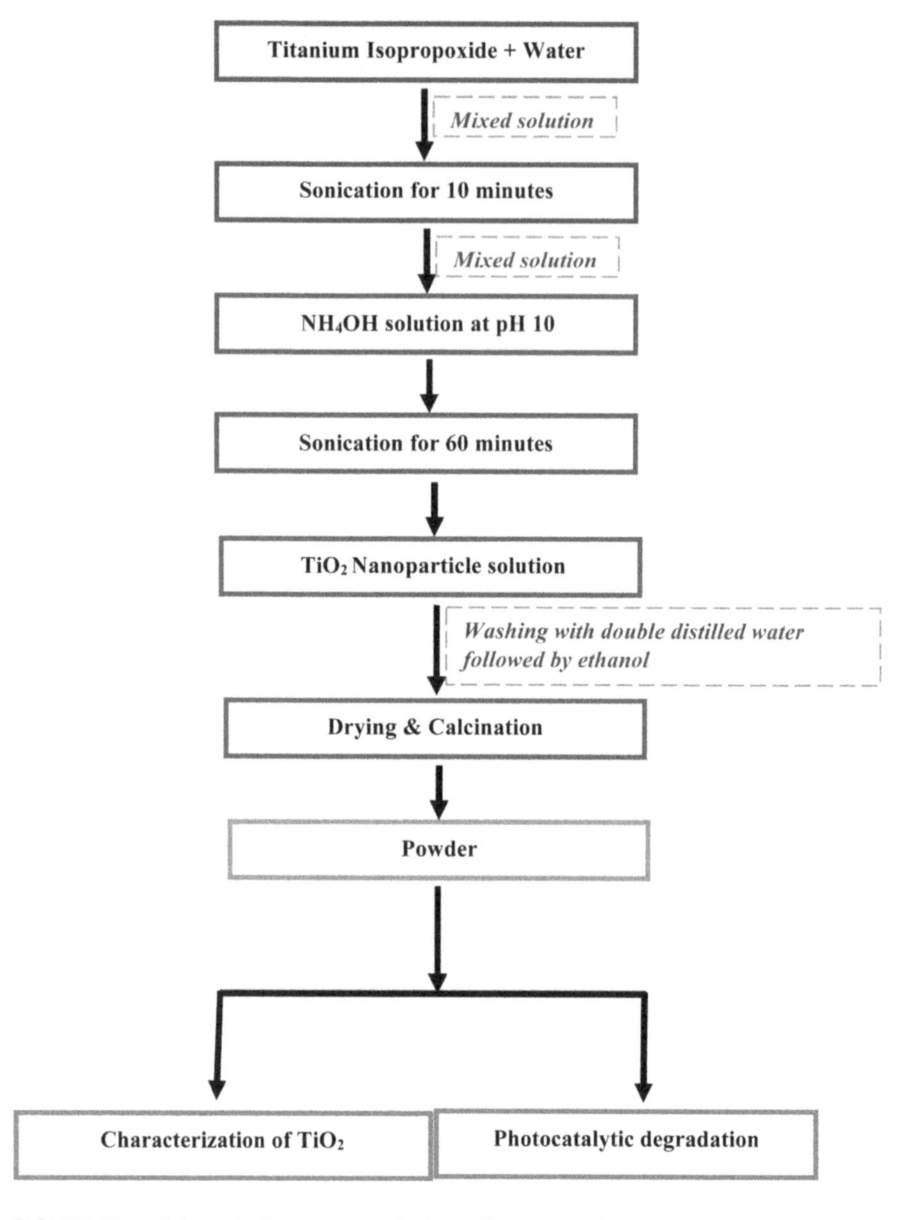

FIGURE 16.1 Schematic diagram of synthesis of TiO_2 nanoparticles by sonochemical method.

16.4.4.2.1 Doping of TiO_2 Nanoparticles

Doping of titanium dioxide has been a significant perspective in band-gap engineering to manipulate the optical property of the photocatalyst. Doping is done to incite bathochromic shift, i.e. lowering of band gap or insertion of intra-band gap that causes more absorption of visible light of the electromagnetic spectrum (see Table 16.1). Doping generates enhanced efficiency of photocatalysts (Carp, 2004).

TABLE 16.1
Doping Percent of Photocatalysts

Doping%	Compound Name		
	TitaniumIsopropoxide	Silver Nitrate	Ferric Nitrate
Molecular weight	284.22	169.87	404
TiO_2: Ag (1%)	0.99 × 284.22 = 281.378 (1.9582 g)	0.01 × 169.87 = 1.6987g	———
TiO_2: Fe (1%)	0.99 × 284.22 = 281.378 (1.9582 g)	———	0.01 × 404 = 4.04 (0.02811 g)
TiO_2: Ag (0.5%): Fe (0.5%)	0.99 × 284.22 = 281.378 (1.9582 g)	0.005 × 169.87 = 0.8493 (0.0059 g)	0.005 × 404 = 2.02 (0.0140 g)

16.4.4.2.2 Single (Ag^+ or Fe^{3+}) Doped TiO_2

The aqueous solution of single doped titanium dioxide was prepared by adding titanium isopropoxide and distilled water in a beaker. The known amount of silver nitrate or ferric nitrate is mixed with TiO_2 suspension for doping. Silver or iron concentrations can be maintained in the ratio of 0.5, 1.0, 2.0 and 5.0% (mole ratio) versus TiO_2. Single doped titanium dioxide can be synthesized by the similar method as described earlier. The product is air dried, calcined and characterized for structure and morphology.

16.4.4.2.3 Double (Ag^+ or Fe^{3+}) Doped TiO_2

The required amount of silver nitrate and ferric nitrate can be mixed with atitanium dioxide solution. The concentration of silver and iron is maintained at 0.5% and 1% molal ratio versus titanium dioxide suspension. To develop double doped titania, the earlier process for the synthesis can be repeated (see Figure 16.2). The prepared product is characterized for structure and morphology.

16.5 BIOLOGICAL METHODS

Novel green techniques employing natural reducing and capping agents to develop nanomaterials of desired structure and characteristic have become a cornerstone for the researchers. Biological synthesis of nanomaterials are simple, safe, economical and nontoxic (Ahmad et al., 2003).Biomolecular extracts viz. enzymes or proteins, amino acids, polysaccharides and vitamins from organisms, when combined with metal ion, forms environmentally friendly yet chemically complex products. Several studies have highlighted the effective synthesis of nanoparticles by microorganisms and plants (Korbekandi et al., 2009; Iravani, 2011). The biological synthesis of nanomaterials can be achieved through different modes of biosynthesis viz. bacteria, actinomycetes, yeasts, fungi, algae and plants. The development of nanoparticles can be assessed by the plant extract composition, metal salt concentration, pH, temperature and contact time, which plays an important role in modifying the characteristic of developed nanomaterials. Biological development of nanoparticles is an innovative approach for the synthesis of nanoparticles (Korbekandi and Iravani, 2012).

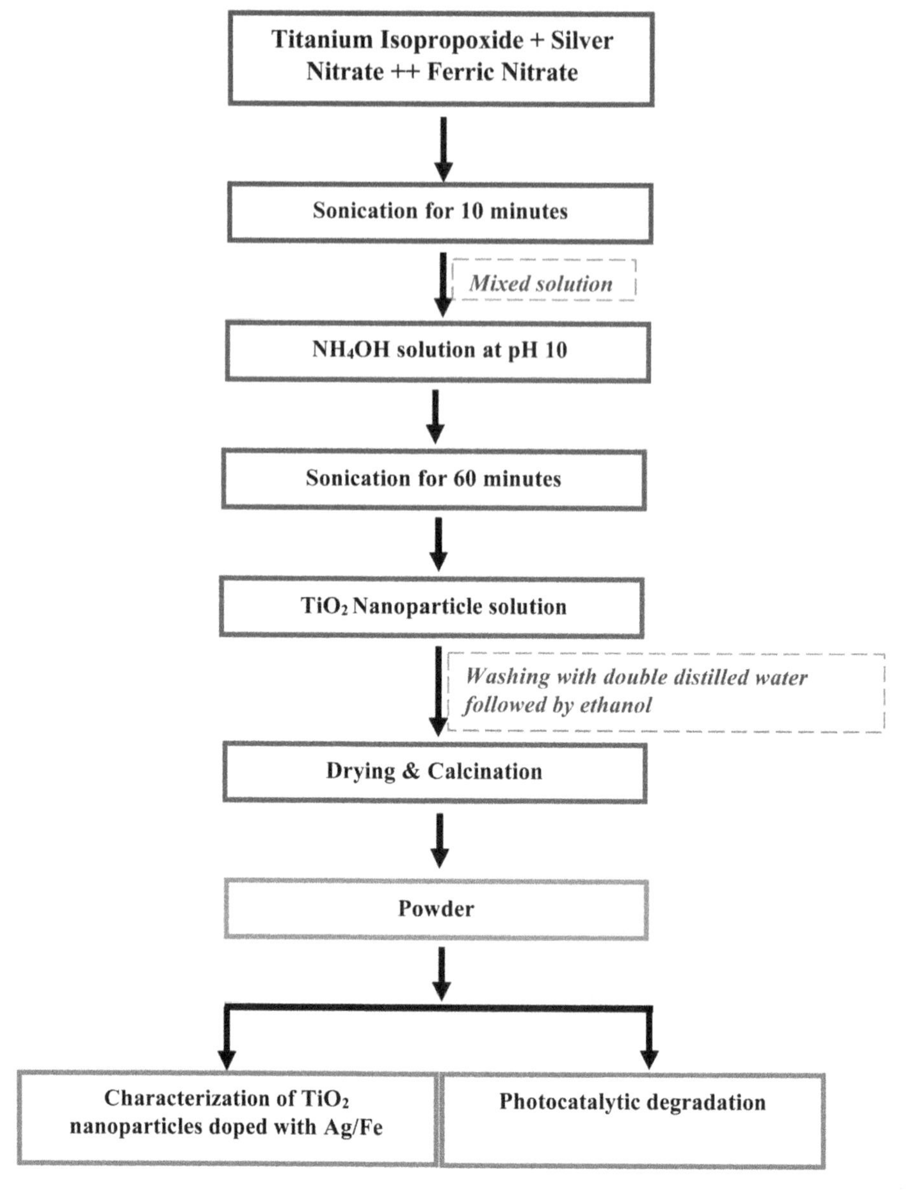

FIGURE 16.2 Schematic diagram of synthesis of Ag/Fe-doped TiO_2 nanoparticles by sonochemical method.

16.5.1 Synthesis of Silver Nanoparticle

Synthesis of silver nanomaterials can be achieved via physicochemical methods which include chemical reduction, electrochemical techniques and photochemical reduction (Babu et al., 2010). The most common physical methods are evaporation-condensation and laser ablation. In this silver nanomaterial synthesis process, stabilizers are incorporated to ensure their stability in aqueous solutions. However, these approaches are costly and involve the utilization of hazardous chemicals (Rai and Duran, 2011). Biological methods ensure the synthesis of silver nanomaterials in safer, economical and nontoxic manner. Plant extracts for the synthesis of silver nanomaterials are beneficial as they are relatively safe and nontoxic. Biological methods using plant-based leaf extracts for silver nanomaterial synthesis can be an innovative approach for the green development of nano-sized particles. Silver nanoparticles are synthesized using leaf extracts of different plants with different concentrations, (i) *Syzygium cumini* (jamun), (ii) *Bryophyllum pinnatum*, (iii) *Sida acuta*, (iv) *Azadirachta indica* (neem) and the precursor, i.e. silver nitrate, and its synthesis involves various steps discussed in the following subsections.

16.5.1.1 Development of Leaf Extract from Plants

The plant leaves are washed thrice with distilled water and kept at room temperature for air drying (Hernández et al., 2015). After drying, the known amount of leaf samples is chopped into fine pieces and 25 g of chopped leaves is added to 100 mL of boiled distilled water and heated for 5 minutes. The sample is allowed to cool for 15 minutes. After that the sample is filtered through Whatman filter paper to get the leaf extract. The fresh leaf extract is used for the synthesis of silver nanoparticles, and the rest of the leaf extract is stored.

16.5.1.2 Development of Silver Nitrate Suspension

A solution of 2 mM $AgNO_3$is prepared by dissolving 0.0339 g of $AgNO_3$in 100 mL of double distilled water. The solution is mixed thoroughly and stored in brown-colored bottle to prevent the autooxidation of silver.

16.5.1.3 Silver Nanoparticle Synthesis

Plant-facilitated synthesis of silver nanomaterials involves the leaf extract and 2 mM silver nitrate solution in the 1:9 ratio, respectively (Mehlhorn, 2016) (see Figure 16.3). A solution of 90 mL silver nitrate is added to 10 mL leaf extract and incubated at room temperature for 24 hours. A color change is observed within first 5 minutes. This process indicates preliminary confirmation for the formation of plant-mediated silver nanoparticles. This shows that the reduction of the metal ions occurs rapidly, and more than 90% of the reduction of silver ions is completed within 24 hours, after addition of the metal ions to the plant extract. Silver nanoparticles settled at the bottom in the form of the gray-colored precipitate. The solution is centrifuged at 8000 rpm for 15 minutes, and the supernatant is discarded and pellets containing silver nanoparticles retrieved. The pellets are washed two times with double distilled water and two times with alcohol to remove cell debris using the same centrifuge techniques. The pellets are taken out of the petridish and kept in an oven to dry at 60°C for 2–3 hours. Thus, silver nanoparticle powder (graycolor) is obtained. Silver nanoparticles exhibit a yellowish brown color in aqueous solution due to the excitation of surface plasma vibrations in silver nanoparticles. Silver nanoparticles are characterized using a UV-visible spectrophotometer, Fourier-transform infrared spectrophotometer, dynamic light scattering, transmission electron microscope, etc.

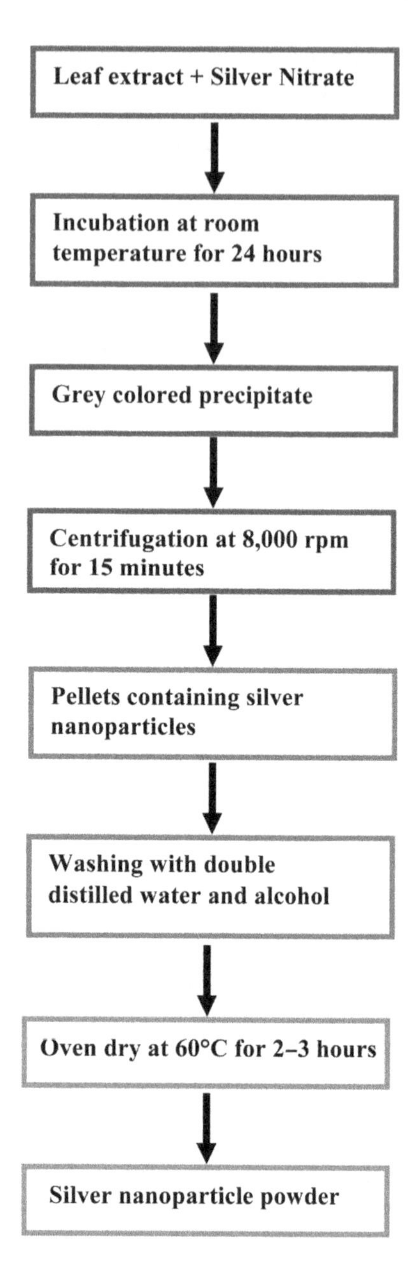

FIGURE 16.3 Schematic diagram of synthesis of silver nanoparticles using plant extract.

16.5.2 Synthesis of Silver Nanoparticle by Bacterial Supernatant

Bacterial supernatant contains various enzymes and metabolites with reducing properties or antioxidants that act on the respective metal salt precursor, which facilitates the reduction of metal salt to metal nanoparticles (Dauthal and Mukhopadhyay, 2016). As the bacterial cell debris is also at a nanoscale, they interfere during the purification step for the nanoparticles. So, the bacterial supernatant is taken for nanoparticles synthesis compared to whole bacterial cells.

16.5.2.1 Isolation of Pure Bacterial Strain

For the isolation of bacteria from any contaminated water/soil sample, a "serial dilution method" is used. A sample of 10 mL is taken in a 100mL beaker and noted as 10^{-1} dilution. From 10^{-1} dilution, 1 mL of water is transferred using a sterile micropipette and added to 9 mL distilled water and noted as 10^{-2} dilution. Likewise dilution is done upto 10^{-5} dilution. Then 0.1 mL of inoculum is taken from 10^{-3} to 10^{-5} dilution and the sample is spread on a petriplate (nutrient agar) to isolate the bacteria. The plates are incubated overnight in the incubator at 37°C. It is the best method for a single cell colony of bacteria. Then the pure colony of bacteria is confirmed by the 16s rDNA technique.

16.5.2.2 Preparation of Supernatant

For the preparation of supernatant, an isolated bacterial strain is cultured in a nutrient broth medium to develop a bacterial biomass for biosynthesis. The bacterial culture is incubated in an orbital shaker at 37°C at 120 rpm (Kannan et al., 2011). The broth is centrifuged at 6000 rpm for 10 minutes. The supernatant is collected for further reaction.

16.5.2.3 Synthesis of Silver Nanoparticles

A bacterial supernatant of 10 mL is mixed with 5 mL solution of 10 mM silver nitrate ($AgNO_3$). Another set of bacterial supernatant without silver nitrate is used as a control. Both the prepared supernatants are incubated at 30°C for 24 hours. The reaction is carried out under closed and dark conditions to prevent photochemical reactions during the experiment. The solution turns into brown from yellow after the incubation. The suspension is centrifuged at 6000 rpm for 10 minutes to obtain silver nanoparticles, and then the obtained product is washed twice with distilled water followed by ethanol and further characterized using advance instruments (see Figure 16.4).

16.6 CHARACTERIZATION OF NANOMATERIALS: ADVANCED INSTRUMENTS

The instruments are the vital part of the methodology of modern science. The realization of the accomplishment in many disciplines such as chemistry, biology, environment, nanotechnology and material science has come about through the application of inventions of modern scientific instruments and techniques, which involve ideas of fresh insights, novel methods and innovative approaches for scientific research. The establishment of an instrumentation facility is the cumulative result of synchronizing the technology involved in physics, chemistry biology, environment and nanoscience/nanotechnology.

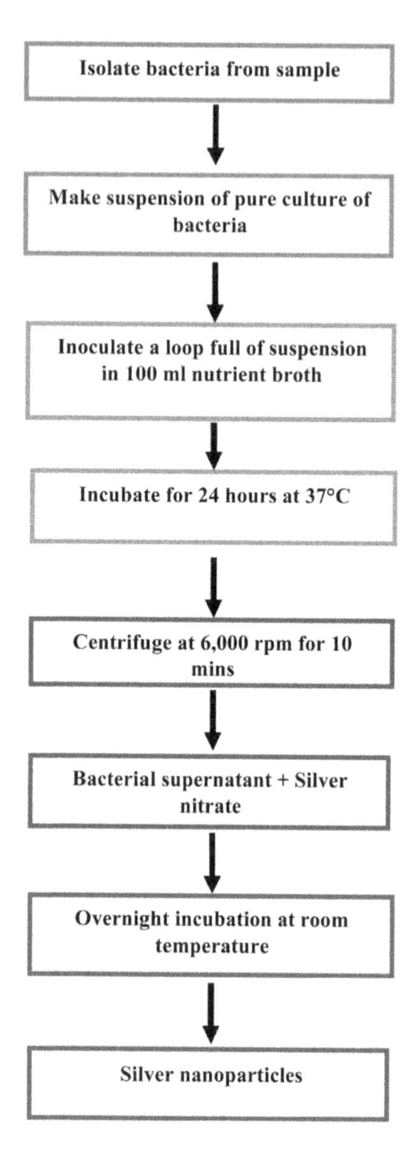

FIGURE 16.4 Schematic diagram of synthesis of silver nanoparticles using bacterial supernatant.

Several instrumental techniques can provide an efficient structure and functional data on nano-sized materials (see Table 16.2). The establishment of analytical techniques, especially for nanoengineering is underway, yet use of new technology results in many questions and unpredictable variables irrespective of the original data. The costs involved in these techniques are also high in terms of time, as well as being labor and capital extensive. Detailed behavior and characterization of manufactured products and their lifespan are essential for both large-scale production and safety.

TABLE 16.2
Instruments for the Characterization of Nanomaterials

S.No.	Instruments	Characterizations
1.	SEM (scanning electron microscope)	Morphology, topography
2.	HRTEM (high-resolution transmissionelectron microscope)	Size, shape, particle size distribution, lattice parameter, crystal symmetry
3.	P-XRD (powder x-ray diffraction)	Crystal structure, phase detection, unitcell dimension
4.	EDAX (energy-dispersive x-rayspectroscopy)	Elemental detection and chemical composition
5.	AFM (atomic force microscope)	Surface topography, lithography
6.	FTIR (Fourier-transform infraredspectroscopy)	Functional group detection
7.	DLS (dynamic light scattering)	Zeta potential, particle size and distribution
8.	BET (Brunauer-Emmett-Teller)	Surface area
9.	NMR (nuclear magnetic resonance)	Structure determination of organic compounds and position of the element
10.	LCMS (liquid chromatography-massspectrometer)	Structure determination of compounds/ intermediate
11.	ESCA (electron spectroscope for chemical analysis)	Surface analysis, quantitative analysis of elements, binding energy of elements, valance state of atoms
12.	ICP (inductively coupled plasma)	Detection of trace metals, material analysis
13.	AAS (atomic absorption spectroscopy)	Determination of metals
14.	GC (gas chromatography)	Separation, identification, qualitative and quantitative analysis of organic compounds
15.	HPLC (high-performance liquid chromatography)	Identification and quantification of a compound, purification of chemical compounds
16.	CHNOS (CHNOS elemental analyzer)	Determination of elemental concentration of carbon, nitrogen, oxygen, sulfur, etc.
17.	UV-VIS (UV-VIS spectrophotometer)	Determination of spectra-based absorption for qualitative and quantitative analysis of organic and inorganic compounds

16.7 NANOMATERIALS AND ENVIRONMENTAL APPLICATIONS

Research and development in nanoscience and technology directs an understanding to create nanomaterials, devices and systems with enhanced efficiency, which gives new dimensions in the field of nanotechnology from multidisciplinary fields viz. physical, chemical and biological (Mandal and Ganguly, 2011). Environmental nanotechnology plays a pivotal role in structuring recent environment science and engineering. A wide array of environmental nanotechnology products already exist for environmental protection. Nanotechnological application for pollution prevention refers to the minimization of raw materials and other resource utilization, as well as reduction of waste generation and use of energy-efficient products (Zhao et al., 2011). The production of nanomaterials and their use have been reported for the beneficial synthesis as a catalyst for energy generation, environmental protection, quality production, agriculture-crop enhancement and advances in wastewater treatment.

The advancements in chemical processes and technology resulted in the generation of abundant levels of contaminants that are above the self-cleansing capability of the environment (Sadraeian and Molaee, 2009). Detection and treatment of contaminants and prevention of new pollutants are among the major challenges of the twenty-first century. Contaminant degradation by the use of the present techniques is not effective and efficient for environmental cleanup. Nanotechnology in the environment can be a novel and innovative approach in the remediation of contaminants to a permissible extent (see Table 16.3). Environmental researchers are focusing on

TABLE 16.3
Nanomaterials and Their Properties

Nanocatalyst	• Speedup of chemical reactions • Enhancement of efficiency of many processes • Controlling emission in industrial processes • Beneficial for fossil fuel industries • Reducing pollution • Useful in material production and chemical industry
Nanocomposite	• Clay-based nanocomposites are beneficial in increasing strength or novel properties and for structural applications • Composite nanomaterials with nanospores useful for synthesis, polymeric optical display substrate material, nanostructure • Clay, arsenic and inorganic nanocomposites have wide environmental applications
Nanotube	• Carbon nanotubes have been used as adsorption materials • Raw materials and other products • Multiwalled nanotubes are used as composites and increase conductivity
Carbon nanotubes (CNTs)	• CNT has high surface area, greater adsorption sites
Nanoscale metal oxide	• High surface area • Greater adsorption sites
Nanofibers with core shell structure	• Reactive core for degradation • Short interval diffusion distance
Nanofilms	• Nanofilms are chemically active • Wear resistant • Highly resistant to erosion
Nanopowders	• Nanopowder is high in density • Nanopowder has potential magnetic, electronic and structural application (Fulekar et al., 2014)
Photocatalyst	
• NanoTiO_2	• Photocatalytic activity initiated in UV and possibly in visible light range • Less toxic, highly stable and economical
• Nano ZnO	• Photocatalytic activity initiated in UV and possibly in visible light range
• Fullerene derivatives	• Photocatalytic activity in solar spectrum • Highly selective (Mehlhorn, 2016)

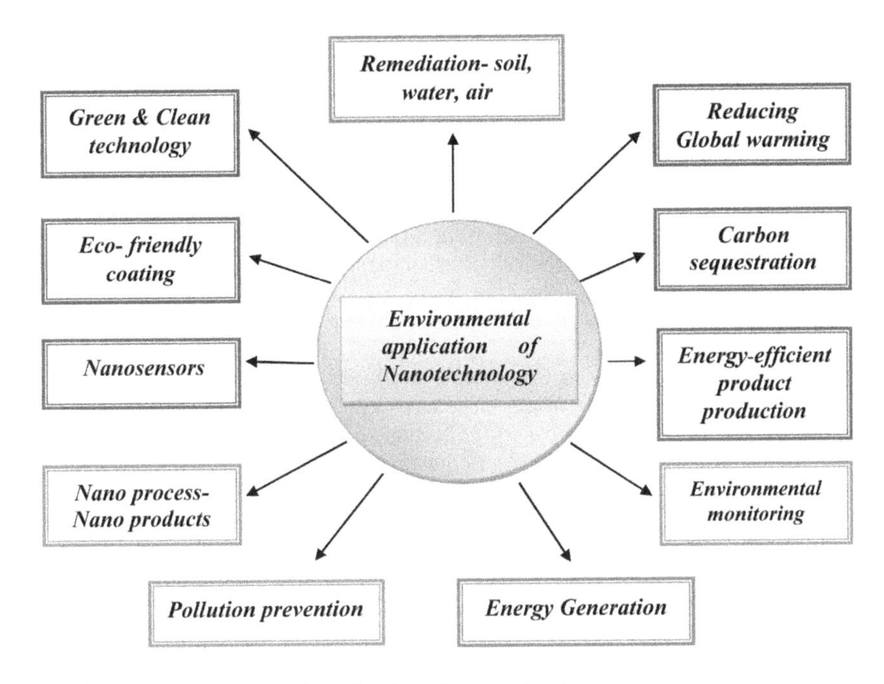

FIGURE 16.5 Environmental application of nanotechnology.

nano-sized materials and manipulating their structure and characteristics across the disciplines of science and technology. Nanoscale by-products of microorganisms such as bacteria and algae as well as weathering of soil can help in the determination of the fate, transfer, transport and availability of environmental contaminants. Nanotechnology in the environment has played an effective role in structuring current environmental science and technology. Down at the nanoscale, the manufacture of innovative, economical and efficient techniques for the detection, catalysis and abatement of contaminants has played a significant role. There is wide array of expectations laid on nanotechnological approaches and techniques for the pollution prevention and abatement to achieve environmental sustainability.

Nanomaterials have a great range of application in various fields and are capable of designing and controlling the structure of products at a nanoscale. It is helpful in making nano-sized materials with novel and enhanced properties. Science involves the modification of individual atoms and molecules, resulting in products with the desired structure and enhanced properties for environmental applications (see Figure 16.5).

Nanotechnology provides efficient nanostructures and devices for wider applications in energy, medicine and safety as well as the protection of the environment and remediation of contaminants. Nanotechnology also promises seamless integration of atoms, photons and biological cells, thus opening new avenues to industry. The literature available in nanotechnology is massive and gives a glimpse into a new world with more opportunity to understand its importance and applications to society. Advancements in research and technology development will generate unique economic, social and environmental benefits.

REFERENCES

Ahmad, A., Mukherjee, P., Senapati, S., Mandal, D., Khan, M.I., & Kumar, R. (2003). Extracellular biosynthesis of silver nanoparticles using the fungus *Fusarium oxysporum*. *Colloids and Surfaces B: Biointerfaces*, *28*, 313–318.

Ahmad, N., & Sharma, S. (2012). Green synthesis of silver nanoparticles using extracts of *Ananas comosus*. *Green and Sustainable Chemistry*, *02*(04), 141–147.

Amin, M. T., Alazba, A. A., & Manzoor, U. (2014). A review of removal of pollutants from water/wastewater using different types of nanomaterials. *Advances in Materials Science and Engineering*, *2014*, 1–24.

Asahi, R. (2001). Visible-light photocatalysis in nitrogen-doped titanium oxides. *Science*, *293*(5528), 269–271.

Babu, V. R., Kim, C., Kim, S., Ahn, C., & Lee, Y. (2010). Development of semi-interpenetrating carbohydrate polymeric hydrogels embedded silver nanoparticles and its facile studies on E. coli. *Carbohydrate Polymers*, *81*(2), 196–202.

Banfield, J.F., & Navrosky, A. (2002). Nanoparticles and the environment. *Reviews in Mineralogy and Geochemistry*, *44*, 6–16.

Bharde, A., Rautaray, D., Bansal, V., Ahmad, A., Sarkar, I., Yusuf, S., & Sastry, M. (2006). Extracellular biosynthesis of magnetite using fungi. *Small*, *2*(1), 135–141.

Carp, O. (2004). Photoinduced reactivity of titanium dioxide. *Progress in Solid State Chemistry*, *32*(1–2), 33–177.

Chakraborty, N., Banerjee, A., Lahiri, S., Panda, A., Ghosh, A. N., & Pal, R. (2008). Biorecovery of gold using cyanobacteria and an eukaryotic alga with special reference to nanogold formation—A novel phenomenon. *Journal of Applied Phycology*, *21*(1), 145–152.

Cheremisinoff, N. P. (ed.) (2002). An overview of water and waste-water treatment. In *Handbook of water and wastewater treatment technologies*. Oxford: Butterworth- Heinemann, 1–61.

Colvin, V. L. (2003). The potential environmental impact of engineered nanomaterials. *Nature Biotechnology*, *21*(10), 1166–1170.

Dauthal, P., & Mukhopadhyay, M. (2016). Noble metal nanoparticles: Plant-mediated synthesis, mechanistic aspects of synthesis, and applications. *Industrial & Engineering Chemistry Research*, *55*(36), 9557–9577.

Diallo, M. S., & Savage, N. (2005). Nanoparticles and water quality. *Journal of Nanoparticle Research*, *7*(4–5), 325–330.

Feynman, R. P. (2011). *Six easy pieces essentials of physics explained by its most brilliant teacher*. New York: Basic Books.

Fulekar, M. H., Pathak, B., & Kale, R. K. (2014). *Environment and sustainable development*. New Delhi: Springer.

Gericke, M., & Pinches, A. (2006). Microbial production of gold nanoparticles. *Gold Bulletin*, *39*(1), 22–28.

Goodsell, D. S. (2004). *Bionanotechnology: Lessons from nature*. Hoboken, NJ: Wiley-Liss.

Goswami, D. Y., Trivedi, D. M., & Block, S. S. (1997). Photocatalytic disinfection of indoor air. *Journal of Solar Energy Engineering*, *119*(1), 92.

Hernández, L. G., Isias, D. A., Guerrero, M. F., Ortega, P. A., & Lechuga, L. G. (2015). Use of extract of *Cupressus goveniana* for synthesis and stabilization of nanoparticles silver. TMS2015 Supplemental Proceedings, 1105–1112.

Hildebrand, H., Mackenzie, K., & Kopinke, F. (2009). Pd/Fe_3O_4 nano-catalysts for selective dehalogenation in wastewater treatment processes—influence of water constituents. *Applied Catalysis B: Environmental*, *91*(1–2), 389–396.

Hillie, T., & Hlophe, M. (2007). Nanotechnology and the challenge of clean water. *Nature Nanotechnology*, *2*(11), 663–664.

Ihara, T. (2003). Visible-light-active titanium oxide photocatalyst realized by an oxygen-deficient structure and by nitrogen doping. *Applied Catalysis B: Environmental*, *42*(4), 403–409.

Iravani, S. (2011). Green synthesis of metal nanoparticles using plants. *Green Chemistry, 13*(10), 2638.

Irie, H., Watanabe, Y., & Hashimoto, K. (2003). Nitrogen-concentration dependence on photocatalytic activity of TiO_2-xNxPowders. *The Journal of Physical Chemistry B, 107*(23), 5483–5486.

Jung, J. H., Oh, H. C., Noh, H. S., Ji, J. H., & Kim, S. S. (2006). Metal nanoparticle generation using a small ceramic heater with a local heating area. *Journal of Aerosol Science, 37*(12), 1662–1670.

Kabashin, A. V., & Meunier, M. (2003). Synthesis of colloidal nanoparticles during femtosecond laser ablation of gold in water. *Journal of Applied Physics, 94*(12), 7941.

Kamat, P. V., & Meisel, D. (2003). Nanoscience opportunities in environmental remediation. *Comptes Rendus Chimie, 6*(8–10), 999–1007.

Kannan, N., Mukunthan, K., & Balaji, S. (2011). A comparative study of morphology, reactivity and stability of synthesized silver nanoparticles using *Bacillus subtilis* and *Catharanthus roseus* (L.) G. Don. *Colloids and Surfaces B: Biointerfaces, 86*(2), 378–383.

Kharissova, O. V., Dias, H. R., Kharisov, B. I., Pérez, B. O., & Pérez, V. M. (2013). The greener synthesis of nanoparticles. *Trends in Biotechnology, 31*(4), 240–248.

Korbekandi, H., & Iravani, S. (2012). Silver nanoparticles. In *The delivery of nanoparticles*, Hashim, A. A. (ed.). London: IntechOpen.

Korbekandi, H., Iravani, S., & Abbasi, S. (2009). Production of nanoparticles using organisms. *Critical Reviews in Biotechnology, 29*(4), 279–306.

Kruis, F. E., Fissan, H., & Rellinghaus, B. (2000). Sintering and evaporation characteristics of gas-phase synthesis of size-selected PbS nanoparticles. *Materials Science and Engineering: B, 69–70*, 329–334.

Kvítek, L., Prucek, R., Panáček, A., Novotný, R., Hrbáč, J., & Zbořil, R. (2005). The influence of complexing agent concentration on particle size in the process of SERS active silver colloid synthesis. *Journal of Materials Chemistry, 15*(10), 1099–1105.

Laval, J., Thomas, D., & Mazeran, P. (2000). Nanobiotechnology and its role in the development of new analytical devices. *The Analyst, 125*(1), 29–33.

Mandal, G., & Ganguly, T. (2011). Applications of nanomaterials in the different fields of photosciences. *Indian Journal of Physics, 85*(8), 1229–1245.

Mehlhorn, Heinz. (2016). *Nanoparticles in the fight against parasites.* Heidelberg/Berlin: Springer.

Merga, G., Wilson, R., Lynn, G., Milosavljevic, B. H., & Meisel, D. (2007). Redox catalysis on "Naked" silver nanoparticles. *The Journal of Physical Chemistry C, 111*(33), 12220–12226.

Nolan, N. T., Seery, M. K., Hinder, S. J., Healy, L. F., & Pillai, S. C. (2010). A systematic study of the effect of silver on the chelation of formic acid to a titanium precursor and the resulting effect on the anatase to rutile transformation of TiO_2. *The Journal of Physical Chemistry C, 114*(30), 13026–13034.

Obare, S. O., & Meyer, G. J. (2004). Nanostructured materials for environmental remediation of organic contaminants in water. *Journal of Environmental Science and Health, Part A, 39*(10), 2549–2582.

Oliveira, M. M., Ugarte, D., Zanchet, D., & Zarbin, A. J. (2005). Influence of synthetic parameters on the size, structure, and stability of dodecanethiol-stabilized silver nanoparticles. *Journal of Colloid and Interface Science, 292*(2), 429–435.

Otto, M., Floyd, M., & Bajpai, S. (2008). Nanotechnology for site remediation. *Remediation, 19*(1), 99–108.

Pandey, J., Khare, R., Kamboj, M., Khare, S., & Singh, R. (2011). Potential of nanotechnology for the treatment of waste water. *Asian Journal of Biochemical and Pharmaceutical Research, 1*(2), 272–282.

Pandey, S., Kumari, M., Singh, S. P., Bhattacharya, A., Mishra, S., Chauhan, P. S., & Mishra, A. (2015). Bioremediation via nanoparticles. *Handbook of research on uncovering new methods for ecosystem management through bioremediation advances in environmental engineering and green technologies*, Hershey PA: IGI Global, 491–515.

Parashar, U.K., Saxena, P.S., & Shrivastava, A. (2009). Bioinspired synthesis of silver nanoparticles. *Digest Journal of Nanomaterials and Biostructures*, *4*(1), 159–166.

Rai, M., & Duran, N. (2011). *Metal nanoparticles in microbiology*, Berlin: Springer.

Rai, M., & Duran, N. (2014). *Metal nanoparticles in microbiology*, Berlin: Springer.

Rai, M., Yadav, A., & Gade, A. (2009). Silver nanoparticles: As a new generation of antimicrobials. *Biotechnology Advances*, *27*, 76–83.

Rickyerby, D., & Morrison, M. (2006). Nanotechnology and the environment: A European perspective. *Science and Technology of Advanced Materials*, *8*, 19–24.

Roco, M.C. (2001). International strategy for nanotechnology research and development. *Journal of Nanoparticle Research*, *3*, 353–360.

Sadraeian, M., & Molaee, Z. (2009). Bioinformatics analyses of *Deinococcus radiodurans* in order to waste clean up. *2009* Second International Conference on Environmental and Computer Science.

Saleh, T. A., & Gupta, V. K. (ed.) (2016). Synthesis, classification, and properties of nanomaterials. *Nanomaterial and polymer membranes*. Amsterdam, Netherlands: Elsevier, 83–133.

Senapati, S., Syed, A., Khan, S., Pasricha, R., Khan, M., Kumar, R., & Ahmad, A. (2014). Extracellular biosynthesis of metal sulfide nanoparticles using the fungus fusarium oxysporum. *Current Nanoscience*, *10*(4), 588–595.

Sharma, N. C., Sahi, S. V., Nath, S., Parsons, J. G., Torresde, J. L., & Pal, T. (2007). Synthesis of plant-mediated gold nanoparticles and catalytic role of biomatrix-embedded nanomaterials. *Environmental Science & Technology*, *41*(14), 5137–5142.

Solairajan, A. S., Alexraj, S., Kumar, P. G., & Rajan, P. V. (2014). Review on nano fabrication and application. *Advanced Materials Research*, *984–985*, 508–513.

Tan, G. (2017). Green synthesis of silver nanoparticles using *Allium cepa* and *Allium sativum* extract: A comparative characterization study. *Journal of Biotechnology*, *256S*, S17–S43.

Tarasenko, N., Butsen, A., Nevar, E., & Savastenko, N. (2006). Synthesis of nanosized particles during laser ablation of gold in water. *Applied Surface Science*, *252*(13), 4439–4444.

The Royal Society & The Royal Academy of Engineering. (2004). *Nanoscience and nano technologies: Opportunities and uncertainties*. Plymouth, UK: Latimer Trend Ltd.

Tiwari, D., Behari, J., & Sen, P. (2008). Application of nanoparticles in waste water treatment. *World Applied Sciences Journal*, *33*, 567–571.

Tsuji, T., Iryo, K., Watanabe, N., & Tsuji, M. (2002). Preparation of silver nanoparticles by laser ablation in solution: Influence of laser wavelength on particle size. *Applied Surface Science*, *202*(1–2), 80–85.

Yuan, G., & Keane, M. A. (2003). Catalyst deactivation during the liquid phase hydrodechlorination of 2,4-dichlorophenol over supported Pd: Influence of the support. *Catalysis Today*, *88*(1–2), 27–36.

Zhang, W-X., & Elliott, D.W. (2006). Applications of iron nanoparticles for groundwater remediation. *Remediation*, *16*(2), 7–21.

Zhang, X. (2011).Application of microorganisms in biosynthesis of nanomaterials—A review. *Wei Sheng Wu Xue Bao.*, *51*(3), 297–304.

Zhao, X., Lv, L., Pan, B., Zhang, W., Zhang, S., & Zhang, Q. (2011). Polymer-supported nanocomposites for environmental application: A review. *Chemical Engineering Journal*, *170*(2–3), 381–394.

17 Environmental Nanotechnology Approaches for the Remediation of Contaminants

*Manviri Rani and Uma Shanker**

CONTENTS

Nanomaterials (NMs) are playing a progressively essential role in providing pioneering and effective solutions to a wide range of environmental issues due to their enhanced properties over traditional materials. Environmental remediation mostly utilizes different technologies such as sorption, advanced oxidation processes

* Corresponding author: Email: shankeru@nitj.ac.in

(AOPs), photocatalysis, filtration, etc., for the elimination of pollutants from the environment (soil, water and air). In recent years, metal-based (nano zerovalent iron [nZVI], Ag, Au, etc.) transition metal oxides and their nanocomposites, carbon-based (carbon nanotubes [CNTs], grapheme oxides, fullerenes, etc.) bionanocomposites, etc., have been extensively reported for eliminating a number of pollutants such as short-chain chlorinated compounds, polycyclic aromatic hydrocarbons (PAHs), organic dyes, pesticides and heavy metals, phenols, etc. The synthesis and application of NMs are understandably prefigured as an environmentally valuable technology; however, the ecological risks associated with their application have only begun to be assessed. This chapter will highlight the successful implications of various transition metal-based NMs for the removal of pesticides, PAHs and organic dyes. The chapter will also throw light on the classification of different nanomaterials used for the removal of environmental contaminants. This chapter will finally conclude with brief description status about the green synthesis of such NMs.

17.1 INTRODUCTION

Air, soil and water pollution by volatile organic compounds, phenols, pesticides, polycyclic aromatic hydrocarbons (PAHs), particulate matters, heavy metals, industrial effluents, sewage and other organic compounds are undoubtedly grave problems that society is facing everyday (Khan and Ghoshal, 2000; Vaseashta et al., 2007). Due to rapid industrialization and urbanization as well as advances in agricultural practices and economic development, a water crisis is intensifying along with deterioration in the quality of water (Alcamo et al., 2007; World Health Organization [WHO], 2015; Shanker et al., 2017a). Hence, management of the environment is necessary to get sustainable and better life. Heavy metal ions (density >6 g/cc) and precarious water pollutants, even at ultralow concentrations, form potentially carcinogenic and mutagenic compounds within the bio-system (Santhosh et al., 2016). Due to high population density and growth of industrial practices, heavy metal ions (Cr^{6+}, Cu^{2+}, Cd^{2+}, Pb^{2+} and Zn^{2+}) were reported having highly recalcitrant contaminants in wastewater and become lethal above allowed concentrations (Rani and Shanker 2018i). In the recent times, pollution by organic substances has become a global threat and its level is increasing continuously due to urbanization, fast development and the varying way of life of people (Shanker et al., 2017b). The high stability toward heat, light and oxidizing agents led to their persistence and accumulation in the environment (Gupta et al., 2011). Accidents related to these pollutants have caused severe damage to the environment (Table 17.1). Among the constituent of wastewater (Table 17.2), synthetic dyes, pesticides, amine, phenol and substituted phenols and PAHs are abundantly present (Shanker et al., 2017c). It has been reported by several researchers that many such pollutants transformed into more toxic by-products (Gupta et al., 2011, 2012a, 2012b, 2012c; Rani, 2012; Shanker et al., 2017c; Rani et al., 2017b; Rani et al., 2018). Such pollutants come to humans via air, food, water, soil and dust.

Innovative technologies are continuously being explored for the removal of pollutants from wastewater (Masciangoli and Zhang, 2003). In this direction, numerous types of materials with advanced properties and novel methodologies have been

TABLE 17.1
Some Important Organic Pollutants Related Accidents Worldwide (Rani, 2012)

Pesticide	Place	Year	Causes	Reference
Parathion	India	1958	Contaminated food due to leakage	Rani (2012)
	India	1962	Inhalation in manufacturing plant	
HCH	India	1963	Contaminated rice	
Endrin	India	1964	Contaminated food	
HCH	India	1963	Contaminated rice	
Endrin	India	1964	Contaminated food	
DDT	India	1965	Contaminated chutney	
Diazinon	India	1968	Contaminated food	
HCH	India	1976	Mixed with wheat	
Endrin	India	1977	Contaminated crabs in rice field	
Aluminum phosphide	India	1983	Contaminated food grain	
Methyl isocyanate	India	1984	Storage tank leakage	
Cartap hydrochloride	India	1988	Factory workers	
Endosulfan	India	1997	Contamination due to aerial spray	
Phorate	India	2001	Spray drift from banana field	
Endosulfan	India	2002	Contaminated wheat flour	
2,3,7,8-Tetrachlorobenzo-10-dioxin (TCDD)	Italy	1976	Air pollution due to poisonous gas	De (2014)
Sarin	Japan	1985-95	Mass poisoning	Morita et al. (1995)
Pesticides	United States	1968-78	Contaminated food	Rani and Shanker (2018i)
Phenol	United States	1974	Accidental spillage	Gupta et al. (2008)
Phenol	India	1999	Accidental overdose of phenol	
Phenol	New Zealand	1980	Absorption of phenol through skin	
2,4-Dinitrophenol	China	2009	Non-oral exposure to workers in a chemical factory	
2,4-Dinitrophenol	United States	1933-38	Poisoning due to weight loss pill	
Phenol	United States	1974	Accidental spillage of 37,900	
PAHs (Lakeview Gusher)	United States	1910	1200 tons of crude oil released	Rani and Shanker (2018i)

(*Continued*)

TABLE 17.1 (*Continued*)
Some Important Organic Pollutants Related Accidents Worldwide (Rani, 2012)

Pesticide	Place	Year	Causes	Reference
PAHs (Kuwaiti oil lakes)	Kuwait	1991	Kuwaiti oil lakes accidental spillage	
PAHs (*Kuwaiti oil fires*)	Kuwait	1991	136,000 tons of crude oil released	
PAHs	S. Korea	2007	MT *Hebei Spirit* oil spill	
PAHs	United States	2010	*Deepwater Horizon* oil spill	
PAHs (Sundarbans oil spill)	Bangladesh	2014	Accidental spillage	
PAHs (Ennore oil spill)	Chennai	2017	Accidental spillage	
PAHs (Lakeview Gusher)	USA	1910	1200 tons of crude oil released	

explored for the removal of toxins from the environment. Complete removal of environmental pollutants is a challenging task, due to the complexity, low boiling point and mild reactivity and persistent nature. Recently use of NMs for the development of new environmental remediation technologies has been trending due to the enhanced properties of such materials over their bulk counterparts due to a high surface to volume ratio (Tratnyek and Johnson, 2006; Rachna et al., 2018, 2019a, 2019b, 2019c; Rani et al., 2019).

Moreover, it has also been established that surface functionalization of NMs is more facile than bulk material, and this will help in target-specific efficient remediation. Further, such deliberate modification of the NMs can offer extra beneficial

TABLE 17.2
Constituents of Wastewater (Henze, 1992)

Type	Components	Effects
Microorganisms	Pathogenic bacteria, virus, etc.	Risk while bathing and eating fish
Organic materials	Oxygen depletion in rivers, lakes and fjords	Eutrophication, aquatic death may contain disease causing microorganisms
Synthetic organic materials	Pesticides, fat, oil and grease, dyes, phenols, amines, PAHs, pharmaceuticals, etc.	Toxic effect, aesthetic inconveniences, bioaccumulation
Nutrients	N, P, ammonium, Ca, Na, Mg, K, etc.	Eutrophication, oxygen depletion, toxic effects
Inorganic materials	Acids, bases, heavy metals (Hg, Pb, Cd, Cr, Cu, Ni)	Corrosion, toxic effect, hardness, aquatic death, bioaccumulation
Radioactivity	Various radioactive elements	Toxic effect, accumulation

features that directly enhance the potential of the material for pollutant removal compared to bulk materials. Several types of NMs, metal-based, metal oxides, nanocomposites and bionanocomposites have been developed in order to get the desired properties, selectivity and stability (Campbell et al., 2015; Ojea-Jiménez et al., 2012; Shah and Imae, 2016; Guerra et al., 2017a, 2017b).

It is also important to mention here that the materials used for the removal of pollutants should not be toxic and not act as another pollutant once employed for remediation. Hence, application of green materials is exceptionally remarkable for the removal of recalcitrant from wastewater. This will not only surge user confidence and acceptance of such technology but will also generate zero waste after their use, as this will offer a safer and green alternative for the complete removal of pollutants. Moreover, these methodologies based on target-specific removal of pollutants are particularly attractive, as they can overawe low competences derived from off-targeting molecules. Thus, diverse studies have been dedicated to the application of the principles of nanotechnology and combining it with chemical and physical alteration of the surface of the materials in an effort to find engineered materials that can deal with many of the challenges involved with the remediation of pollutants (Kamat and Meisel, 2003; Pandey and Fulekar, 2012).

Low cost, stability, target-specific capture, facile synthesis, green chemistry, safety, biodegradability and reusability are some of the crucial challenges that should be addressed while developing novel NMs for environmental remediation. Regardless of the effective benefits of the NMs discussed earlier, few are inherently unstable under normal circumstances; hence, their generation needs superior methods for preparation at the nanoscale. Supplementary efforts are required to avoid agglomeration, enhance monodispersity and surge stability of synthesized nanoparticles. Another important challenge in the removal process is the possible toxicity of metallic nanoparticles along with by-products formation and retrieval costs from the remediation site. In view of this, superior NMs or its combination with other compound with no toxicity either in material or in its metabolites with effective capability for the complete removal of contaminants is the need of present time.

17.2 ENVIRONMENTAL CONCERN OF POLLUTANTS

Growing populations, rapid industrialization, modernization of developments and agronomic and domestic wastes are continuously deteriorating the quality of water and soil around and are hence major concerns around the globe. Synthetic dyes, pesticides, aromatic amines and phenols and emerging PAHs are extensively entering water bodies from untreated discharge. These recalcitrant contaminants have toxicity in the form of secondary waste, are persistent and sometimes metabolites that in some cases are more toxic than parental compound (Rani, 2012). Azo dyes and benzidine are highly cancer-causing and even explosive (Shanker et al., 2017b). The untreated discharge to water bodies must be prohibited as it might contain carcinogens (Jassal et al., 2015a). The European Commission has disqualified several noxious azo dyes used in the leather industry in view of such concerns. Presently, China is producing 40–45% of the world's total dye consumption

(Franssen et al., 2010). It has been reported that 1000–3000 m3 of water is let out after processing around 12–20 tonnes of textiles per day (Pagga and Brown, 1986; Kdasi et al., 2004). Another huge problem is the use of pesticides (dispersed off: 85–90%); its estimated use was 5.5×10^8 kg in the United States and 2.59×10^9 kg globally during 1995 (Rani and Shanker, 2018i). Regardless of stern conventions, priority hazardous substances (EC, 2008) are still found in rivers and seafood, indicating their long persistence or current use (Kannan et al., 1997; Navarro et al., 2010). There have been several cases of pesticide poisoning reported in developing countries like India, and by virtue of that many farmers lose their lives every year, e.g., in 1997–2002, several farmers died due to endosulfan poisoning (Rao et al., 2005).

Untreated PAHs are discharged to the water streams at the rate of 80,000 tons/year and it is documented ubiquitously that it is present in the environment as carcinogens and mutagens (IARC, 1998; Wright, 2002). It has been reported that 46–90% of individual PAHs are emitted by motor vehicles in the cities (Dubowsky et al., 1999). Indoor emission contributes to ~16% in the United States, 29% in Sweden and 33% in Poland (Li et al., 2008a). Individuals spend 80–93% of their time indoors and inhale PAHs (Brunekreef et al., 2005). Oil spills in coastal regions are the main reason for the increase in PAH pollution and cause the loss of various marine organisms (Rani and Shanker, 2019). A total of 17 unsubstituted PAHs have been identified as priority carcinogens by United States Environmental Protection Agency (U.S. EPA) (Fu et al., 2018).

Aromatic phenols and amines are also major organic constituents ordinarily found in wastewater (range: 1–100 mg/L) (Said et al., 2013; Pradeep et al., 2014). The U.S. EPA has determined that exposure to phenol in drinking water at a concentration of 4 mg/L for up to 10 days is not expected to cause any adverse effects in a child (Jadhav and Vanjara, 2004). Phenols are dangerous for the life of aquatic bodies at 9–25 mg/L (Veeresh et al., 2004; El-Ashtoukhy et al., 2013). Consequently, phenols are listed as specific priority pollutants released by the EPA (Babich and Davis, 1981; Singh et al., 2013). Bisphenol A is another common pollutant found in wastewater due to its extensive use and bulk production as plastic antioxidant. Bisphenol A (BPA) can affect marine creatures and distress physiological functions even at picogram concentrations (Suzuki et al., 2004; Kang et al., 2006; Negri-Cesi, 2015). The annual growth rate was found to be 4.6% from 2013 to 2019 owing to worldwide demand of around 6.5 million tons (Im and Loffler, 2016). A literature survey reveals that heavy metals have atomic weights between 63.5 and 200.6 and density greater than 5 g/cm^3 (Srivastava and Majumder, 2008; Barakat, 2011; Fu and Wang, 2011). The fast industrialization during recent years has significantly contributed to heavy metals released into the environment. The pollution of water due to the discharge of heavy metals from various industrial sources (metal plating facilities, battery manufacturing, fertilizer, mining, paper and pesticides, metallurgical, fossil fuel, tannery) into the environment has been instigating disquiet around the globe (Marques et al., 2000). Heavy metals, owing to their nonbiodegradable nature, tend to accumulate in living organisms unlike organic contaminants, and this leads to treatment of industrial wastewaters including Pb, Cr, Cd, Hg, As, Ni, Cu and Zn. There are various symptoms of the metals toxicity like high blood pressure, vascular occlusion speech disorders, sleep disabilities, fatigue, aggressive behavior, poor concentration, irritability, mood swings, depression, increased allergic reactions, autoimmune diseases and memory loss (Qu et al., 2013). Heavy metals such as Se, Zn and Mg might dislocate the humanoid cellular enzymes, while cadmium, lead, arsenic and mercury are known to be harmful to the human body. Although the human body needs a few heavy metals (manganese, iron, chromium, copper and zinc), large amounts of such metals might be lethal (Rao et al., 2006; Nordberg et al., 2007; Qu et al., 2013).

Keeping these facts in view, it may be concluded that there has been an increasing concern about the removal of noxious pollutants (wide use, toxicity and ubiquitous presence) and toxic metabolites prevalent in the environment. Moreover, these pollutants may also be produced in the environment by degradation of several biocalcitrant substances. Among the remediation techniques (including variety of different materials), sorption (absorption and adsorption), chemical reactions, photocatalysis, advanced oxidation processes and filtration have been used (Figure 17.1)

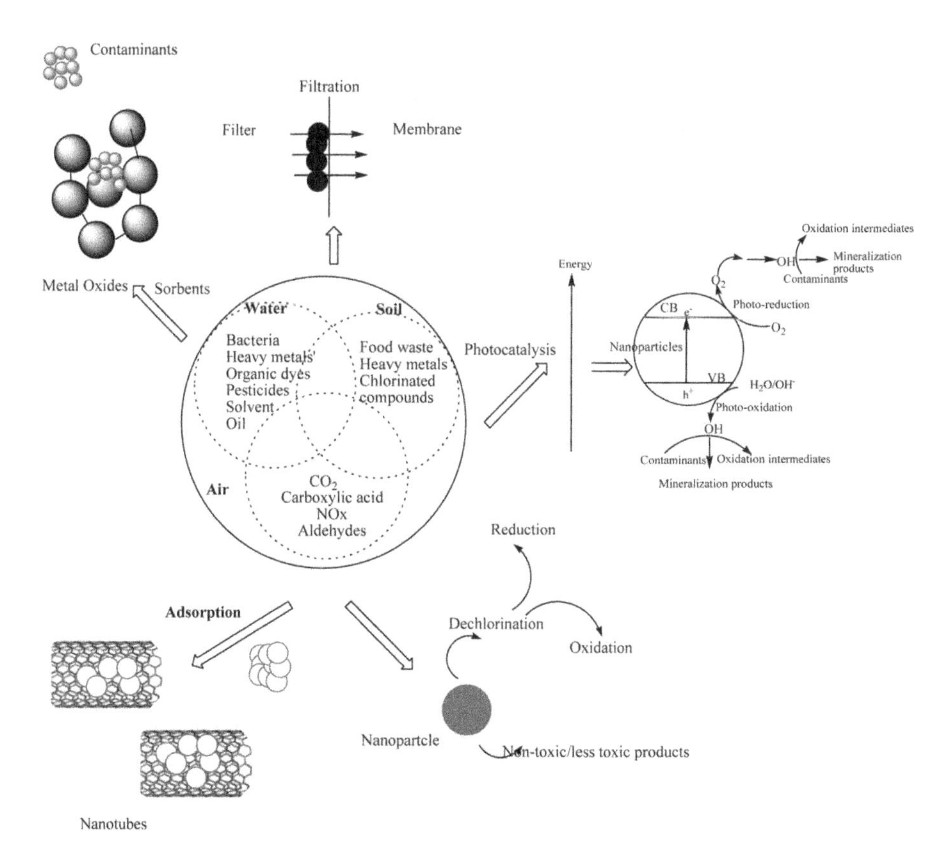

FIGURE 17.1 Various approaches for environmental remediation.

(Masciangoli and Zhang, 2003; Mauter and Elimelech, 2008; Sharma et al., 2009a; Gui et al., 2010; Zhao et al., 2011a; Tong et al., 2012; Ng et al., 2013). A number of reviews indicating the use of TiO_2, ZnO and iron-based NMs are available. A number of reviews have been published on nanoparticles (NPs), surface modification and functionalized NM (silica-based, carbonaceous and metal-based) for water treatments and environmental industries application (Shanker et al., 2017a; Darwish and Mohammadi, 2018; Rachna et al., 2019a). It was concluded that functionalization can help to overcome the limitations of bared NPs and provide favorable future for environmental remediation such as removal of pollutants like PAHs, dyes, heavy metals, etc., from wastewater. Chemical approaches are better for their diversity and ability to decorate NMs with a large number of functionalities (Darwish and Mohammadi, 2018). Engineered NMs are categorized either based on chemical backbone (inorganic, carbon-based and polymer) or on modification ways (doped, functional, mixed). Carbon-based functionalized materials are highly used as very promising adsorbents for treating organic and inorganic wastewater pollutants. Jun et al. (2018) reviewed functionalized CNT and graphene for organic pollutant removal (organic dye and phenol). Nanoadsorbents-based polymer nanocomposites are more beneficial for environmental sustainability in view of their low energy consumption, large adsorption capacity, simple operation, high stability, selective sorption, recycling and biodegradation (Singh and Ambika, 2018). Vishwakarma (2018) discussed the impact of engineered NMs like metal-based nanoparticles (Ag, Fe and Zn), metal oxide (TiO_2, ZnO and Fe_2O_3), CNTs and nanocomposites for industrial and public health. This chapter aims at reviewing the different synthesis methods for functionalization of various nanoparticles for their application in environmental remediation. This chapter also presents the future challenges, perspectives and directions in the area of NMs functionalization and their utilization.

17.3 CONVENTIONAL TECHNIQUES EMPLOYED

Various techniques like chemical precipitation, ion exchange, membrane filtration, electrochemical methods, solvent extraction, coagulation-flocculation, membrane process, flotation and adsorption have been employed for the remediation of pollutants. Among them, adsorption was reported best in terms of its effectiveness and low-cost as metal-bounded adsorbents are compact and strongly bonded. A brief discussion about the few effective techniques follows.

17.3.1 Oxidation

Chemical oxidation was reported to be a highly efficient methodology for the remediation of a variety of persistent organic contaminants. The oxidation methods were found less effective for treatment of inorganic pollutants such as heavy metals and dissolved minerals and salts, while a highly efficient methodology for the remediation of new and emerging pollutants from wastewater. Chlorine, ozone and hydrogen peroxide were commonly used as oxidizing agents for wet oxidation in air, while in electrochemical oxidation an electrical current is employed to achieve a redox reaction. Several threats have been encountered with the use of chlorine, like its extreme toxicity even at low concentrations, corrosive nature and hazardous by-products that require supplementary treatment. Moreover, use of ozone requires a lot of electricity, with similar drawbacks like chlorine. For certain recalcitrants at high concentrations in wastewater, electrochemical oxidation was an inefficient and unsuccessful removal technology.

17.3.2 Photocatalysis

This methodology is highly efficient in various forms like UV photolysis and TiO_2-catalyzed UV photolysis for degradation of various organic pollutants from wastewater Rachna et al. (2019a, 2019b). It is important to mention that low concentrations of volatile organic compounds (VOCs) are hard to degrade by photocatalysis alone, and hence this technology generally is combined with other removal methods. Photocatalysis is frequently restricted by the clarity of the water being treated, as the UV irradiation must be able to penetrate. Due to the high energy intake, UV photolysis can have a high life-cycle impact. UV lamps also entail repeated cleaning, and their replacement increases labor costs. The presence of hard water as well as co-contaminates can make photocatalysis less effective.

17.3.3 Fenton/Photo-Fenton

The Fenton process uses hydrogen peroxide and an iron catalyst to oxidize pollutants from contaminated wastewater. The involvement of UV irradiation to this process will enhance the rate of oxidant production and hence the efficiency of the treatment. Fenton and photo-Fenton reactions have been found to be highly effective in the removal of a variety of halogenated organics, pesticides, herbicides, most nonhalogenated organics, dyes and pharmaceuticals and personal care products (PCPs). Fenton and photo-Fenton processes suffer many limitations similar to hydrogen peroxide oxidation and UV photolysis. During the Fenton reagent process, the handling of extremely reactive materials is required.

17.3.4 Adsorption

Numerous adsorption methods are found effective for the removal of most contaminants, i.e. organic and inorganic. Most such adsorption methods utilize carbon-based materials to capture contaminant molecules inside its mesoporous structure. Raw materials for activated carbon are low-cost, while high energy is required for producing quality activated carbon for substantial environmental impacts.

In activated carbon treatment, the material was found reusable after the removal of contaminants, while other adsorption media are often expensive. Adsorption only removes the contaminant and does not degrade it which ultimately produces a harmful waste stream, which further requires treatment and finally enhances the cost.

17.3.5 Filtration

Several filtration techniques like sand filtration and membrane filtration (microfiltration [MF], ultrafiltration [UF] and reverse osmosis) have been reported to be efficient for the removal of a variety of pollutants from water. The filtration process involved size-dependent trapping of pollutant molecules between pore spaces within the filter. The filtration method alone is generally unable to eradicate new emerging toxicants, heavy metal ions and other dissolved ions from water, while membrane, UF and reverse osmosis (RO) are all comparatively more efficient for water purification. Energy inputs are required for equipment to flow the water across the membrane, and it frequently needs the specific conditions to be maintained in order to prevent fouling (a common occurrence in filtration). Generally, backwashing is required to avoid blockage, and punitive chemicals are being used to clean the membrane for reuse and to improve performance. RO also necessitates remineralization of the water and pH adjustments that ultimately increase the costs.

17.3.6 Biological Treatment

Biological treatment was a common process for the removal of several organics, but failed to remove emerging contaminants like halogenated organics or certain nonhalogenated organics, though anaerobic processes can be technically effective if slow. Biological treatment alone was reported ineffective for the removal of organic contaminants from water, but once combined with filtration showed excellent performance, mainly in the removal of heavy. The effectiveness of biological treatment is dependent on activity of microorganisms, water composition, loading rate, the type of media, the temperature and extent of aeration during the treatment of wastewater. Biological treatment is vulnerable to seasonal changes, shock, the existence of spare nutrients and other contaminants existing in the water. Fouling and filter clogging are regular issues in bioreactors and biofilters that decline the performance of the technology over time.

17.4 NANOMATERIAL-BASED REMEDIATION TECHNIQUES FOR ORGANIC POLLUTANTS

17.4.1 Inorganic Nanomaterials

17.4.1.1 Metal- and Metal Oxide–Based Nanomaterials

Numerous metal-based NMs, particularly transition metal based, have been employed for the remediation of numerous contaminants due to their enhanced properties and flexibility of applications toward contaminants in aqueous media. The majority of studies have been on metal-based NMs and are dedicated to the removal of heavy metals and chlorinated organic pollutants from wastewater. Such NMs exhibit high adsorption of contaminants with fast kinetics (Karn et al., 2009; Das et al., 2015). Table 17.3 depicts the several metal-based NMs employed for various environmental remediation.

During the last few decades numerous efforts have been made to synthesize shape controlled, stable, uniformly dispersed NMs via different methods such as hydrothermal, coprecipitation and thermal decomposition and/or reduction (Cushing et al., 2004; Willis et al., 2005). Silver nanoparticles (Ag NPs) with particle size less than 10 nm are employed as a water disinfectant because of their substantial antibacterial, antifungal

TABLE 17.3
Metal-Based Nanomaterials and Applications in Environmental Remediation of Contaminants

Material	Application	Reference
Ag NPs/Ag ions	Water disinfectant—*Escherichia Coli*	Chou et al., 2015; Gupta and Silver 1998; Bosetti et al., 2002; Xiu et al., 2012; Pal et al., 2007
TiO_2 NPs	Water disinfectant, soil—MS-2 phage, *E. coli*, hepatitis B virus, aromatic hydrocarbons, biological nitrogen, phenanthrene	Cho et al., 2005; Alizadeh et al., 2013; Da Silva et al., 2016; Gu et al., 2012, Li et al., 2014
Metal-doped TiO_2	Water contaminants—2-chlorophenol, endotoxin, *E. coli*, rhodamine B, *Staphylococcus aureus*	Bao et al., 2011; Sreeja et al., 2016
Titanate nanotubes	Gaseous—nitric oxide	Chen et al., 2013
Binary mixed oxide	Water—methylene blue dye	Rasalingam et al., 2014
Iron based	Water—heavy metals, chlorinated organic solvents	Hooshyar et al., 2013
Bimetallic NPs	Water, soil—chlorinated and brominated contaminants	Lien and Zhang 2005

and antiviral properties and high toxicity toward *E. coli* and *Pseudomonas aeruginosa* (Gupta and Silver, 1998; Bosetti et al., 2002).

Nanoparticles	Dye	Remarks	Reference
Nanocrystalline TiO_2/activated carbon composite	Chromotrope 2R	Effective photodegradation under UV irradiation with 80-activated carbon-TiO_2 calcined at 450°C. Initial quantum yield = 1.01%	Wang et al. (2007)
Intercalated CdS into titanate (0.08 g)	Congo Red	Complete removal in 15 minutes	Sehati and Entezari (2016)
Silver-TiO_2 nanocomposites	Eosin Y	100% degradation with 50% of mineralization in 160 minutes	Alfaro et al. (2011)
Cr-doped titanium oxide	Methyl orange	Nearly 90% of the dye was degraded using 5% mol Cr^{3+}-doped TiO_2 nanoparticles in a time period of around 350 minutes	Hamadanian et al. (2014)
Dye sensitized TiO_2	Reactive red 120	Visible light; photocatalytic and sonophotocatalytic degradation	Kavitha and Palanisamy (2011)
Cr(VI), phenol and humic acid	TiO_2	Photocatalytic reduction	Malakutian and Mansuri (2015)
Mesoporous TiO_2	Methylene blue and *E. coli*	Photocatalytic activation	Kim et al. (2009)
TiO_2	Methylene blue	Photocatalytic degradation	Lakshmi et al. (1995)
TiO_2-P25/activated carbon	Phenol	Photocatalytic degradation	Lam et al. (2010)
TiO_2	2-Chlorophenol	Photocatalytic	Lin et al. (2006)

Particles of (11–23 nm) Ag NPs showed inferior bactericidal activity (Gogoi et al., 2006). Ag NPs coupled with other materials like metal oxides and polymers showed improved efficiency of the resulting nanocomposite.

Titanium oxides nanoparticles (TiO_2 NPs) are the most explored material for environmental remediation (wastewater treatment, air purification, self-cleaning, photocatalyst) singly or in combination with other materials due to their low-cost, nontoxicity, semiconducting, photocatalytic, electronic and gas-sensing nature (Li et al., 2008b). TiO_2 NPs generate highly reactive species like hydroxyl, peroxy radicals that act as disinfectant for several microorganisms like fungi, bacteria, viruses and algae (Cho et al., 2005; Zan et al., 2007). TiO_2 exhibits excellent photocatalytic activity under UV irradiation, but is limited in sunlight. In order to improve its activity under sunlight, TiO_2 is often doped with other materials like transition metal/metal oxides, metal hexacyanoferrates, etc. Figure 17.2 describes the photocatalytic mechanism of TiO_2 doped with zinc hexacyanoferrates nanomaterials and its application in degradation of carcinogenic PAHs.

TiO_2 was further mixed with various metal oxides and implemented for the removal of various contaminants such as organic dyes, pesticides, heavy metals, phenols, PAHs, etc. Such composites have reported high potential in smaller-scale industrial wastewater treatment systems. Nonetheless, there are some challenges when using this class of NPs for the remediation of environmental contaminants.

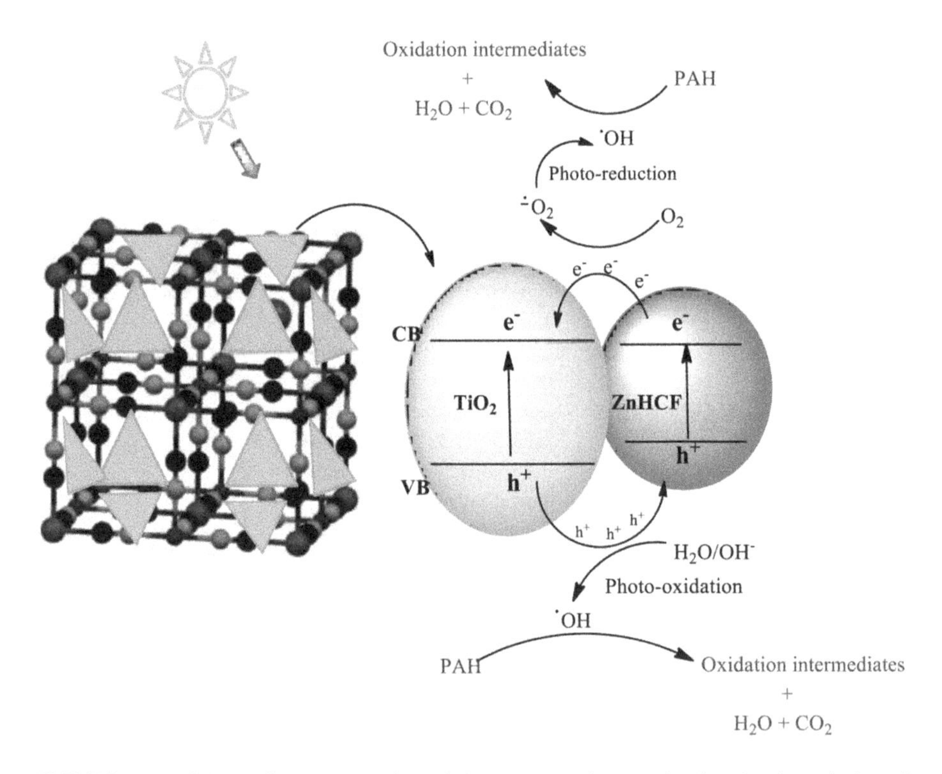

FIGURE 17.2 Schematic representation of charge separation mechanism for degradation of PAHs over photocatalysts.

TABLE 17.4
Degradation of Various Phenols by Nanomaterials

Target	Nanoparticles	Brief Summary	Mechanism	Reference
Phenol	Pr-doped TiO_2 (1 g/L)	Complete degradation in aqueous solutions after 2 hours irradiation; 0.072 mol% of Pr(III)	Photocatalytic degradation	Chiou et al. (2008)
Phenol	Bismuth-doped TiO_2 nanotubes composite	Efficiency: bismuth-doped TiO_2 (by a factor of 4) > undoped TiO_2 rate was 5.2 times faster due to OH• and superoxides	Photocatalytic degradation (Visible light)	
2,4-DNP	Nanosized particles of TiO_2	At a pH of 8, 70% removal within 7 hours	Photocatalytic degradation (UV)	Lin et al. (2008)
2,4-DNP	A bamboo-inspired hierarchical nanoarchitecture of $Ag/CuO/TiO_2$ nanotube array	Under simulated solar light irradiation, the 2,4-dinitrophenol (2,4-DNP) photocatalytic degradation rate over $Ag/CuO/TiO_2$ was about 2.0, 1.5, and 1.2 times that over TiO_2 nanotubes, CuO/TiO_2 and Ag/TiO_2, respectively.	Photocatalytic degradation	Zhang et al. (2005)
3-Aminophenol	Aqueous TiO_2 suspensions	TiO_2 P25 Degussa as the photocatalyst	Photodegradation	Sanna et al. (2016)
Nitrophenols	Aqueous titanium dioxide dispersion		Photocatalytic degradation	

There are many challenges with the implication of TiO_2 NPs—for example, aggregation is one of the key concerns since it can considerably affect the activity of the material, and subsequently reduce the benefit of TiO_2 NPs for the process. The details of the use of TiO_2 nanoparticles used for the degradation of various pollutants are given in Tables 17.4 and 17.5.

It has also been reported that iron nanoparticles with core-shell structure (Fe0; zerovalent) involving the core and mixed valent (Fe(II) and Fe(III)) oxides developing the shell were employed for the remediation of environmental contaminants. These were found effective for chlorinated compounds and heavy metals that can be reduced via electron donation from the zerovalent iron core. Moreover, the shell portion of the NPs facilitate the remediation of pollutants, like heavy metals with higher standard reduction potential (E°) than of the Fe^{2+}/Fe couple (Lien and Zhang, 2005; Li and Zhang, 2006; Wu and Ritchie, 2006; Wang et al., 2009; Nagpal et al., 2010; Zhang et al., 2010a; Kharisov et al., 2012), Ni (Tee et al., 2005; Wu and Ritchie, 2006; Fang et al., 2011; Xie et al., 2014), or Cu (Zheng et al., 2009; Zhu et al., 2010).

TABLE 17.5
List of Various OCs Pesticides Degraded Using TiO_2 Nanoparticles

Type	Pesticide (mg/L)	Nanoparticles	Brief Summary	Mechanism	Reference
OCs	α-, β-, γ-, δ-BHC, dicofol and cypermethrin (20 μg of each)	TiO_2-coated films (2.24 mg/cm^2)	Within 10 minutes; 32%, 44%, 0.2%, 9%, 31% and 0.4%, respectively, remained	Photocatalytic degradation (UV)	Yu et al. (2007)
	Dicofol	TiO_2 (0.25 mg/mL)	Completely degraded in 2 hours, rate constant (k) = 0.167 minute^{-1}	Photocatalytic degradation (UV)	Yu et al. (2008)
	Aldrin (5)	TiO_2 (25 mg)	Complete degradation into dieldrin, chlordene, and 12-hydroxy-dieldrin in water	Photodegradation	Bandala et al. (2002)
	Lindane (5×10^{-4} mmol/L)	Degussa P-25, anatase TiO_2, N-doped TiO_2	Effectively degraded (100%)under visible light, k = 0.099 minute^{-1}, 0.117 minute^{-1}, 0.007 minute^{-1}	Photodegradation	Senthilnathan and Philip (2010)
OPs	Terbufos (5)	TiO_2 (50 mg)	~99% degradation within 90 minutes, rate constant, k = 0.0854 minute^{-1}	Photocatalysis	Wu et al. (2009)
	Monocrotophos, dichlorv (20)	TiO_2-zeolite (50 mg)	100%	Photocatalysis	Gomez et al. (2015)
	Acephate (10)	TiO_2 (4 g/L)	Complete mineralization, rate constant (0.6 mmol/(min)	Photocatalysis	Han et al. (2009)
	Phoxim (20)	La-doped TiO_2 (0.5 g/L)	Complete mineralization after 4 hours	Photocatalysis	Deka and Sinha (2015)
	Monocrotophos, quinalphos (20 each)	ZnO/TiO_2 (0.1 gm)	84.2% and 96%, respectively, k = 0.00567 minute^{-1} and 0.00199 minute^{-1}, respectively	Photocatalysis	Kaur et al. (2013)
	Malathion (10 ppm)	ZnO, TiO_2, Au/ZnO, (10^{-4}, 3×10^{-5}, 10^{-5} M)	30.0% degradation within 30 minutes	Photocatalysis	Fouad and Mohamed (2011)

	Malathion (12)	TiO_2 (anatase)/WO_3 (2 wt.%) (125 or 250 mg)	Complete degradation after 2 hour	Photocatalysis	Ramos-Delgadoa et al. (2013)
	Chlorpyrifos (5)	$CoFe_2O_4$@TiO_2-reduced grapheme (0.4 g/L)	89.9 % in 60 minutes	Photocatalysis	Gupta et al. (2015)
	Parathion methyl (PM) (100 μL of 10,000 mg/L)	Titania-iron mixed oxides (50 mg in 400 μL)	Highest degradation (<70%) with Ti:Fe ratio 0.25:1and 1:0.25	SN2	Henych et al. (2015)
	PM (100 μL of 10,000 mg/L)	Titania-ceria mixed oxides (50 mg in 400 μL)	Degradation ability: Nano-ceria> nano-titania; highest degradation at Ti:Ce 2:8 and 1:1 molar ratio	SN2	Henych et al. (2016)
Miscellenous	Pyridazinone pestcides: Chloridazon (10)	Au/TiO_2 (10^{-4}, 3×10^{-5}, 10^{-5} M)	50% in 30 minutes	Photocatalysis	Fouad and Mohamed (2011)
	Chloridazon (125 mL; 0.18 mM) and metribuzin (125 mL of 0.30 mM)	TiO_2 (1 g/L)	Acidic condition: chloridazon > metribuzin Alkaline condition: metribuzin > chloridazon	Photocatalysis	Khan et al. (2012)
	Neonicotinoids pestcides: Thiamethoxam (3.4 mM)	TiO_2 (1 g/L)	TiO_2 P25 is more efficient photocatalyst than UV100 and PC500	Photocatalysis	Mir et al. (2013)
	Acetamiprid	TiO_2	More efficiently degraded in alkaline pH and all electron acceptors enhanced the degradation rate	Photocatalysis	Khan et al. (2010)
	Imidacloprid, thiamethoxam, clothianidin	Immobilized TiO_2	Mineralized within 2 hours with $k = 0.035$, 0.019 and 0.021 $minute^{-1}$, respectively	Photocatalysis (UV)	Zabar et al. (2012)
	Imidacloprid, isoproctum, phosphamidon (1.14×10^{-4} M each)	5 wt% TiO_2 supported on porous nanosilica (3 g/L)	At neutral pH, degradation within 90, 240, and 120 minutes, respectively	Photocatalysis (UV)	Sharma et al. (2009b)

17.5 SILICA-BASED NANOMATERIALS

Mesoporous silica materials have received great attention because of the high surface to volume ratio, easy surface modification, large pore volumes and tunable pore size in environmental remediation (Tsai et al., 2016). Due to their attractive performance as adsorbents, a number of studies have reported the application of such materials for contaminant remediation in the gas phase (Leal et al., 1995; Satyapal et al., 2001; Huang et al., 2003; Xu et al., 2003; Franchi et al., 2005; Kim et al., 2005; Mandal and Bandyopadhyay, 2005; Harlick and Sayari, 2006; Knofel et al., 2007; Son et al., 2008; Afkhami et al., 2010; Brigante et al., 2016). The hydroxyl groups on the surface of silica materials are crucial for further surface modification, gas adsorption and other surface phenomena such as wetting. Grafting of functional groups onto the pore walls is also a well-known strategy to design new adsorbents and catalysts (Huang et al., 2003).

17.5.1 Carbon-Based Nanomaterials/Carbonaceous

The unique physical, chemical and electronic properties of elemental carbon and its mutable hybridization states such as fullerene C60, fullerene C540, single-walled nanotubes, multi-walled nanotubes and graphene compared to metal-based NMs account for the better implication to environmental remediation (Mauter and Elimelech, 2008). In a variety of investigations determining the suitability of carbon nanotubes and graphene for environmental remediation applications, it has been reported that surface treatments, activation or functionalization of the pristine carbon material is first required. Multi- and single-walled carbon nanotubes (MWCNTs and SWCNTs) have been the subject of many studies. The high surface area enables high adsorption properties of such materials, making them useful for the removal of pollutants from air and large volumes of aqueous solution (Theron et al., 2008; Ren et al., 2011; Di Paola et al., 2012; Kharisov et al., 2014). Moreover, carbon-based NMs also utilize irradiation like UV and sunlight to act as photocatalysts and eradicate pollutants.

Carbon-based NMs exhibited excellent adsorption capacities for different classes of organic compounds such as PAHs, VOCs, herbicides and industrial dyes in both synthetic and natural waters [266, 316–321]. It was also reported that these NMs are easily functionalized in order to improve their performance. Figure 17.3 represents various ways to functionalize carbon-based NMs.

Under UV irradiation, photons of energy greater than or equal to the band gap of the nanotubes encourage the formation of valence band holes (h+) and conduction band electrons (e−). The holes are responsible for the creation of hydroxyl radicals that take part in the oxidation of chlorinated organic compounds. The electrons form superoxide radicals participates in the reduction of heavy metal pollutants. Numerous reports are available for the use of graphene to synthesize photocatalytic nanocomposites [123–126]. Graphene composites with TiO_2 NPs show improved photocatalytic activity compared to bare TiO_2 NPs due to an increase in conductivity

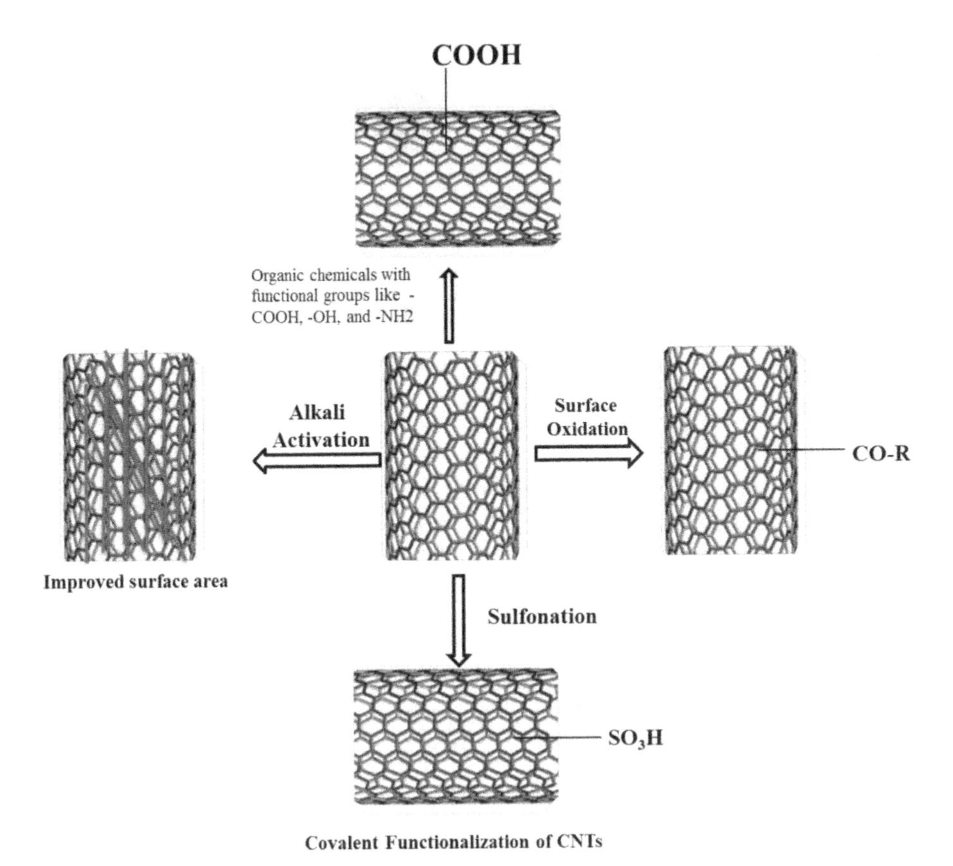

FIGURE 17.3 Pictorial representation of covalent functionalization of CNTs.

(Figure 17.4) (Zhang et al., 2010b; Liu et al., 2011; Yang et al., 2013; Chowdhury and Balasubramanian, 2014). Figure 17.3 summarizes some of the different types of CNTs and their applications in environmental remediation.

Numerous oxygen-containing functional groups—like carboxylic acids, epoxides and hydroxyls—are present on the carbon surface of graphene oxides and facilitate acid-base interactions with basic contaminant gases, such as ammonia (Seredych and Bandosz, 2007; Wang et al., 2013). Moreover, ZnO-graphene and CdS-graphene composites exhibit photocatalytic competences to water pollutants (Zhang et al., 2010a, 2013; Liu et al., 2011). The functionalization of graphene with other materials, such as metal oxides, improved the material's applicability for pollutant degradation. The other explored carbon-based NMs are MWCNTs and SWCNTs with hexagonal configuration (i.e., one nanotube surrounded by six others), forming porous and tube-like structures. Adsorption on external sites attains equilibrium faster than adsorption on internal sites due to the direct exposure of the external sites to the contaminating material. MWCNTs generally do not exist in bundles, except when specific methods of synthesis are used to construct such configurations. Aggregated

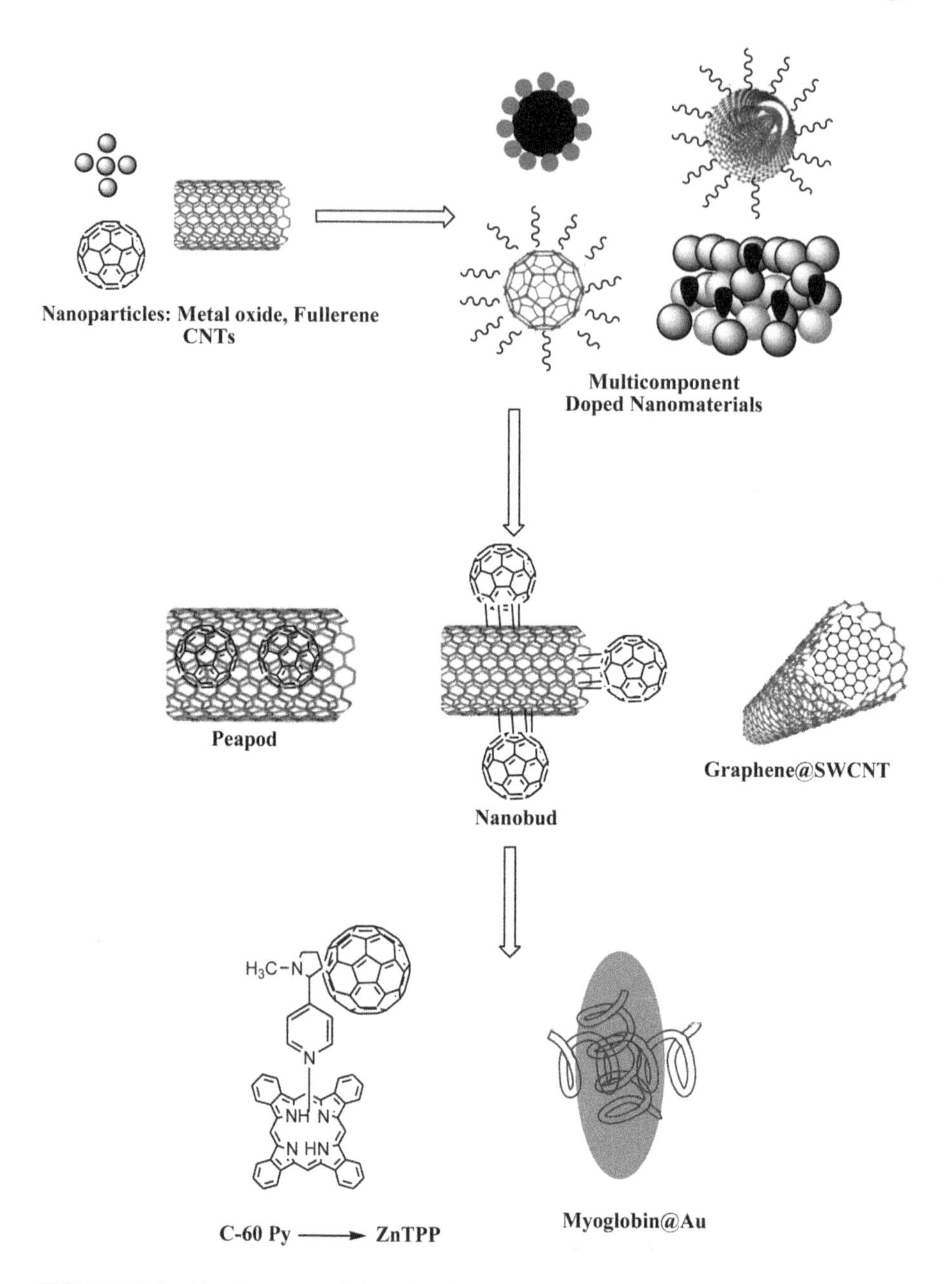

FIGURE 17.4 Graphene materials and their use in environmental remediation.

pores were shown to be more expressively responsible for the adsorption properties of such materials than the less available inner pores. The other parameter that can influence the adsorption capacity of CNTs is the oxygen content. Several chemicals—such as HNO_3, $KMnO_4$, H_2O_2, NaOCl, H_2SO_4, KOH and NaOH—can be utilized to oxidize CNTs. It was finally recommended that in order to enhance their efficiency, carbon-based NMs must be functionalized or coated with other materials having suitable functional groups. Hence, such functionalized materials having manifold features into one template to surge the desired performance.

17.6 POLYMER-BASED NANOMATERIALS

NMs with large surface area to volume ratio contribute to higher reactivity with concomitant improved performance in environment remediation, but at the same time the fast agglomeration, nonspecificity and low stability restrict the use of such technologies. In view of this and in order to enhance stability of NMs, the use of a host materials that act as a matrix to other types of materials have been extensively used for remediation of environmental contaminants (Zuttel et al., 2002; Zhao et al., 2011a).

Polymers are mostly used for the detection and removal of contaminant chemicals (e.g., manganese, nitrate, iron, arsenic, heavy metals, etc.), gases (e.g., CO, SO_2, NOx), organic pollutants (e.g., aliphatic and aromatic hydrocarbons, pharmaceuticals or VOCs) and a wide array of biologics (e.g., bacteria, parasites, viruses, etc.). Polymeric hosts (e.g., surfactants, emulsifiers, stabilizing agents and surface functionalized ligands) are often employed to enhance stability and overcome some of the limitations of pristine NPs as well as to impart other desirable properties such as enhanced mechanical strength, thermal stability, durability and recyclability of the material in question. Amphiphilic polyurethane (APU) NPs have been developed for the remediation of polycyclic aromatic hydrocarbons (PAHs) from soils, thus validating the hypothesis that organic NPs can be engineered with desired properties (Tungittiplakorn et al., 2004). The hydrophilic surface of the nanoparticles promotes mobility in the soil, while the hydrophobic interior of the material confers affinity for the hydrophobic organic contaminants. APU NPs removed phenanthrene from contaminated aquifer sand (i.e., 80% recovery). An analysis of different formulations indicated that the APU nanoparticle affinity for phenanthrene increased when the size of the hydrophobic backbone was also increased. Furthermore, increasing the number of ionic groups on the precursor chain contributed to a reduction in APU particle aggregation in the presence of polyvalent cations (Tungittiplakorn et al., 2004). While the application of these materials in the environment could be beneficial for contaminant remediation, there is no report on the biodegradability of such materials, which contributes to concerns regarding their fate after application.

Poly(amidoamine) or dendrimers (PAMAM) have been utilized in wastewater remediation for water samples contaminated with metal ions such as Cu^{2+}. Those dendritic nanopolymers contain functional groups such as primary amines, carboxylates, and hydroxamates that are able to encapsulate a broad range of solutes in water, including cations (e.g., Cu^{2+}, Ag^{+}, Au^{+}, Fe^{2+}, Fe^{3+}, Ni^{2+}, Zn^{2+} and U^{6+}) (Diallo et al., 2005). They are used as chelating agents and ultrafilters to bind with metal ions, thus facilitating water purification. These materials have also been used as antibacterial/antivirus agents (Brinkmann et al., 1999). The key feature of the dendritic nanopolymers is that they have a lower tendency to pass through the pores of UF membranes compared to linear polymers of similar molecular weight due to their lower polydispersity and globular shape. Therefore, they have been employed to improve UF and MF processes for the recovery of dissolved ions from aqueous solutions. First, a solution of functionalized dendritic nanopolymers is mixed with contaminated water and then the bound nanopolymer contaminants are transferred to UF or MF units to recover the clean water. They can be separated from each other by changing the acidity (pH) of the solution and then the recovered concentrated solution of contaminants is collected for disposal and the nanopolymers may be recycled (Biricova and Laznickova, 2009). In a different approach, our group has described the use of functionalized biodegradable and nontoxic polymeric NMs for target-specific capture of VOCs (Campbell et al., 2015; Guerra et al., 2017a, 2018). Biodegradability is an important and desired feature as it eliminates concerns regarding the fate of the materials after their application. Nontoxicity is another important issue that should be considered when designing new environmental remediation technologies. In our work, the incorporation of amine functional groups from poly(ethyleneimine) (PEI) onto the surface of self-assembled poly(lactic acid)-poly(ethylene oxide) (PLA-PEG) polymeric nanoparticles allowed for the capture of specific VOCs bearing aldehyde and carboxylic acid functional groups by means of condensation (i.e., imine formation) and acid/base reactions, respectively. Thus, the target-specific reactivity of the modified NPs allowed for the rapid and selective capture of desired contaminants in the gas phase (i.e., up to 98% removal). While some technologies rely on physical interactions that could potentially be easily dissociated, this technology is based on the formation of covalent or ionic bonds between the amine functional groups and the targeted aldehyde and/or carboxylic moieties, which may contribute to a more specific and robust capture technology.

Through headspace analysis via gas chromatography, we demonstrated that the amine-functionalized NPs were able to significantly reduce aldehydes (by greater than 69% depending on the analyte structure) and carboxylic acid vapors (by greater than 76%). Aldehydes and carboxylic acids of different chain lengths, linear and branched, were efficiently captured by the PEI functionalized nanoparticle. The NPs were also efficient in the capture of aldehyde and carboxylic acids even in the presence of comparable or more volatile non-targeted vapors, thus exhibiting a selective and targeted capture characteristic (Campbell et al., 2015). In a later iteration of the strategy, we successfully modified cellulose nanocrystals (CNCs) with PEI for the efficient capture of aldehyde VOCs (Guerra et al., 2018).

In a similar context, polymer-supported nanocomposites (PNCs) consist of materials that utilize a polymer as a host material on which NPs are either interspersed or coated on top. This material combines the desirable properties of both the polymer host (i.e., exquisite mechanical strength) and the NPs (i.e., high reactivity, arising from their large surface area). Various polymers have also been used in the fabrication of membranes that incorporate metal and metal oxide NPs for environmental remediation applications (Xu and Bhattacharyya, 2005; Kim and Van Der Bruggen, 2010; Ng et al., 2013; Trujillo-Reyes et al., 2014). In polymeric nanocomposites, polymers are typically used as host materials and other constituents of the composite, such as NPs, are responsible for the contaminant remediation (Zhao et al., 2011b). However, functionalized polymeric NMs have also been described as the main agent responsible for the remediation. Polymer NMs can be used to design technologies exhibiting target-specific capture of compounds from gaseous mixtures in order to prevent off-target fouling that could otherwise contribute to a decrease in the material's performance (Ng et al., 2013).

Polymer/inorganic hybrid NMs have also been widely investigated in environmental applications and notably studied for the adsorptive removal of various toxic metal ions, dyes andmicroorganisms from water/wastewater streams. They exhibit high stability in terms of chemical and thermal properties. Further, these hybrid materials also demonstrate a high capacity for and selective sorption of heavy metals form aqueous media. They have been fabricated through different strategies such as sol-gel processes, self-assembly techniques and assembling of nanobuilding blocks, gathering the features of both materials.

Different studies have shown the importance of the integrated hybrid nanoparticles. For example, Khare et al. (2016) have designed chitosan-based carbon nanofibers (CNFs) incorporated with iron oxide nanoparticles along with polyvinyl alcohol nanocomposite films. This unique combination of materials revealed an efficient adsorption capacity of Cr^{6+} from water and showed a high capacity for metal uptake (~80 mg/g of chitosan/iron-CNF composite). Mittal et al. (2015) have reported that impregnation of SiO_2 NPs in acrylamide hydrogel improves the monolayer adsorption capacity of acrylamide hydrogel to remove 1408.67 mg/g of cationic dyes (MB) (efficiency = 96%) with the hydrogel nanocomposite dose of 0.2 g/L. They also fabricated SiO2 NPs and gum karaya grafted with poly(acrylic acid-co-acrylamide) nanocomposite containing hydrogel for removal of methylene blue (MB) by adsorption from aqueous solutions. These data demonstrated that the grafted copolymerized gum karaya hydrogel could be used as an eco-friendly and efficient adsorbent for the removal of MB dye from industrial wastewater.

17.7 UTILIZATION OF GREEN SYNTHESIZED FNMS IN REMEDIATION OF ENVIRONMENT

In order to avoid use of toxic solvents such as dichlorometane, dimethylformamide, hydrazine, $NaBH_4$, etc., green routes employing environmentally benign and renewable materials are preferred. Currently, a lot of researchers worldwide are working on synthesis of nanoparticles using by employing sunlight, or plant-based surfactants or microorganisms and water or combination of those (Rani and Shanker, 2019) reviewed green strategies for the preparation of metallic nanoparticles applied for water purification. The use of enzymes, vitamins, monosaccharides, polysaccharides and biodegradable polymers including microwave-assisted synthesis were discussed in detail.

Figure 17.5 represents the various green methodologies used for synthesis of NMs. It is more reliable, sustainable and bioinspired bottom-up approach for NMs employed in remediation of. Green synthesis of NMs has several benefits over traditional methods such as cheaper, efficiency, ecofriendly, safe, reusable, easy couture of the size, morphology, shape, and surface functionality, generation of relative stability, remarkable biocompatibility and biodegradability (Li et al., 2011). There are some examples highlighting the use of green synthesized NMs in removal of various organic pollutants. Till date their most of application is reported on dye and waited for exploration on several contaminants. Shanker et al. (Jassal et al., 2015a, 2015b,

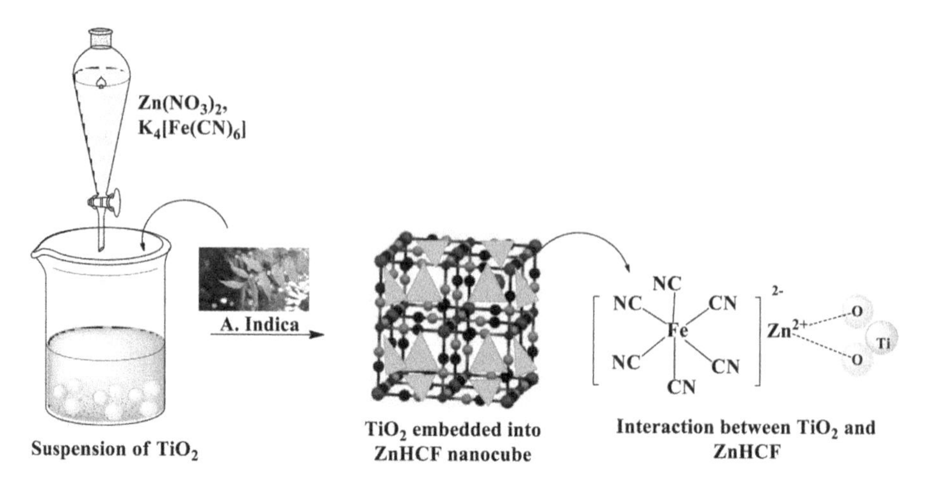

FIGURE 17.5 Possible reaction mechanism for the formation of TiO_2@ZnHCF nanocubes in the presence of leaf extract of *A. indica* as capping agent.

2015c, 2016a, 2016b; Shanker et al., 2016a, 2016b, 2017a, 2017b, 2017c; Rani and Shanker, 2018a–i; Rani and Shanker, 2017; Rani et al., 2017b, 2017b) used several metal hexacyanoferrates, bimetallic oxides, cobaltates nanostructures and transition metal oxide for photo-degradation of dyes and other organic pollutants. Sunlight and natural surfactant *(Aegle marmelos, Sapindus mukorossi, Azadirachta indica)*, respectively, were used for green synthesis.

Sunlight assisted green synthesized transition metal oxide (ZnO, CuO, Co_3O_4, NiO and Cr_2O_3) nanoparticles were able to photocatalytically degrade the 80% to 88% of the dyes in a short period of 180 minute depending on the sizes of respective nanoparticles (Shanker et al., 2016a). Potassium metal hexacyanoferrate KMHCF nanoparticles obtained with aid of *A. marmelos* degraded almost 95% of organic colorants (Jassal et al., 2016a). Using Morinda tinctoria leaf extract silver nanoparticles were synthesized and effectively degraded 95% of dye in 72 hours under sunlight irradiation. In addition to this, Polygonum Hydropiper was also used as a biogenic source for synthesizing silver nanoparticles with high catalytic efficiency in degrading MB completely within thirteen minutes. Yeast (Saccharomyces cerevisiae) extract was also utilized for synthesizing silver nanoparticles which were effectively applied for degrading ~90% of MB within 6 hours. A high degradation of 98% for Orange II was reported using green synthesized bimetallic Fe/Pd nanoparticles, whereas, only 16% of the dye was degraded using Fe nanoparticles. Green synthesized copper oxide nanoparticles and ZnO/Graphene Oxide (GO) nanocomposite were also used for decolorization of the dye.

Green tea or starch fabricated iron–based FNPs have been estimated for photocatalysis. Compared to bared ones, green synthesized FNPs showed superior activities: bimetallic Fe/Pd > Fe and Fe_2O_3@ZnHCF nanocubes > ZnHCF > Fe_2O_3 naoparticles (Rachna et al., 2018). Better catalytic efficiency of green tea-iron iron nanoparticles than Fe nanoparticles fabricated by borohydride reduction further supported this fact. RGO-based plasmonic nanohybrids (Ag-RGO, Au-RGO and Ag/Au-RGO) synthesized using extracts of baker's yeast showed exceptional photothermal transformation adeptness for waste desalination and purification under visible-light. Moreover, Ag-RGO synthesized by means of aqueous extracts of *Psidium guajava* leaves showed detection capability for MB even at very low concentrations. The almond shell was applied to conjugate the Ag-almond nanocomposites obtained from *Ruta graveolens* sleeves (which acts as both reducing and stabilizing agents) was able to degrade 4-nitrophenol, rhodamine B (RhB) and MB. Nanocomposites of GO with crystalline cellulose displayed astonishing adsorption capacity for MB (2630 mg/g). Besides, cellulose-Ag nanocomposites have unlimited talent for commercial applications (catalysis, antibacterial, sensor and environment). Silver nanoparticles with *Amaranthus gangeticus* could able to remove more than 50% of Congo red dye in 15 minutes. Dyes could be photocatalytic removed by biosynthesized nanocomposites of ZnO and copper oxide with Graphene oxide. In addition, supermagnetic $MgFe_2O_4@\gamma\text{-}Al_2O_3$ FNPs were used for remediation of several reactive dyes.

Green synthesis is commonly used for metal oxides but limited for bimetallic and higher forms. Green functionalized of ZnO with Ce ion or Ag for photocatalytic degradation of organic pollutants. Rani and Shanker fabricated bimetallic oxides (below 50 nm) such as NiCuO nanorods, $CuCr_2O_4$ nanoflowers and $NiCrO_3$ nanospheres via use of *A. marmelos* leaf extract consisting of terpenoids, alkaloids and phenylpropanoids as phytochemicals for controlling the morphology of FNPs.

Green synthesized ZnO/GO nanocomposite had better efficiency for Brilliant blue-R and other dyes than that of annealed one (90.64% degradation). Table 17.4 represents various NMs used for the degradation of various organic dyes. Recently Rani and Shanker (2018i) reported facile green synthesis of various metal oxides functionalized with PMMA and their use for the degradation of MB dye. It was also revealed from reusability analysis of nanocomposites for MB degradation that it was recyclable ten times without significant loss in the activity. Photodegradation of Eriochrome Black T using Fe^{3+} and Pt^{4+} impregnated TiO_2 nanostructures was reported.

A few studies recently reported on the use of green synthesized nanoparticles implemented for treatment of PAHs and other pollutants. Biosynthesized nanoparticles ZnO showed good potential in photodegradation of anthracene 96% for the into non-toxic 9,10-anthraquinone. The potential of green synthesized metal hexacyanoferrate MHCF nanoparticles was tested for the treatment of toxic PAHs, namely, anthracene, phenanthrene chrysene, fluorine, and benzo (a) pyrene in water as well as in soil under dark exposure, UV and sunlight. The selection of PAHs was based on different rings. Using metal hexacyanoferrate nanoparticles different toxic PAHs like anthracene > phenanthrene > fluorine > chrysene > BaP were degraded to good extent in water (70–93%) as well as in soil (68–84%) at neutral pH under sunlight exposure.

The order of degradation was found to be dependent on the molecular weight, size, structure and aromaticity of the PAHs. Small and non-toxic by-products such as malealdehyde, 4-oxobut-2-enoic acid, and *o*-xylene were identified. This indicates the positive aspects of using synthesized nanoparticles in removal of PAHs. Working mechanism of metal hexacyanoferrate nanoparticles indicated the initial adsorption followed by photo-degradation process occurring over their surface (Figures 17.6a, b and 17.7).

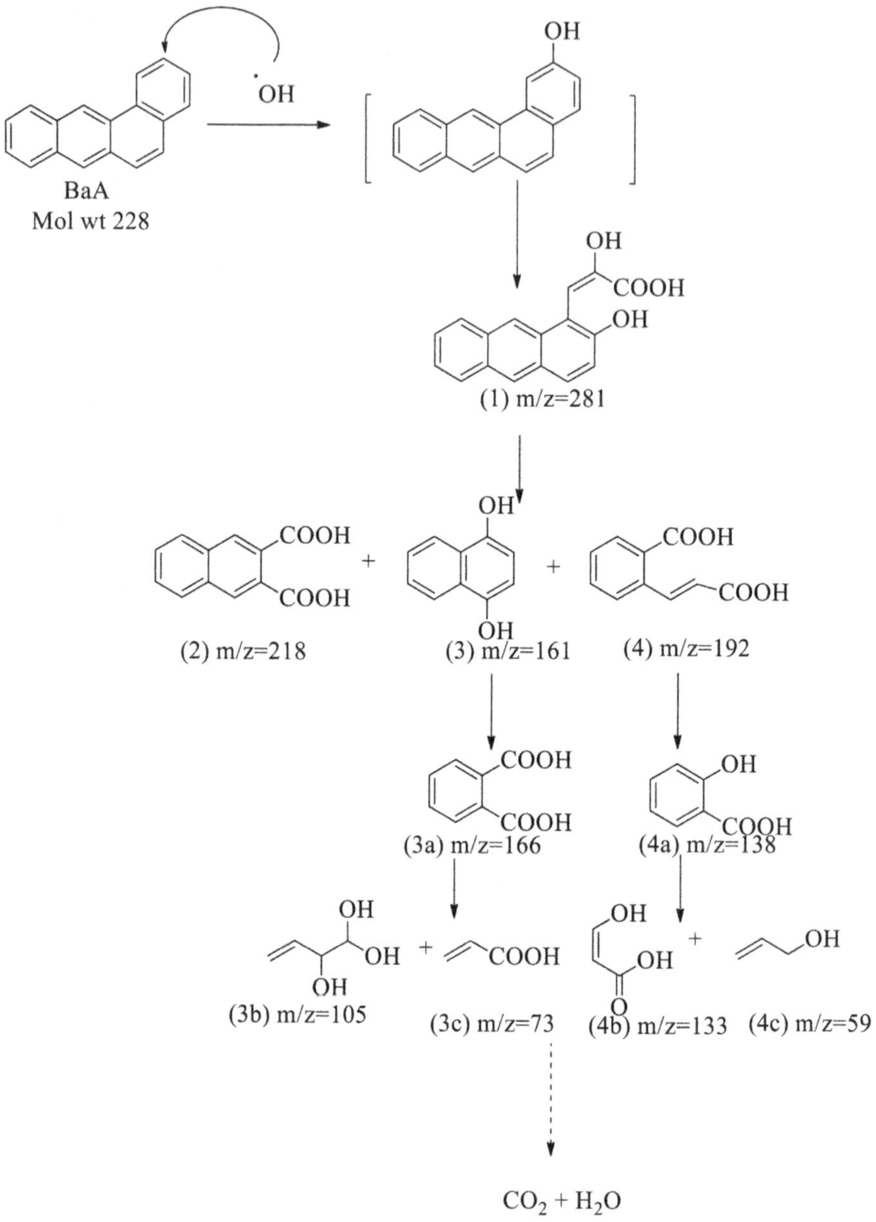

BaA
Mol wt 228

$\cdot OH$

(1) m/z=281

(2) m/z=218

(3) m/z=161

(4) m/z=192

(3a) m/z=166

(4a) m/z=138

(3b) m/z=105

(3c) m/z=73

(4b) m/z=133

(4c) m/z=59

$CO_2 + H_2O$

(a)

FIGURE 17.6 Proposed degradation pathway for the degradation of (a) BaA and (b) BaP over ZnHCF@ZnO nanocubes (concentration: 2 mg/L; pH:~7; catalyst dose: 25 mg; natural sunlight).

(Continued)

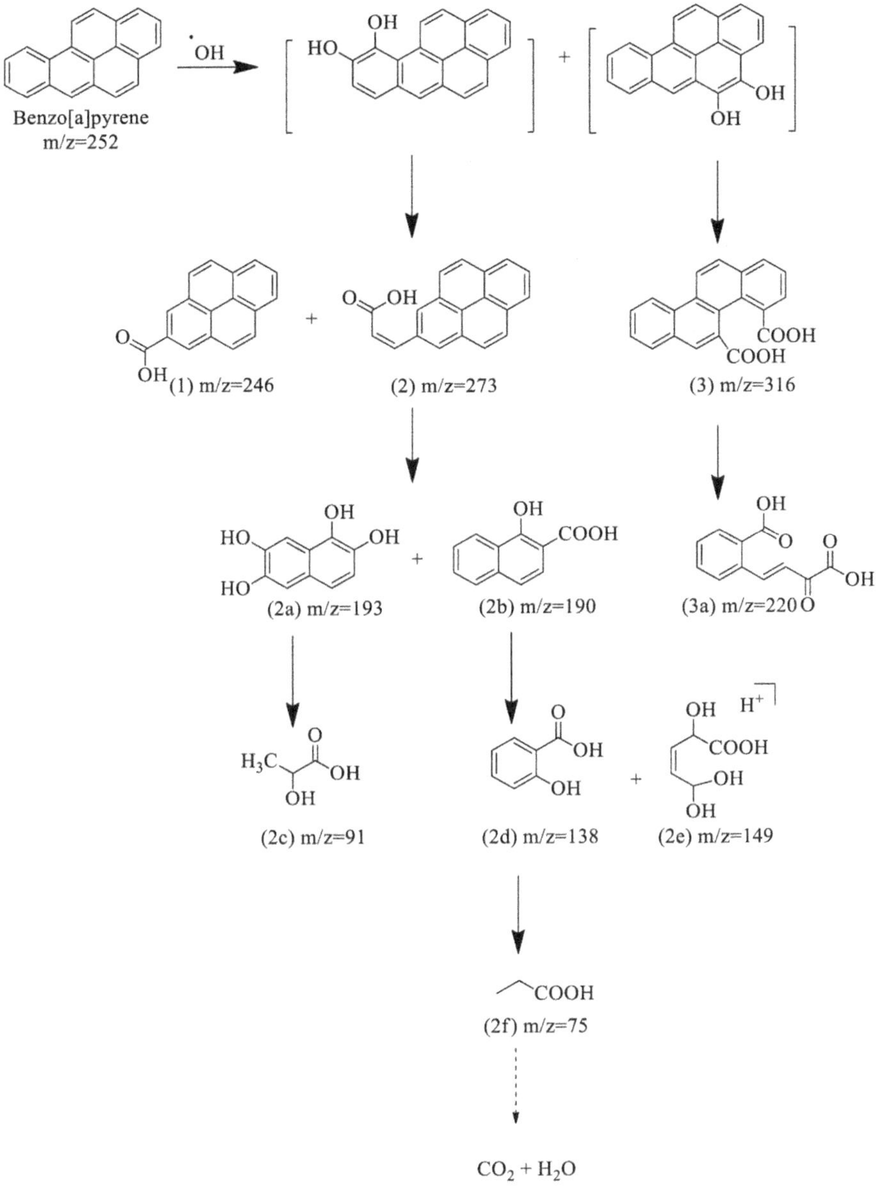

Benzo[a]pyrene m/z=252

$\cdot OH$

(1) m/z=246 + (2) m/z=273

(3) m/z=316

(2a) m/z=193 + (2b) m/z=190

(3a) m/z=220

(2c) m/z=91

(2d) m/z=138 + (2e) m/z=149

(2f) m/z=75

$CO_2 + H_2O$

(b)

FIGURE 17.6 *(Continued)*

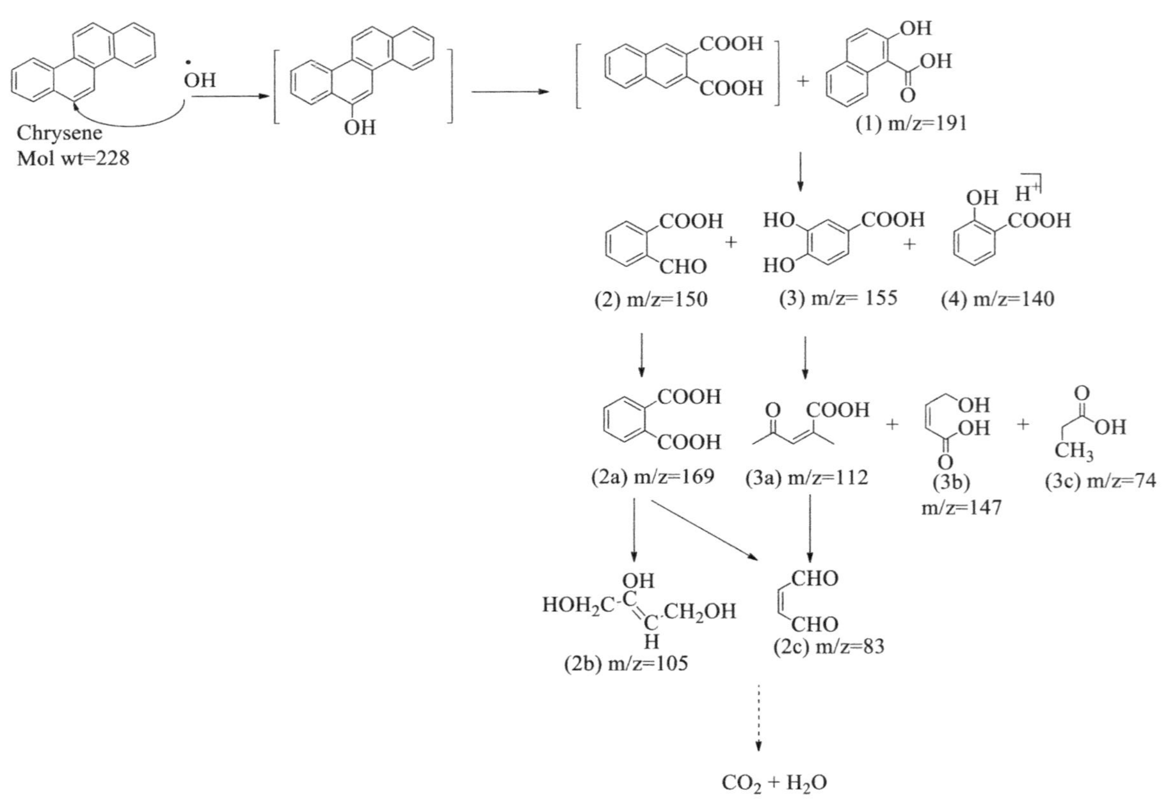

FIGURE 17.7 Proposed degradation pathway for the degradation of chrysene over Fe2O3@ZnHCF nanocubes (concentration; 2 mg L-1; pH~7; catalyst dose: 25 mg; natural sunlight).

Moreover, like FNPs of metal oxide with ZnHCF, mixtures of metal oxides are less expensive as synthesized from cheap process and reusable (n = 10) with enhancement of charge separation. Diverse potentials were observed by different researchers for FNPs fabricated from a green source containing different phytochemicals surfactants. Due to this, FNPs of different size and morphology were obtained. Hence, in view of this, commercialization of green synthesized FNPs should be promoted.

17.8 SUMMARY, CONCLUSION AND RECOMMENDATIONS

The application of nanotechnology in water purification and environmental remediation has considerable potential, as demonstrated by several studies. Traditional treatment technologies do not always offer the most cost-effective solution for removing several common pollutants, and in particular are not cost-effective for the removal of pollutants present at low concentrations. In addition, many of these techniques are already stretched to their limits and may be unable to meet increasingly stringent water quality standards. Conventional treatment methods are often energy-intensive while generating a considerable amount of sludge and hazardous wastes. Nano-based techniques may thus become very important, especially for the removal of emerging pollutants and low levels of contaminants. In contrast to traditional technologies, the effectiveness of many nano-based methods can be improved via particle modification, and their cost may be lowered by industrial-scale production and the development of synthetic methods that consider cheaper feedstock and less energy. In addition, many nano-based technologies allowed for reuse often require less space, can offer reduced toxic intermediate formation and are adaptable to support the emerging concept of water treatment decentralization (Li et al., 2008c). Many nano-based treatment technologies have been shown to perform well as complements to or substitutes for traditional treatment technologies (Rani et al. 2017a; Rani and Shanker, 2018i). Release of engineered NM to the environment may occur when used for cleanup or water treatment, and their implications to humans and the environment are not well understood. Release may occur through dissolution of metals, desorption from composite materials or by the intentional introduction to a polluted environmental media (Wang et al., 2011; Miralles et al., 2012). C-NMs and insoluble metallic NMs (such as TiO_2) are likely to be persistent in the environment since they are not easily biodegraded (Shanker et al. 2016b; Rani et al., 2018i). Several studies have predicted that many of these persistent NMs will aggregate (homoaggregation or heteroaggregation) in the natural environment and most likely settle out (Shanker et al., 2016b), leading to increased exposure of organisms in the sediment phase. The fate of soluble ENMs will be influenced by their dissolution rate and potential. Toxic ion shedding and ROS production have been implicated as two major toxicity pathways for Me/MeO ENMs, and toxicity to a variety of aquatic and terrestrial organisms has been shown. Release of surface coatings of MPs into water may cause concern, although anionic surfactants can reduce contact between MPs and bacteria, resulting in reduced toxicity. Many of these concerns can be addressed by safer technology design (Liu et al., 2011), surface modification to improve biocompatibility and increased efforts on understanding the impacts of ENMs.

In order to further materialize the potential of nano-based methods, research needs to focus on developing nanoparticle synthesis methods that are less energy-intensive and require cheap feedstock. In addition, approaches that can provide fast and reliable prediction of ENM/nanocomposite toxicity (e.g., high throughput approaches that use key ENM physicochemical properties or single cells should be further researched and developed.

This is important for predicting the toxicology of nano-based technologies without testing each ENM/composite. Systematic studies of the fate of nanoparticles in the environment, which can be extrapolated to predict the fate of similar particles, need to be intensified since studying each nanoparticle/nanocomposite is onerous and almost impossible. This review submits that nanotechnology is emerging as a promising alternative to traditional methods of water treatment and pollution remediation.

17.9 CONCLUSION

Inorganic, carbonaceous and polymeric NMs are among the different types of materials that can be successfully employed for a variety of environmental remediation applications. Selecting the best NM to mitigate a particular pollutant in a specific environmental context requires a full analysis of the type of contaminant to be removed, the accessibility to the remediation site, the amount of material needed to implement efficient remediation and whether it is advantageous to recover the remediation NM (recycling). In that each material has its own advantages and issues related to its applicability, we provided here an overall perspective of some NMs that have been utilized in the context of environmental remediation.

Although a variety of studies have been undertaken to investigate the use of nanotechnology, concerns regarding the application of nanotechnology for environmental remediation purposes have yet to be addressed. Also, while many studies do demonstrate efficacy in laboratory settings, more research is necessary in order to fully understand how nanotechnology can significantly affect the remediation of environmental contaminants in real case scenarios (e.g., the remediation of contaminated water, soil, and air from industrial processes). Also, while the mechanisms through which the different nanotechnologies are applied are well known, what happens to these materials after they have been applied for contaminant capture or degradation is underexplored. Even though the recyclability of some materials have been described, it appears that at some point the efficacy of these materials declines, which makes them no longer useful. Therefore, research is necessary to elucidate the fate of these materials after introduction to the environment for remediation purposes in order to avoid the possibility of these materials becoming themselves a source of environmental contamination. These challenges must be overcome to realize the full potential of NMs for environmental applications. Nevertheless, nanotechnology provides a wealth of strategies that can be leveraged to address environmental contamination.

REFERENCES

Afkhami, A., Saber-Tehrani, M., Bagheri, H. Simultaneous removal of heavy-metal ions in wastewater samples using nano-alumina modified with 2, 4-dinitrophenylhydrazine. J. Hazard. Mater. **2010**, 181, 836–844.

Alcamo, J., Flörke, M., Märker, M. Future long-term changes in global water resources driven by socio-economic and climatic changes. Hydrol. Sci. J. **2007**, 52, 247–75.

Alfaro, S. O., Rodríguez-González, V., Zaldívar-Cadena, A. A., Lee, S. W. Sonochemical deposition of silver-TiO_2 nanocomposites onto foamed waste-glass: Evaluation of Eosin Y decomposition under sunlight irradiation. Catal. Today. **2011**, 166, 166–171.

Alizadeh, M. F., Aminzadeh, B., Vahidi, H. Degradation of petroleum aromatic hydrocarbons using TiO_2 nanopowder film. Environ. Technol. **2013**, 34, 1183–1190.

Babich, H., Davis, D. L. Phenol: A review of environmental and health risks. Regul. Toxicol. Pharmacol. **1981**, 1, 90–101.

Bandala, E. R., Gelover, S., Leal, M. T. Arancibia-Bulnes, C., Jimenez, A., Estrada, C. A. Solar photocatalytic degradation of Aldrin. Catal. Today. **2002**, 76, 189–199.

Bao, Q., Zhang, D., Qi, P. Synthesis and characterization of silver nanoparticle and graphene oxide nanosheet composites as a bactericidal agent for water disinfection. J. Colloid Interface Sci. **2011**, 360, 463–470.

Barakat, M. A. New trends in removing heavy metals from industrial wastewater. Arab. J. Chem. **2011**, 4, 361–377.

Biricova, V., Laznickova, A. Dendrimers: A Analytical characterization and applications. Bioorg. Chem. **2009**, 37, 185–192.

Bosetti, M., Masse, A., Tobin, E., Cannas, M. Silver coated materials for external fixation devices: In vitro biocompatibility and genotoxicity. Biomaterials. **2002**, 23, 887–892.

Brigante, M., Pecini, E., Avena, M. Magnetic mesoporous silica for water remediation: Synthesis, characterization and application as adsorbent of molecules and ions of environmental concern. Micropor. Mesopor. Mater. **2016**, 230, 1–10.

Brinkmann, N., Giebel, D., Lohmer, G., Reetz, M. T., Kragl, U. Allylic substitution with dendritic palladium catalysts in a continuously operating membrane reactor. J. Catal. **1999**, 183, 163–168.

Brunekreef, B., Janssen N. A., de Hartog J. J., Oldenwening M., Meliefste K., Hoek G., Lanki T. Personal, indoor and outdoor exposures of PM25 and its components for groups of cardiovascular patients in Amsterdam and Helsinki. Res. Rep. Health Eff. Inst. **2005**, 127, 1–70.

Campbell, M. L., Guerra, F. D., Dhulekar, J., Alexis, F., Whitehead, D. C. Target-specific capture of environmentally relevant gaseous Aldehydes and Carboxylic Acids with functional Nanoparticles. Chem. A Eur. J. **2015**, 21, 14834–14842.

Chen, X., Cen, C., Tang, Z., Zeng, W., Chen, D., Fang, P., Chen, Z. The key role of pH value in the synthesis of titanate nanotubes-loaded manganese oxides as a superior catalyst for the selective catalytic reduction of NO with NH3. J. Nanomater **2013**, 871528

Chiou, C. H., Wu, C. Y., Juang, R. S. Influence of operating parameters on photocatalytic degradation of phenol in UV/TiO_2 process. Chem. Eng. J. **2008**, 139, 322–329.

Cho, M., Chung, H., Choi, W., Yoon, J. Different inactivation behaviors of MS-2 phage and *Escherichia coli* in TiO_2 photocatalytic disinfection. Appl. Environ. Microbiol. **2005**, 71, 270–275.

Chou, K.-S., Lu, Y.-C., Lee, H.-H. Effect of alkaline ion on the mechanism and kinetics of chemical reduction of silver. Mater. Chem. Phys. **2005**, 94, 429–433.

Chowdhury, S., Balasubramanian, R. Recent advances in the use of graphene-family nanoadsorbents for removal of toxic pollutants from wastewater. Adv. Colloid Interface Sci. **2014**, 204, 35–56.

Cushing, B. L., Kolesnichenko, V. L., O'Connor, C. J. Recent advances in the liquid-phase syntheses of inorganic nanoparticles. Chem. Rev. **2004**, 104, 3893–3946.
Da Silva, M. B., Abrantes, N., Nogueira, V., Gonçalves, F., Pereira, R. TiO_2 nanoparticles for the remediation of eutrophic shallow freshwater systems: Efficiency and impacts on aquatic biota under a microcosm xperiment. Aquat. Toxicol. **2016**, 178, 58–71.
Dàna, E., Sayari, A. Optimization of copper removal efficiency by adsorption on amine-modified SBA-15: Experimental design methodology. Chem. Eng. J. **2011**b, 167, 91–98.
Darwish, M., Mohammadi, A. Functionalized nanomaterial for environmental techniques. Nanotechnol. Environ. Sci. **2018**, 9, 315–350.
Das, S., Sen, B., Debnath, N. Recent trends in nanomaterials applications in environmental monitoring and remediation. Environ. Sci. Pollut. Res. **2015**, 22, 18333–18344.
De A. K. Environmental Chemistry (7th ed.). New Age International, India, **2014**.
Deka, A. C., Sinha, S. K. Mycogenic silver nanoparticle biosynthesis and its pesticide degradation potentials. Int. J. Technol. Enhanc. Emerg. Eng. Res. **2015**, 3, 108–113.
Di Paola, A., García-López, E., Marcì, G., Palmisano, L. A survey of photocatalytic materials for environmental remediation. J. Hazard. Mater. **2012**, 211–212, 3–29.
Diallo, M. S., Christie, S., Swaminathan, P., Johnson, J. H., Goddard, W. A. Dendrimer enhanced ultrafiltration.1. Recovery of Cu(II) from aqueous solutions using PAMAM dendrimers with ethylene diamine core and terminal NH2 groups. Environ. Sci. Technol. **2005**, 39, 1366–1377.
Dubowsky, S. D. et al. The contribution of traffic to indoor concentrations of polycyclic aromatic hydrocarbons. J. Expo. Anal. Environ. Epidemiol. **1999**, 9, 312–321.
El-Ashtoukhy, S. Z., El-Taweel, Y. A., Abdelwahab, O., Nassef, E. M. Treatment of petrochemical wastewater containing phenolic compounds by electrocoagulation using a fixed bed electrochemical reactor. Int. J. Electrochem. Sci. **2013**, 8, 1534–1540.
European Commission (EC). Priority substances and certain other pollutants according to annex II of directive 2008/105/EC, **2008** http://ec.europa.eu/environment/water/water-framework/priority_substances.html.
Fang, Z., Qiu, X., Chen, J., Qiu, X. Debromination of polybrominated diphenyl ethers by Ni/Fe bimetallic nanoparticles: Influencing factors, kinetics, and mechanism. J. Hazard. Mater. **2011**, 185, 958–969.
Fouad, D. M., Mohamed, M. B. Photodegradation of chloridazon using coreshell magnetic nanocompsites. J. Nanotechnol. **2011**, 2011, 1–7.
Franchi, R. S., Harlick, P. J. E., Sayari, A. Applications of pore-expanded mesoporous silica. 2. Development of a high-capacity, water-tolerant adsorbent for CO2. Ind. Eng. Chem. Res. **2005**, 44, 8007–8013.
Franssen, M. C. R., Kircher, M., Wohlgemuth, R. Industrial Biotechnology in the chemical and Pharmaceutical Industries, Industrial Biotechnology Sustainable Growth and Economic Success. Weinheim, Germany: Wiley-VCH Verlag GmbH & Co, **2010**.
Fu, F., Wang, Q. Removal of heavy metal ions from wastewaters: A review. J. Environ. Manage. **2011**, 92, 407–418.
Fu, J., Kyzas, G. Z., Cai, Z., Deliyanni, E. A., Liu, W., Zhao, D. Photocatalytic degradation of phenanthrene by graphite oxide-TiO_2-$Sr(OH)_2$/$SrCO_3$ nanocomposite under solar irradiation: Effects of water quality parameters and predictive modelling. Chem. Eng. J. **2018**, 335, 290–300.
Gogoi, S. K., Gopinath, P., Paul, A., Ramesh, A., Ghosh, S. S., Chattopadhyay, A. Green fluorescent protein-expressing escherichia coli as a model system for investigating the antimicrobial activities of silver nanoparticles. Langmuir **2006**, 22, 9322–9328.
Gomez, S., Marchena, C. L., Renzini, M. S., Pizzio, L., Pierella, L. In situ generated TiO_2 over zeolitic supports as reusable photocatalysts for the degradation of dichlorvos. Appl. Catal. B Environ. **2015**, 162, 167–173.

Gu, J., Dong, D., Kong, L., Zheng, Y., Li, X. Photocatalytic degradation of phenanthrene on soil surfaces in he presence of nanometer anatase TiO_2 under UV-light. J. Environ. Sci. **2012**, 24, 2122–2126.

Guerra, F. D., Campbell, M. L., Whitehead, D. C., Alexis, F. Capture of aldehyde VOCs using a series of amine-functionalized cellulose nanocrystals. Chemistry Select. **2018**, 3, 5495–5501.

Guerra, F. D., Campbell, M. L., Whitehead, D. C., Alexis, F. Tunable properties of functional nanoparticles for efficient capture of VOCs. Chemistry Select. **2017**a, 2, 9889–9894.

Guerra, F. D., Smith, G. D., Alexis, F., Whitehead, D. C. A survey of VOC emissions from rendering plants. Aerosol Air Qual. Res. **2017**b, 17, 209–217.

Gui, X., Wei, J., Wang, K., Cao, A., Zhu, H., Jia, Y., Shu, Q., Wu, D. Carbon nanotube sponges. Adv. Mater. **2010**, 22, 617–621.

Gupta, A., Silver, S. Molecular genetics: Silver as a biocide: Will resistance become a problem? Nat. Biotechnol. **1998**, 16, 888–893.

Gupta, B., Rani, M., Kumar, R. Degradation of thiram in water, soil and plants: A study by high-performance liquid chromatography. Biomed. Chromatogr. **2012**a, 26, 69–75.

Gupta, B., Rani, M., Kumar, R., Dureja, P. Decay profile and metabolic pathways of quinalphos in water, soil and plants. Chemosphere. **2011**, 85(5), 710–716. https://doi.org/10.1016/j.chemosphere.2011.05.059.

Gupta, B., Rani, M., Kumar, R., Dureja, P. Identification of degradation products of thiram in water, soil and plants using LC-MS technique. J. Environ. Sci. Health B. **2012**b, 47(8), 823–831. https://doi.org/10.1080/03601234.2012.676487.

Gupta, B., Rani, M., Kumar, R., Dureja, P. In vitro and in vivo studies on degradation of quinalphos in rats. J. Hazard Mat. **2012**c, 213–214, 285–291.

Gupta, S., Ashrith, G., Chandra, D., Gupta, A. K., Finkel, K.W., Guntupalli, J.S. Acute phenol poisoning: A life-threatening hazard of chronic pain relief. Clinical Toxicol. **2008**, 46, 250–253.

Gupta, S. S., Chakraborty, I., Maliyekkal, S. M., Mark, T. A., Pandey, D. K., Das, S. K., Pradeep, T. Simultaneous dehalogenation and removal of persistent halocarbon pesticides from waste water using grapheme nanocomposites: A case study of lindane. Sustain. Chem. Eng. **2015**, 3, 1155–1163.

Hamadanian, M., Ali, S. S., Mehra, A. M., Jabbari, V. Photocatalyst Cr-doped titanium oxide nanoparticles: Fabrication, characterization, and investigation of the effect of doping on methyl orange dye degradation. Mat. Sci. Semicond. Process. **2014**, 21, 161–166.

Han, S. T., Jing, L., Xi, H. L., Xu, D. N., Zuo, Y., Zhang, J. H. Photocatalytic decomposition of acephate in irradiated TiO_2 suspensions. J. Hazard Mater. **2009**, 163, 1165–1172.

Harlick, P. J. E., Sayari, A. Applications of pore-expanded mesoporous silicas. 3. Triamine silane grafting for enhanced CO_2 adsorption. Ind. Eng. Chem. Res. **2006**, 45, 3248–3255.

Henych, J., Janos, P., Kormunda, M., Tolasz, J., Stengl, V. Reactive adsorption of toxic organophosphates parathion methyl and DMMP on nanostructured Ti/Ce oxides and their composites. Arab. J. Chem. **2016**. http://dx.doi.org/10.1016/j.arabjc.2016.06.002.

Henych, J., Stengl, V., Slusna, M., Grygar, T. M., Janos, P., Kuran, P., Stastny M. Degradation of organophosphorus pesticide parathion methyl on nanostructured titania–iron mixed oxides. Appl Surf Sci. **2015**, 344, 9–16.

Henze, M. Characterization of wastewater for modelling of activated sludge processes. Water Sci. Tech. **1992**, 25, 1–15.

Hooshyar, Z., Rezanejade Bardajee, G., Ghayeb, Y. Sonication enhanced removal of nickel and cobalt ions from polluted water using an iron based sorbent. J. Chem. **2013**, 786954.

Huang, H. Y., Yang, R. T., Chinn, D., Munson, C. L. Amine-grafted MCM-48 and silica xerogel as superior sorbents for acidic gas removal from natural gas. Ind. Eng. Chem. Res. **2003**, 42, 2427–2433.

IARC. Polynuclear aromatic compounds, part I, chemical, environmental and experimental data. IARC Monographs on the Evaluation of Carcinogenic Risks to Humans, Volume 32. **1998**.

Im, J., Loffler, F. E. Fate of bisphenol a in terrestrial and aquatic environments. Environ. Sci Technol. **2016**, 50, 8403–8416.

Jadhav, D. N., Vanjara, A. K. Removal of phenol from wastewater using sawdust, polymerized sawdust and sawdust carbon. Indian J. Chem. Technol. **2004**, 11, 35–41.

Jassal, V., Shanker, U., Gahlot, S. Green synthesis of some iron oxide nanoparticles and their interaction with 2-amino, 3-amino and 4-aminopyridines. Mater. Today Proc. **2016**a, 3, 1874–1882.

Jassal, V., Shanker, U., Gahlot, S., Kaith, B. S., Kamaluddin, Md. A. I., Samuel, P. *Sapindus mukorossi* mediated green synthesis of some manganese oxide nanoparticles interaction with aromatic amines. Applied Physics A. **2015**a, 122, 271–282.

Jassal, V., Shanker, U., Kaith, B. S. *Aeglemarmelos* mediated green synthesis of different nano-structured metal hexacyanoferrates: Activity against photodegradation of harmful organic dyes. Scientifica. **2016**b, 1–13. http://dx.doi.org/10.1155/2016/2715026.

Jassal, V., Shanker, U., Kaith, B. S., Shankar, S. Green synthesis of potassium zinc hexacyanoferrate nanocubes and their potential application in photocatalytic degradation of organic dyes. RSC Adv. **2015**b, 5, 26141–26149.

Jassal, V., Shanker, U., Shankar, S. Synthesis, characterization and applications of nano-structured metal hexacyanoferrates. J. Environ. Anal. Chem. **2015**c, 2, 1000128–1000141.

Jun, L. Y., Mubarak, N. M., Yee, M. J., Yon, L. S., Bing, C. H., Khalid, M., Abdullah, E. C. An overview of functionalised carbon nanomaterial for organic pollutant removal. J. Indus. Eng. Chem. **2018**, 77, 909–917.

Kamat, P. V., Meisel, D. Nanoscience opportunities in environmental remediation. C. R. Chim. **2003**, 6, 999–1007.

Kang, J. H., Kondo, F., Katayama, Y. Human exposure to bisphenol A. Toxicology. **2006**, 226, 79–89.

Kannan, K., Tanabe, S., Giesy, J. P., Tatsukawa, R. Organochlorine pesticides and polychlorinated biphenyls in foodstuffs from Asian and oceanian countries. Rev. Environ. Contam. Toxicol. **1997**, 152, 1–55.

Karn, B., Kuiken, T., Otto, M. Nanotechnology and in situ remediation: A review of the benefits and potential risks. Environ. Health Perspect. **2009**, 117, 1813–1831.

Kaur, P., Bansal, P., Sud, D. Heterostructured nanophotocatalysts for degradation of organophosphate pesticides from aqueous streams. J. Korean Chem. Soc. **2013**, 57(3), 382–388.

Kavitha, S. K., Palanisamy, P. N. Photocatalytic and sonophotocatalytic degradation of reactive red 120 using dye sensitized TiO_2 under visible light. Int. J. Civil Environ. Eng. **2011**, 3, 1–14.

Kdasi, A., Idris, A., Saed, K., Guan, C. Treatment of textile wastewater by advanced oxidation processes—A review. Global Nest Int. J. **2004**, 6, 222–230.

Khan, A., Haque, M. M., Mir, N. A., Muneer, M., Boxall, C. Heterogeneous photocatalysed degradation of an insecticide derivative acetamiprid in aqueous suspensions of semiconductor. Desalination. **2010**, 261, 169–174.

Khan, A., Mir, N. A., Faisal, M., Muneer, M. Titanium dioxide mediated photcatalysed degradation of two herbicide derivatives chloridazon and metribuzin in aqueous suspensions. Int. J. Chem. Eng. **2012**, 2012, 1–8.

Khan, F. I., Ghoshal, A. K. Removal of volatile organic compounds from polluted air. J. Loss Prev. Process Ind. **2000**, 13, 527–545.

Khare, P., Yadav, A., Ramkumar, J., Verma, N. Microchannel-embedded metal–carbon–polymer nanocomposite as a novel support for chitosan for efficient removal of hexavalent chromium from water under dynamic conditions. Chem. Eng. J. **2016**, 293, 44–54.

Kharisov, B. I., Rasika Dias, H. V., Kharissova, O. V. Nanotechnology-based remediation of petroleum impurities from water. J. Pet. Sci. Eng. **2014**, 122, 705–718.

Kharisov, B. I., Rasika Dias, H. V., Kharissova, O. V., Manuel Jiménez-Pérez, V., Olvera Pérez, B., Muñoz Flores, B. Iron-containing nanomaterials: Synthesis, properties, and environmental applications. RSC Adv. **2012**, 2, 9325–9358.

Kim, E. Y., Kim D. S., Ahn B. T. Synthesis of mesoporous TiO_2 and its application to photocatalytic activation of Methylene Blue and E. coli, Bull. Korean Chem. Soc., **2009**, 30, 193–196.

Kim, S., Ida, J., Guliants, V. V., Lin, J. Y. S. Tailoring pore properties of MCM-48 silica for selective adsorption of CO_2. J. Phys. Chem. B. **2005**, 109, 6287–6293.

Knofel, C., Descarpentries, J., Benzaouia, A., Zelenak, V., Mornet, S., Llewellyn, P. L., Hornebecq, V. Functionalised micro-/mesoporous silica for the adsorption of carbon dioxide. Micropor. Mesopor. Mater. **2007**, 99, 79–85.

Lakshmi, S., Renganathan, R., Fujita, S. Study on TiO_2- mediated photocatalytic degradation of methylene blue. J. Photochem. Photobiol. A: Chem. **1995**, 88(2–3), 163–167.

Lam, S. M., Sin, J. C., Mohamed, A. R. Parameter effect on photocatalytic degradation of phenol using TiO_2-P25/activated carbon (AC), Korean J. Chem. Eng., **2010**, 27, 1109–1116.

Leal, O., Bolívar, C., Ovalles, C., García, J. J., Espidel, Y. Reversible adsorption of carbon dioxide on amine surface-bonded silica gel. Inorg. Chim. Acta. **1995**, 240, 183–189.

Li, D., Cui, F., Zhao, Z., Liu, D., Xu, Y., Li, H., Yang, X. The impact of titanium dioxide nanoparticles n biological nitrogen removal from wastewater and bacterial community shifts in activated sludge. Iodegradation **2014**, 25, 167–177.

Li, Q., Mahendra, S., Lyon, D. Y., Brunet, L., Liga, M. V., Li, D., Alvarez, P. J. J. Antimicrobial nanomaterials for water disinfection and microbial control: Potential applications and implications. Water Res. **2008**a, 42, 4591–4602.

Li, X. Q., Zhang, W. X. Iron nanoparticles: The core–shell structure and unique properties for Ni(II) sequestration. Langmuir. **2006**, 22, 4638–4642.

Li, Y., Zhang, P., Du, Q., Peng, X., Liu, T., Wang, Z., Xia, Y., Zhang, W., Wang, K., Zhu, H., et al. Adsorption of fluoride from aqueous solution by graphene. J. Colloid Interface Sci. **2011**, 363, 348–354.

Li, Z., et al. Concentration and profile of 22 urinary polycyclic aromatic hydrocarbon metabolites in the US population. Environ. Res. **2008**b, 107, 320–33.

Lien, H.-L., Zhang, W. Hydrodechlorination of chlorinated ethanes by nanoscale Pd/Fe bimetallic particles. J. Environ. Eng. **2005**, 131, 4–10.

Lin, H., Huang, C. P., Li, W., Ni, C., Shah S. I., Tseng, Y. H. Size dependency of nanocrystalline TiO_2 on its optical property and photocatalytic reactivity exemplified by 2-chlorophenol. Appl. Catal., B, **2006**, 68, 1–11.

Lin, L., Qiu, P., Cao, X., Jin, L. Colloidal silver nanoparticles modified electrode and its application to the electroanalysis of Cytochrome c. Electrochim. Acta. **2008c**, 53, 5368–5372.

Liu, X., Pan, L., Lv, T., Lu, T., Zhu, G., Sun, Z., Sun, C. Microwave-assisted synthesis of ZnO-graphene composite for photocatalytic reduction of Cr(vi). Catal. Sci. Technol. **2011**, 1, 1189–1193.

Malakutian M., Mansuri F., Hexavalent chromium removal by titanium dioxide photocatalytic reduction and the effect of phenol and humic acid on its removal efficiency, Int. J. Environ. Health Eng. **2015**, 4, 1–8.

Mandal, B. P., Bandyopadhyay, S. S. Simultaneous absorption of carbon dioxide and hydrogen sulfide into aqueous blends of 2-amino-2-methyl-1-propanol and diethanolamine. Chem. Eng. Sci. **2005**, 60, 6438–6451.

Marques, P. A. S. S., Rosa, M. F., Pinheiro, H. M. pH effects on the removal of Cu2+, Cd2+ and Pb2+ from aqueous solution by waste brewery waste. Bioprocess Eng. **2000**, 23,135–141.

Masciangoli, T., Zhang, W. Environmental technologies. Environ. Sci. Technol. **2003**, 37, 102–108.

Mauter, M., Elimelech, M. Environmental applications of carbon-based nanomaterials. Environ. Sci. Technol. **2008**, 42, 5843–5859.

Mir, N. A., Khan, A., Muneer, M., Vijayalakhsmi, S. Photocatalytic degradation of a widely used insecticide Thiamethoxam in aqueous 2 suspension of TiO_2: Adsorption, kinetics, product analysisandtoxicityassessment. Sci. Total Environ. **2013**, 458, 388–398.

Miralles, P., Church, T. L., Harris A. T. Toxicity, uptake, and translocation of engineered nanomaterials in vascular plants. Environ. Sci. Technol. **2012**, 46, 9224–9239

Mittal, H., Maity, A., Ray, S. S. Synthesis of co-polymer-grafted gum karaya and silica hybrid organic–inorganic hydrogel nanocomposite for the highly effective removal of methylene blue. Chem. Eng. J. **2015**, 279, 166–179.

Morita, H., Yanagisawa, N., Nakajima, T., Shimizu, M., Hirabayashi, H., Okudera, H., Nohara, M., Midorikawa, Y., Mimura, S. Sarin poisoning in Matsumoto, Japan. Lancet. **1995**, 346, 290–293.

Nagpal, V., Bokare, A. D., Chikate, R. C., Rode, C. V., Paknikar, K. M. Reductive dechlorination of -hexachlorocyclohexane using Fe-Pd bimetallic nanoparticles. J. Hazard. Mater. **2010**, 175, 680–687.

Navarro, A., Tauler, R., Lacorte, S., Barcelo, D. Occurrence and transport of pesticides and alkylphenols in water samples along the Ebro River Basin. J Hydrol. **2010**, 383, 18–29.

Negri-Cesi, P. Bisphenol A interaction with brain development and functions. Res. Int. J. **2015**, 6, 1–12.

Ng, L. Y., Mohammad, A. W., Leo, C. P., Hilal, N. Polymeric membranes incorporated with metal/metal oxide nanoparticles: A comprehensive review. Desalination. **2013**, 308, 15–33.

Nordberg, G. F., Flower, B., Nordberg, M., Friberg, L. Handbook on the Toxicology of Metals (3rd ed.). Cambridge, MA: Academic Press, **2007**.

Ojea-Jiménez, I., López, X., Arbiol, J., Puntes, V. Citrate-coated gold nanoparticles as smart scavengers for mercury(II) removal from polluted waters. ACS Nano. **2012**, 6, 2253–2260.

Pagga, U, Brown, D. The degradation of dyestuffs: Part II behaviour of dyestuffs in aerobic biodegradation tests. Chemosphere. **1986**, 15, 479–491.

Pal, S., Tak, Y.K., Song, J.M. Does the antibacterial activity of silver nanoparticles depend on the shape of the nanoparticle? A study of the gram-negative bacterium Escherichia coli. Appl. Environ. Microbiol. **2007**, 73, 712–1720.

Pandey, B., Fulekar, M. H. Nanotechnology: Remediation technologies to clean up the environmental pollutants. Res. J. Chem. Sci. **2012**, 2, 90–96.

Pradeep, N. V., Anupama, S., Arun Kumar, J. M., Vidyashree, K. G., Ankitha, K., Lakshmi, P., Pooja, J. Treatment of sugar industry wastewater in anaerobic down flow stationary fixed film (DSFF) reactor. Sugar Tech. **2014**, 16, 9–18.

Qu, X., Alvarez, P. J. J., Li, Q. Applications of nanotechnology in water and wastewater treatment, Water Res. **2013**, 47(12), 3931–3946.

Rachna, Rani M., Shanker, U. Sunlight mediated improved photocatalytic degradation of carcinogenic benz[a]anthracene and benzo[a]pyrene by zinc oxide encapsulated hexacyanoferrate nanocomposite. J. Photochem. Photobiol. A Chem. **2019**b, 381, 111861.

Rachna, Rani, M., Shanker, U. Degradation of tricyclic polyaromatic hydrocarbons in water, soil and river sediment with a novel TiO_2 based heterogeneous nanocomposite. J Environ. Manage. **2019**a, 248, 109340–109351.

Rachna, Rani, M., Shanker, U. Enhanced photocatalytic degradation of chrysene by Fe2O3@ ZnHCF nanocubes. Chem. Eng. J. **2018**, 348, 754–764.

Rachna, Rani, M., Shanker, U. Sunlight active ZnO@FeHCF nanocomposite for the degradation of bisphenol A and nonylphenol. J. Environ. Chem. Eng. **2019**c, 7, 103153–103170.

Ramos-Delgadoa, N. A., Gracia-Pinilla, M. A., Maya-Trevinoa, L., Hinojosa-Reyesa, L., Guzman-Mara, J. L., Hernandez-Ramirez, A. Solar photocatalytic activity of TiO_2 modified with WO_3 on the degradation of an organophosphorus pesticide. J. Hazard. Mater. **2013**, 263, 36–44.

Rani, M. 2012. Studies on decay profiles of quinalphos and thiram pesticides. PhD Thesis, Indian Institute of Technology Roorkee, Roorkee, Uttarakhand, India, **2012**, 1, 5.

Rani, M., Rachna, Shanker, U. Metal hexacyanoferrates nanoparticles mediated degradation of carcinogenic aromatic amines. Environ. Nanotechnol. Monit. Manag. **2018**, 10, 36–50.

Rani, M., Rachna, Shanker, U. Mineralization of carcinogenic anthracene and phenanthrene by sunlight active bimetallic oxides nanocomposites. J. Colloid Interface Sci. **2019**, 555, 676–688.

Rani, M., Shanker, U. Advanced treatment technologies. In Handbook of Environmental Materials Management, ed. C. M. Hussain. Cham: Springer International Publishing AG, **2018**h. https://doi.org/10.1007/978-3-319-58538-3_33-1.

Rani, M., Shanker, U. Degradation of traditional and new emerging pesticides in water by nanomaterials: Recent trends and future recommendations. Int. J. Environ. Sci. Technol. **2018**i, 15, 1347–1380.

Rani, M., Shanker, U. Effective adsorption and enhanced degradation of various pesticides from aqueous solution by Prussian blue nanorods. J. Env. Chem. Eng. **2018**c, 6, 1512–1521.

Rani, M., Shanker, U. Green solvents in chemical reactions. In Industrial Applications of Green Solvents. Millersville, PA: Materials Research Forum LLC, **2019**.

Rani, M., Shanker, U. Insight in to the degradation of bisphenol A by doped ZnO@ ZnHCF nanocubes: High photocatalytic performance. J. Colloid. Interfac. Sci. **2018**a, 530, 16–28.

Rani, M., Shanker, U. Photocatalytic degradation of toxic phenols from water using bimetallic metal oxide nanostructures. Colloids Surf. A. **2018**e, 553, 546–561.

Rani, M., Shanker, U. Promoting sun light-induced photocatalytic degradation of toxic phenols by efficient and stable double metal cyanide nanocubes. Environ. Sci. Pollut. Res. **2018**f, 25, 23764–23779.

Rani, M., Shanker, U. Remediation of polycyclic aromatic hydrocarbons using nanomaterials. In The Handbook Green Adsorbents for Pollutant Removal, eds. G. Crini, E. Lichtfouse. London: Springer Nature, **2018**g, pp. 343–387.

Rani, M., Shanker, U. Removal of carcinogenic aromatic amines by metal hexacyanoferrate nanocubes synthesized via green process. J. Env. Chem. Eng. **2017**, 5, 5298–5311.

Rani, M., Shanker, U. Removal of chlorpyriphos, thaimethoxam and tebuconazole from water using green synthesized metal hexacyanoferrates nanoparticles. Environ. Sci. Pollut. Res. **2018**d, 25, 10878–10893.

Rani, M., Shanker, U. Sun-light driven rapid photocatalytic degradation of methylene blue by poly (methyl methacrylate)/metal oxide nanocomposites. Colloids Surf. **2018**b, 559, 136–147.

Rani, M., Shanker, U., Chaurasia, A. Catalytic potential of laccase immobilized on transition metal oxides nanomaterials: Degradation of alizarin red S dye. J. Environ. Chem. Eng. **2017**a, 5(3), 2730–2739.

Rani, M., Shanker, U., Jassal, V. Recent strategies for removal and degradation of persistent and toxic organochlorine pesticides using nanoparticles: A review. J. Environ. Manag. **2017**b, 190, 208–222.

Rao, C. H. S., Venkateswarlu, V., Surender, T., Eddleston, M., Buckley, N. A. Pesticide poisoning in South India—Opportunities for prevention and improved medical management. Trop. Med. Int. Health. **2005**, 10, 581–588.

Rao, M. M., Ramesh, A., Rao, G. P. C., Seshaiah, K. Removal of copper and cadmium from the aqueous solutions by activated carbon derived from Ceiba pentandra hulls. J. Hazard. Mater. B. **2006**, 129, 123–129.

Rasalingam, S., Peng, R., Koodali, R. T. Removal of hazardous pollutants from wastewaters: Applications of TiO_2-SiO_2 mixed oxide materials. J. Nanomater. **2014**, 617405.

Ren, X., Chen, C., Nagatsu, M., Wang, X. Carbon nanotubes as adsorbents in environmental pollution management: A review. Chem. Eng. J. **2011**, 170, 395–410.

Said, M., Ahmad, A., Wahab, M. A. Removal of phenol during ultrafiltration of Palm oil mill effluent (POME): Effect of pH, ionic strength, pressure and temperature. Der Pharma Chem. **2013**, 5, 190–200.

Sanna, V., Pala, N., Alzaric, V., Nuvolic, D., Mauro, C. ZnO nanoparticles with high degradation efficiency of organic dyes under sunlight irradiation. Mater. Lett. **2016**, 162, 257–260.

Santhosh, C., Velmurugan, V., Jacob, G., Jeong, S. K., Grace, A. N., Bhatnagar, A. Role of nanomaterials in water treatment applications: A review. Chem. Eng. J. **2016**, 306, 1116–1137.

Satyapal, S., Filburn, T., Trela, J., Strange, J. Performance and properties of a solid amine sorbent for carbon dioxide removal in space life support applications. Energ. Fuel. **2001**, 15, 250–255.

Sehati, S., Entezari, M. H. Sono-intercalation of CdS nanoparticles into the layers of titanate facilitates the sunlight degradation of Congo red. J. Colloid Interface Sci. **2016**, 462, 130–139.

Senthilnathan, J., Philip, L. Removal of mixed pesticides from drinking water system using surfactant-assisted nano-TiO_2, Water. Air. Soil Pollut. **2010**, 210, 143–154.

Seredych, M., Bandosz, T. J. Removal of ammonia by graphite oxide via its intercalation and reactive adsorption. Carbon. **2007**, 45, 2130–2132.

Shah, K. J., Imae, T. Selective gas capture ability of gas-adsorbent-incorporated cellulose nanofiber films. Biomacromolecules. **2016**, 17, 1653–1661.

Shanker, U., Jassal, V., Rani, M. Catalytic removal of organic colorants from water using some transition metal oxide nanoparticles synthesized under sunlight. RSC Adv. **2016**a, 6, 94989–94999.

Shanker, U., Jassal, V., Rani, M. Degradation of toxic PAHs in water and soil using potassium zinc hexacyanoferrate nanocubes. J. Environ. Manag. **2017**a, 204, 337–348.

Shanker, U., Jassal, V., Rani, M. Green synthesis of iron hexacyanoferrate nanoparticles: Potential candidate for the degradation of toxic PAHs. J. Environ. Chem. Eng. **2017**b, 5, 4108–4120.

Shanker, U., Jassal, V., Rani, M., Kaith, B. S. Towards green synthesis of nanoparticles: From bio-assisted sources to benign solvents. A review. Int. J. Env. Anal. Chem. **2016**b, 96, 801–835.

Shanker, U., Rani, M., Jassal, V. Degradation of hazardous organic dyes in water by nanomaterials. Environ. Chem. Lett. **2017**c, 15(4), 623–642. http://dx.doi.org/10.1007/s10311-017-0650-2.

Sharma, M. V. P., Sadanandam, G., Ratnamala, A., Kumari, V. D., Subrahmanyam, M. An efficient and novel porous nanosilica supported TiO_2 photocatalyst for pesticide degradation using solar light. J. Hazard. Mater. **2009**a, 171, 626–633.

Sharma, Y. C., Srivastava, V., Singh, V. K., Kaul, S. N., Weng, C. H. Nano-adsorbents for the removal of metallic pollutants from water and wastewater. Environ. Technol. **2009**b, 30, 583–609.

Singh, A., Kumar, V., Srivastava, J. N. Assessment of bioremediation of oil and phenol contents in refinery waste water via bacterial consortium. Pet. Environ. Biotechnol. **2013**, 4, 1–11.

Son, W.-J., Choi, J.-S., Ahn, W.-S. Adsorptive removal of carbon dioxide using polyethyleneimine-loaded mesoporous silica materials. Micropor. Mesopor. Mater. **2008**, 113, 31–40.

Sreeja, S., Vidya Shetty, K. Microbial disinfection of water with endotoxin degradation by photocatalysis using Ag@TiO_2 core shell nanoparticles. Environ. Sci. Pollut. Res. **2016**, 23, 18154–18164.

Srivastava, N. K., Majumder, C. B. Novel biofiltration methods for the treatment of heavy metals from industrial wastewater. J. Hazard. Mater. **2008**, 151, 1–8.

Suzuki, T., Nakagawa, Y., Takano, I., Yaguchi, K., Yasuda, K. Environmental fate of bisphenol A and its biological metabolitesin river water and their xenoestrogenic activity. Environ. Sci. Technol. **2004**, 38, 2389–2396.

Tee, Y. H., Grulke, E., Bhattacharyya, D. Role of Ni/Fe nanoparticle composition on the degradation of trichloroethylene from water. Ind. Eng. Chem. Res. **2005**, 44, 7062–7070.

Theron, J., Walker, J. A., Cloete, T. E. Nanotechnology and water treatment: Applications and emerging opportunities. Crit. Rev. Microbiol. **2008**, 34, 43–69.

Tong, H., Ouyang, S., Bi, Y., Umezawa, N., Oshikiri, M., Ye, J. Nano-photocatalytic materials: Possibilities and challenges. Adv. Mater. **2012**, 24, 229–251.

Tratnyek, P. G., Johnson, R. L. Nanotechnologies for environmental cleanup. Nano Today. **2006**, 1, 44–48.

Trujillo-Reyes, J., Peralta-Videa, J. R., Gardea-Torresdey, J. L. Supported and unsupported nanomaterials for water and soil remediation: Are they a useful solution for worldwide pollution? J. Hazard. Mater. **2014**, 280, 487–503.

Tsai, C. H., Chang, W. C., Saikia, D., Wu, C. E., Kao, H. M. Functionalization of cubic mesoporous silica SBA-16 with carboxylic acid via one-pot synthesis route for effective removal of cationic dyes. J. Hazard. Mater. **2016**, 309, 236–248.

Tungittiplakorn, W., Lion, L. W., Cohen, C., Kim, J. Y. Engineered polymeric nanoparticles for soil remediation. Environ. Sci. Technol. **2004**, 38, 1605–1610.

Vaseashta, A., Vaclavikova, M., Vaseashta, S., Gallios, G., Roy, P., Pummakarnchana, O. Nanostructures in environmental pollution detection, monitoring, and remediation. Sci. Technol. Adv. Mater. **2007**, 8, 47–59.

Veeresh, G. S., Kumar, P., Mehrotra, K. Treatment of phenol and cresols in upflow anaerobic sludge blanket (UASB) process: A review. Water Res. **2004**, 39, 154–165.

Wang, S., Sun, H., Ang, H. M., Tadé, M. O. Adsorptive remediation of environmental pollutants using novel graphene-based nanomaterials. Chem. Eng. J. **2013**, 226, 336–347.

Wang H. H., Kou, X. M., Pei, Z. G., Xiao, J.Q., Shan, X. Q., Xing, B. S. Physiological effects of magnetite (Fe_3O_4) nanoparticles on perennial ryegrass (Lolium perenne L.) and pumpkin (Cucurbita mixta) plants. Nanotoxicology **2011**, 5, 30–42.

Wang, W., Silva, C. G., Faria, J. L. Photocatalytic degradation of Chromotrope 2R using nanocrystalline TiO_2/activated-carbon composite catalysts. Appl. Catal. B Environ. **2007**, 70, 470–478.

Wang, X., Chen, C., Chang, Y., Liu, H. Dechlorination of chlorinated methanes by Pd/Fe bimetallic nanoparticles. J. Hazard. Mater. **2009**, 161, 815–823.

Willis, A. L., Turro, N. J., O'Brien, S. Spectroscopic characterization of the surface of iron oxide nanocrystals. Chem. Mater. **2005**, 17, 5970–5975.

World Health Organization (WHO). (**2015**). World health statistics 2015. World Health Organization. https://apps.who.int/iris/handle/10665/170250.

Wright, D. A. **2002**. P. Welbourn Environmental toxicology. Cambridge University Press, Cambridge.

Wu, L., Ritchie, S. M. C. Removal of trichloroethylene from water by cellulose acetate supported bimetallic Ni/Fe nanoparticles. Chemosphere. **2006**, 63, 285–292.

Wu, R. J., Chen, C. C., Chen, M. H., Lua, C. S. Titanium dioxide mediated heterogeneous photocatalytic degradation of terbufos: Parameter study and reaction pathways. J. Hazard. Mater. **2009**, 162, 945–953.

Xie, Y., Fang, Z., Cheng, W., Tsang, P. E., Zhao, D. Remediation of polybrominated diphenyl ethers in soil using Ni/Fe bimetallic nanoparticles: Influencing factors, kinetics and mechanism. Sci. Total Environ. **2014**, 485–486, 363–370.

Xiu, Z., Zhang, Q., Puppala, H. L., Colvin, V. L., Alvarez, P. J. J. Negligible particle-specific antibacterial activity f silver nanoparticles. Nano Lett. **2012**, 12, 4271–4275.

Xu, J., Bhattacharyya, D. Membrane-based bimetallic nanoparticles for environmental remediation: Synthesis and reactive properties. Environ. Prog. **2005**, 24, 358–366.

Xu, X., Song, C., Andresen, J. M., Miller, B. G., Scaroni, A. W. Preparation and characterization of novel CO_2 "molecular basket" adsorbents based on polymer-modified mesoporous molecular sieve MCM-41. Micropor. Mesopor. Mater. **2003**, 62, 29–45.

Yang, M., Zhang, N., Xu, Y. Synthesis of fullerene-, carbon nanotube-, and graphene-TiO_2 nanocomposite photocatalysts for selective oxidation: A comparative study. ACS Appl. Mater. Interfaces. **2013**, 5, 1156–1164.

Yu, B., Zeng, J., Gong, L., Yang, X. Q., Zhang, L., Chen, X. Photocatalytic degradation investigation of dicofol. Chin. Sci. Bull. **2008**, 53, 27–32.

Yu, B., Zeng, J., Gong, L., Zhang, M., Zhang, L., Chen, X. Investigation of the photocatalytic degradation of organochlorine pesticides on a nano-TiO_2 coated film. Talanta. **2007**, 72, 1667–1674.

Zabar, R., Komel, T., Fabjan, J., Kralj, M., Trebse, P. Photocatalytic degradation with immobilised TiO_2 of three selected neonicotinoid insecticides: Imidacloprid, thiamethoxam and clothianidin. Chemosphere. **2012**, 89(3), 293–301.

Zan, L., Fa, W., Peng, T., Gong, Z. Photocatalysis effect of nanometer TiO_2 and TiO_2-coated ceramic plate on hepatitis B virus. J. Photochem. Photobiol. B Biol. **2007**, 86, 165–169.

Zhang, N., Yang, M.-Q., Tang, Z.-R., Xu, Y.-J. CdS-graphene nanocomposites as visible light photocatalyst for redox reactions in water: A green route for selective transformation and environmental remediation. J. Catal. **2013**, 303, 60–69.

Zhang, X., Wang, Y., Li, G. Mesoporous silica nanoparticles in drug delivery and biomedical applications. J. Mol. Catal. A Chem. **2005**, 237, 199–215.

Zhang, Y., Tang, Z.-R., Fu, X., Xu, Y.-J. TiO_2 graphene nanocomposites for gas-phase photocatalytic degradation of volatile aromatic pollutant: Is TiO_2 graphene truly different from other TiO_2 carbon composite materials? ACS Nano. **2010**a, 4, 7303–7314.

Zhang, Z., Shen, Q., Cissoko, N., Wo, J., Xu, X. Catalytic dechlorination of 2, 4-dichlorophenol by Pd/Fe bimetallic nanoparticles in the presence of humic acid. J. Hazard. Mater. **2010**b, 182, 252–258.

Zhao, X., Lv, L., Pan, B., Zhang, W., Zhang, S., Zhang, Q. Polymer-supported nanocomposites for environmental application: A review. Chem. Eng. J. **2011**a, 170, 381–394.

Zhao, Y., Gao, Q., Tang, T., Xu, Y., Wu, D. Effective NH2-grafting on mesoporous SBA-15 surface for adsorption of heavy metal ions. Mater. Lett. **2011**b, 65, 1045–1047.

Zheng, Z., Yuan, S., Liu, Y., Lu, X., Wan, J., Wu, X., Chen, J. Reductive dechlorination of hexachlorobenzene by Cu/Fe bimetal in the presence of nonionic surfactant. J. Hazard. Mater. **2009**, 170, 895–901.

Zhu, N., Luan, H., Yuan, S., Chen, J., Wu, X., Wang, L. Effective dechlorination of HCB by nanoscale Cu/Fe particles. J. Hazard. Mater. **2010**, 176, 1101–1105.

Zuttel, A., Mauron, P., Kiyobayashi, T., Emmenegger, C., Schlapbach, L. Hydrogen storage in carbon nanostructures. Int. J. Hydrog. Energy. **2002**, 27, 203–212.

Index

Note: *Italicized* page numbers refer to figures, **bold** page numbers refer to tables

B

D

O

P

Q

R

S